Clinical Understanding of the Human Gut Microbiome

Mark Pimentel • Ruchi Mathur
Gillian M. Barlow

Editors

Clinical Understanding of the Human Gut Microbiome

Springer

Editors
Mark Pimentel
Medically Associated Science and Technology (MAST) Program, Cedars-Sinai Medical Center
West Hollywood, CA, USA

Karsh Division of Gastroenterology and Hepatology, Department of Medicine
Cedars-Sinai Medical Center
Los Angeles, CA, USA

Ruchi Mathur
Medically Associated Science and Technology (MAST) Program, Cedars-Sinai Medical Center
West Hollywood, CA, USA

Division of Endocrinology, Diabetes, and Metabolism, Department of Medicine
Cedars-Sinai Medical Center
Los Angeles, CA, USA

Gillian M. Barlow
Medically Associated Science and Technology (MAST) Program, Cedars-Sinai Medical Center
West Hollywood, CA, USA

ISBN 978-3-031-46711-0 ISBN 978-3-031-46712-7 (eBook)
https://doi.org/10.1007/978-3-031-46712-7

This Springer imprint is published by the registered company Springer Nature Switzerland AG
The registered company address is: Gewerbestrasse 11, 6330 Cham, Switzerland

Paper in this product is recyclable

We dedicate this work to our families, colleagues, and friends.
We are grateful for their support throughout this journey.
We are also grateful for the microorganisms who made the wine, beer, and cheese we consumed throughout this journey.
"We cannot fathom the marvelous complexity of an organic being; but on the hypothesis here advanced this complexity is much increased. Each living creature must be looked at as a microcosm—a little universe, formed of a host of self-propagating organisms, inconceivably minute and as numerous as the stars in heaven."
Charles Darwin

Introduction

This book is written with the goal of making the breadth and complexity of gut microbiome research accessible to physicians and clinicians, so that they can then make the best decisions in the care of their patients. The gut microbiome world is complex, and the sheer speed with which discoveries are being made regarding the roles of gut microbiota in any given clinical area (e.g., diabetes, functional gastrointestinal disorders, colorectal cancer, and neurological disorders from the autism spectrum to Parkinson's disease, to name but a few) is incredible. Moreover, this research now includes not just gut bacteria, but also yeast and other fungi, archaea, parasites, and viruses, as well as transcriptomics, proteomics, and metabolomics, such that physicians and clinicians, and indeed other researchers, can have a hard time keeping pace. In writing this book, our goal was to digest down recent advances in as many areas of gut microbiome research as possible, so that improved understanding for physicians can ultimately translate into benefits for their patients.

There are some important considerations that must be taken into account in reading this, or any other, book on the microbiome. Firstly, it is important to recognize that we do not yet know what constitutes a "normal" or "healthy" gut microbiome, or rather, there is no one pattern of gut microbial composition that we can define as "normal." "Normal" gut microbial compositions will vary depending on geographical location and diet, and of course are highly individual. Moreover, it is now understood that the gut microbiome also changes with age, and can in turn influence aging, and exhibits differences between males and females—this has been termed the "microgenderome." As a result, it is challenging to define dysbiosis, i.e., an imbalance in the balance of gut microbial populations, and indeed the term dysbiosis has fallen out of favor somewhat. However, we can say that greater diversity in microbial populations is beneficial, and many disease states are characterized by reductions in microbial diversity. Another caveat is that the majority of studies of the "gut" microbiome are performed using stool samples which, although readily obtained, are not representative of the entire human gastrointestinal tract. As discussed in Chaps. 5 and 10, the small intestine plays central roles in the processes of nutrient absorption and metabolism, and differs from the large intestine in pH, transit time, microbial load, and viscosity of the luminal contents. It is therefore critical to sample and study small intestinal microbial populations directly, particularly for diseases and conditions such as diabetes and obesity. Lastly, although much touted, it should be noted that no probiotic, prebiotic, or synbiotic is as yet approved for the treatment of any gastrointestinal diseases or conditions, nor are they subject to the same regulatory oversight or scrutiny as "conventional" drugs.

The future of gut microbiome research will of necessity not be limited to continuing to characterize the bacterial, fungal, archaeal, and viral populations present in the "normal" gut vs. various disease states, although much work remains to be done to understand the nonbacterial components of the microbiome, particularly in the metabolically relevant small intestine. It will be critical to expand our understanding of how the gut microbiota and their products and metabolites influence and regulate the immune system, how they influence neurological pathways and processing, and potential impacts on other body systems as well as other microbiomes (oral, skin, pulmonary, vaginal, etc.). We also need to expand our understanding of how lifestyle choices affect the gut microbiome—we have made inroads in understanding the effects of diet, but are now beginning to include the effects of food additives, smoking, and environmental pollutants, to name but a few. It is also well established that antibiotic use affects the gut microbiome, and that early and repeated administration particularly during childhood can have long-term consequences, but there is work to be done for other drugs as well as prebiotics and other over-the-counter supplements, and even personal care products. Lastly, it is now emerging that even factors such as pet ownership can influence our microbiota, and exploring these avenues will doubtless yield fascinating results. With regard to manipulating and treating the microbiome, we in our group frame this as a three-pronged approach: (1) drugs for bugs, evolving from the classic approaches to treating pathogenic microbes, but targeting pathways/cell components unique to causal microbes without affecting commensal and other beneficial microbes; (2) drugs from bugs, i.e., using microbial metabolites and products to develop targeted therapeutics, as well as associated diagnostics/theranostics as appropriate; and (3) bugs as drugs, i.e., giving or engineering particular microbes (rather than transplanting a whole microbiome, as is done in fecal microbiota therapy (FMT)) to treat a disease or condition. Utilizing all of these approaches, or possibly combinations thereof, will allow us to take the vast breadth of knowledge we have already gained and which researchers are continuing to generate and to translate these exciting discoveries into tools to improve patients' lives.

Contents

Editors and Contributors

About the Editors

Mark Pimentel is the Executive Director of the Medically Associated Science and Technology (MAST) Program at Cedars-Sinai Medical Center in Los Angeles, California, and also serves as Professor of Medicine at the Geffen School of Medicine and Cedars-Sinai Medical Center. He completed 3 years of an undergraduate degree in honors microbiology and biochemistry at the University of Manitoba, Canada, followed by his medical degree and his BSc (Med) from the University of Manitoba Health Sciences Center, where he also completed a residency in internal medicine. His medical training includes a fellowship in gastroenterology at UCLA.

Dr. Pimentel is a clinician and researcher who has conducted numerous basic, translational, and clinical studies exploring the role of the gut microbiome in human health and disease. Before the term "microbiome" was coined, he worked to determine the role of bacteria in the pathophysiology of IBS. Based on his identification that small intestinal bacterial overgrowth (SIBO) contributes significantly to IBS symptoms, he pioneered the use of antibiotics, particularly rifaximin, in treating IBS. Recognizing links between acute gastroenteritis and the development of IBS, he demonstrated that antibodies produced against the bacterial toxin CdtB cross-react with the host protein vinculin, and developed the world's first diagnostic blood test for IBS. This work was crucial to proving that IBS is an organic, microbiome-based disease. Recently, he has focused on the specific roles of the small bowel microbiome in human disease, including identifying small bowel microbial populations that are altered in SIBO, and characterizing the distinct microbiome profiles along the entire length of the gastrointestinal tract.

His work has been published in the *New England Journal of Medicine*, *Annals of Internal Medicine*, *American Journal of Physiology*, *American Journal of Medicine*, *American Journal of Gastroenterology*, and *Digestive Diseases and Sciences*, among others. He is a diplomate of the American Board of Internal Medicine (Gastroenterology) and a fellow of the Royal College of Physicians and Surgeons of Canada.

Ruchi Mathur is the Director of Clinical Diabetes and Director of Clinical Research and Operations for the MAST Program at Cedars-Sinai Medical Center in Los Angeles, California. She also serves as Professor of Medicine at the Geffen School of Medicine and Cedars-Sinai Medical Center. She received her medical degree from the University of Manitoba, Canada. She completed an internal medicine residency at the Health Sciences Center at the University of Manitoba. She fulfilled a clinical endocrinology and metabolism fellowship at Cedars-Sinai Medical Center, followed by a NIH/NRSA-awarded research year at Cedars-Sinai Medical Center.

Dr. Mathur is an academic Endocrinologist, with a specific interest in the relationship between the gastrointestinal tract and the endocrine system. This includes a focus on polycystic ovary syndrome (PCOS), prediabetes, diabetes, and obesity, and she published the first paper confirming a clinical association with PCOS and IBS, which noted an increase in body mass index (BMI) in women with IBS. Since then, her research has focused on exploring the roles of the gut microbiome, and particularly the small bowel microbiome, in metabolic conditions such as obesity and diabetes. In addition to this work, she has also continued to explore the potential contribution of androgen production by gut microbes to the elevated hormone levels seen in conditions such as PCOS.

Dr. Mathur's work has been published in the *Journal of Clinical Endocrinology and Metabolism, Cell Reports*, *Digestive Disease Sciences*, *Frontiers in Microbiology*, *Journal of Hormone Research*, and *Diabetes Care*. She also coauthored the textbook *Davidson's Diabetes Mellitus: Diagnosis and Treatment* and contributed chapters to *Hacker and Moore's Essentials of Obstetrics and Gynecology, 5th Edition* and *D'Saia's Obstetrics and Gynecology Women's Health Review*. Dr. Mathur is a fellow of the Royal College of Physicians and Surgeons of Canada and belongs to academic societies including the American Association of Clinical Endocrinologists, Androgen Excess Society, American Diabetes Association, Endocrine Society, and Women in Endocrinology.

Gillian M. Barlow is a Project Scientist with the MAST Program at Cedars-Sinai Medical Center, where she also serves as an Assistant Professor of Medicine. She is a highly experienced molecular biologist, with over 20 years of experience. Her research interests include the roles of the gut microbiome in metabolism and weight gain, and in gastrointestinal conditions such as irritable bowel syndrome. Dr. Barlow obtained her BSc from Dublin City University in Ireland, where she majored in Genetics and Immunology, and subsequently earned her PhD in Molecular Biology from University College Dublin. Prior to joining the MAST Program, she was a Research Scientist in the Division of Molecular Neurogenetics (Dept. of Pediatrics) at Cedars-Sinai Medical Center, where she worked on the molecular genetics of Down syndrome, and later worked in the Division of Reproductive Endocrinology and Infertility (Dept. of Obstetrics and Gynecology) where she worked on female infertility and premature ovarian failure. Dr. Barlow has held grants from the American Heart Association and has been a co-investigator on grants from the National Institute of Child Health and Human Development.

Contributors

Gillian M. Barlow Medically Associated Science and Technology (MAST) Program, Cedars-Sinai Medical Center, West Hollywood, CA, USA

Aadra P. Bhatt Division of Gastroenterology and Hepatology, Department of Medicine, Center for Gastrointestinal Biology and Disease, Chapel Hill, NC, USA

Lineberger Comprehensive Cancer Center, University of North Carolina at Chapel Hill, Chapel Hill, NC, USA

Ian M. Carroll Department of Nutrition, University of North Carolina at Chapel Hill, Chapel Hill, NC, USA

Shreya Celly Medically Associated Science and Technology (MAST) Program, Cedars-Sinai Medical Center, West Hollywood, CA, USA

Gerard Clarke APC Microbiome Ireland, University College Cork, Cork, Ireland

Department of Psychiatry and Neurobehavioural Science, University College Cork, Cork, Ireland

John F. Cryan APC Microbiome Ireland, University College Cork, Cork, Ireland

Department of Anatomy and Neuroscience, University College Cork, Cork, Ireland

John A. Damianos Department of Gastroenterology and Hepatology, Mayo Clinic, Rochester, MN, USA

Michael Dougherty Division of Gastroenterology and Hepatology, Department of Medicine, University of North Carolina at Chapel Hill, Chapel Hill, NC, USA

Julia L. Drewes Division of Infectious Diseases, Department of Medicine, Johns Hopkins University School of Medicine, Baltimore, MD, USA

Marie Fogarty Department of Biological Sciences, Durham Technical Community College, Durham, NC, USA

Yi Ge Department of Nutrition, University of North Carolina at Chapel Hill, Chapel Hill, NC, USA

David Goldenberg Division of Gastroenterology, Department of Medicine, Stanford University, Palo Alto, CA, USA

Soo Chan Lee South Texas Center for Emerging Infectious Diseases (STCEID), Department of Molecular Microbiology and Immunology, University of Texas, San Antonio, San Antonio, TX, USA

Jintong Liu Department of Nutrition, University of North Carolina at Chapel Hill, Chapel Hill, NC, USA

Caoimhe M. K. Lynch APC Microbiome Ireland, University College Cork, Cork, Ireland

Department of Anatomy and Neuroscience, University College Cork, Cork, Ireland

Ruchi Mathur Medically Associated Science and Technology (MAST) Program, Cedars-Sinai Medical Center, West Hollywood, CA, USA

Division of Endocrinology, Diabetes, and Metabolism, Department of Medicine, Cedars-Sinai Medical Center, Los Angeles, CA, USA

Gil Y. Melmed Karsh Division of Gastroenterology and Hepatology, Department of Medicine, Cedars-Sinai Medical Center, Los Angeles, CA, USA

Mark Morrison University of Queensland Frazer Institute, Faculty of Medicine, and NHMRC Centre for Digestive Health, University of Queensland, Saint Lucia, QLD, Australia

Nida Murtaza University of Queensland Frazer Institute, Faculty of Medicine, University of Queensland, Saint Lucia, QLD, Australia

Renay Ngobeni-Nyambi Department of Environmental, Water and Earth Sciences, Tshwane University of Technology, Arcadia Campus, Pretoria, South Africa

Kenneth J. O'Riordan APC Microbiome Ireland, University College Cork, Cork, Ireland

Mark Pimentel Medically Associated Science and Technology (MAST) Program, Cedars-Sinai Medical Center, West Hollywood, CA, USA

Karsh Division of Gastroenterology and Hepatology, Department of Medicine, Cedars-Sinai Medical Center, Los Angeles, CA, USA

Ali Rezaie Medically Associated Science and Technology (MAST) Program, Cedars-Sinai Medical Center, West Hollywood, CA, USA

Karsh Division of Gastroenterology and Hepatology, Department of Medicine, Cedars-Sinai Medical Center, Los Angeles, CA, USA

Mark S. Riddle University of Nevada, Reno School of Medicine, Reno, NV, USA

Eric D. Shah Department of Internal Medicine, University of Michigan, Ann Arbor, MI, USA

Will Takakura Medically Associated Science and Technology (MAST) Program, Cedars-Sinai Medical Center, West Hollywood, CA, USA

Department of Medicine, Cedars-Sinai Medical Center, Los Angeles, CA, USA

Nicholas J. Talley School of Medicine and Public Health, and NHMRC Centre for Digestive Health, College of Health, Medicine and Wellbeing, University of Newcastle, Callaghan, NSW, Australia

The Importance of the Microbiome in the Gut

Mark Morrison, Nida Murtaza, and Nicholas J. Talley

Highlights

- The globally rising incidence and prevalence of digestive diseases and disorders emphasizes the gut microbiota remains a relatively poorly characterized but clinically relevant modulator of digestive health.
- The gut microbiota is most dynamic early in life, impacted by a myriad of factors including mode of delivery, the duration of breastmilk and/or milk formula feeding, post-weaning dietary patterns and antibiotic use.
- In later life, depletion in the relative abundance of Gram positive, butyrate-producing bacteria is widely accepted to be a hallmark sign of digestive diseases and disorders. Particularly, *Faecalibacterium prausnitzii* has emerged as an important microbial biomarker of digestive health, with demonstrated anti-inflammatory effects.
- Faecal microbiota transplantation is now an established highly efficacious therapy for chronic *Clostridioides difficile* infection, and with promise for the treatment of inflammatory bowel disease. Research has also revealed the gut microbiota can serve as a key modulator of tumourigenesis and the efficacy of some cancer immunotherapies.
- Translation of "gut microbiome" research into medicine requires an ongoing commitment to its transition into "gold standard" clinical studies and trials, emphasizing baseline measurements, longitudinal assessment of subject response(s) and key measures of diet and gut motility.

M. Morrison
University of Queensland Frazer Institute, Faculty of Medicine, and NHMRC Centre for Digestive Health, University of Queensland, Saint Lucia, QLD, Australia
e-mail: m.morrison1@uq.edu.au

N. Murtaza
University of Queensland Frazer Institute, Faculty of Medicine, University of Queensland, Saint Lucia, QLD, Australia

N. J. Talley (✉)
School of Medicine and Public Health, and NHMRC Centre for Digestive Health, College of Health, Medicine and Wellbeing, University of Newcastle, Callaghan, NSW, Australia
e-mail: nicholas.talley@newcastle.edu.au

1 Introduction

The recognition that stool has "medicinal" properties dates back to early Chinese medicine [1, 2], which predates Leeuwenhoek's discovery that microbes colonize and inhabit various sites of the human body in the 1600s [3]. From the nineteenth Century, the application of Koch's postulates have enabled a sustained trajectory of advances in the diagnosis and treatment of infectious diseases, as exemplified by Marshall's direct application of these postulates, as part of his collaborations with Warren to establish the role of *Helicobacter pylori* as the causative organism for gastritis and peptic ulcer [4]. Theodore Escherich was inspired by Koch during his research into neonatal gut microbiology in the late 1800s, which included the isolation and characterization of *Eschericha coli* and other intestinal bacteria [5]. In his tribute to some of the pioneering scientists of gut microbiota research, Savage [6] refers to Escherich's recognition that although the "intestinal microflora" appears to be brought about by a conglomeration of random events and coincidences, these communities play an important and deterministic role in digestion, health, and disease. However, studies of the "intestinal microflora" were largely overlooked in lieu of the "germ theory" of medical microbiology until the mid-twentieth century, when the collective works of a relatively small number of scientific teams across the globe reinvigorated the field, enabled by their development and use of rodent models and the advent of techniques for the cultivation of fastidious anaerobic bacteria [6]. Since that time, our recognition of the goods and services that the "commensal" bacteria provide, both independently and as members of a larger community, via the immune, neuroendocrine, and

M. Pimentel et al. (eds.), *Clinical Understanding of the Human Gut Microbiome*, https://doi.org/10.1007/978-3-031-46712-7_1

other systems which are central to host physiology and metabolism, has grown dramatically [7–10]. In particular the use of germ-free (i.e. microbiologically naïve) and/or gnotobiotic (defined mixtures of microbes colonizing the gastrointestinal tract) rodent animal models has expanded dramatically during the last decade, in no small part due to the ground-breaking studies of Gordon and colleagues with both specific microbes [11, 12] and later, mouse and human faecal microbiota transplants (FMT) demonstrating transference of an "obese" or "lean" phenotype to naïve mice [13–15]. Similar approaches have advanced our awareness and understanding of the roles provided by commensal bacteria in resistance to pathogen colonization and the impacts of antibiotics on this process (e.g., [16, 17]), and the use of germ-free rodent animal models to explore the cause-consequence relationships of the gut microbiota in health and disease is rapidly expanding (e.g., [18]). However, and despite the potential associated with using germ-free rodent animal models, caveats and pitfalls also exist. In that context, the systematic review and analysis by Walter et al. [19] not only queries and provides some rationale to explain the (surprisingly) high success rate in the transferance of phenotype via faecal microbiota transfer to germ-free rodent animal models, but also provides strategies and approaches that can enhance the rigour and conclusions that can be drawn from such studies.

The expansion of the study of the "intestinal microflora" has also been driven by the steady and rapid expansion in the scale of its assessment and description, driven in large part by advances in molecular and genomic biology via PCR and DNA/RNA sequencing technologies, as well as the use of various computational and statistical approaches, collectively referred to as bioinformatics, to characterize, compare, and contrast these datasets. Two new terms to the (preferred) exclusion of "microflora" have become commonplace in the literature: "microbiota" in holistic reference and recognition of the biologically extant life forms resident in these communities; and "microbiome" which typically refers to the collective genomic content recovered from the microbiota. Quigley [20] has provided a recent overview of the basic concepts and principles of microbiome research from a clinician's perspective. The salient feature of these approaches are that they can be used in a culture-independent way, not reliant on what first grows in the laboratory, but on the ability to extract microbial DNA from real-world specimens (tissue, stool, sputum, blood, etc.). These technological advances initially supported the use of individual genes such as those encoding ribosomal RNA (rRNA) to produce taxonomy-based "16S rRNA profiles" and sequencing of the entire metagenome of these communities via "shotgun metagenomic sequencing" [21]. The sustained trajectory of technological advances in nucleic acid sequencing platforms [22] includes enhancements in time- and cost-effective enhancements in sequencing depth, which have further supported the extension of these approaches to augment these datasets beyond the prokaryotes (i.e. bacteria and some methane-producing Archaea) to include yeast and fungi (the mycobiome), protists (simple distinct eukaryotic organisms), as well as their respective viromes [23–26].

While the term "microbiome" is still commonly used to refer to the collective genome-based content (and its analysis) of the microbial communities of different habitats [20], a concerted effort has been made to revise this definition to emphasize the biomic (rather than -omic) elements of the term "microbiome" [27]. For simplicity, we define microbiome research as the holistic study of the interactions and processes inherent to a <u>microbial community</u> (micro-), placed in context with the physiochemical attributes and/or biological inputs of their <u>surrounding environment</u> (-biome). To that end, and following on from the US-led Human Microbiome Project [28] and EU-led Metagenomics of the Human Intestinal Tract (MetaHIT, [29]), the National Institutes of Health have supported several clinically-led and oriented integrated human microbiome projects (iHMP) that have also utilized other analytical platforms for protein, metabolite and immune cell analyses to produce a "multi-omic" assessment of the host-microbe interactions relating to pregnancy and birth, inflammatory bowel disease, and (pre)diabetes [30], and similar activities have been supported in parallel across the European Union [31].

The intent of this Chapter is to provide some specific examples of the importance of microbiome research to address some of the long-held concepts and key knowledge gaps pertaining to the importance of the gut microbiota in health and disease. To that end, we will consider how innovations in tissue sampling, the analysis and integration of microbiome-derived datasets with other objective measures of clinical and psychometric assessments routinely collected in clinical studies, including faecal microbiota transplantation, are offering new insights and opportunities to establish the cause-or-consequence relationships between our gut microbiota with health and disease.

2 The Importance of the Gut Microbiome in Early Life

The microbial world is thought to establish its interrelationships with the human body *in utero* via the placenta. While the colonization of the placenta *per se* remains controversial [32, 33] the developing foetus interacts with the maternal microbiota via their metabolic outputs, such as short chain fatty acids (SCFAs) (e.g., [34]). The mode of delivery (vaginal birth vs. cesarean delivery), isolation from the mother during the pueperal period, and/or early feeding practices (e.g. breastfeeding vs. formula) are well known to help shape the gut microbiota in early life [35]. Natural delivery (ND) provides the first fecal microbiota transplant but a granular understanding of the dynamics of microbial colonization, persistence, and ecological succession in the neonatal period has remained elusive. The application of high throughput sequencing technologies with both paired stool samples from mothers and their babies, as well as longitudinal studies of almost 600 full-term newborns and infants by Shao et al. [36] has now revealed the consequences of caesarean delivery (CD) on the stool microbiota during the neonatal period, and the persistence of these communities into infancy. Most notably, lineages of genus *Bacteroides* are depleted and Gram positive and Gram negative "opportunistic pathogens" are more prevalent in CD infants compared to ND infants. Interestingly, the stool microbiota of those ND infants from mothers receiving antibiotics prophylactically during childbirth and/or ND infants that are formula-fed also bore similarities with CD infants. The translational potential of these findings is reinforced by the recent pilot study by Korpela et al. [37] that showed full-term CD infants rapidly develop a microbiota more closely resembling that of ND infants, and particularly lineages of *Bacteroides* and *Bifidobacterium,* when provided with maternal faecal microbiota added to the breastmilk first consumed by the newborn.

In summary then, there are changes in the maternal gut microbiome throughout pregnancy, and the transmission (or not) of the maternal microbiota during and after child delivery in both full- or pre-term pregnancies, have now been better characterized. There is also evidence that at least some of the maternal faecal microbiota can be transferred and persist in full-term infants via their post-natal oral ingestion in breast milk. This knowledge offers new opportunities to overcome the microbial "dysbiosis" associated with pre-term birth, mode of delivery, and antibiotic use in early life, and to restore an ecological succession that promotes gut homeostasis and immune system development in these infants.

3 The Mucosa-Associated Microbiome in Disorders of Gut-Brain Interaction (DGBI) and Inflammatory Bowel Diseases (IBD)

The advent of microbiome research has seen an extensive literature of the links between the gut microbiome and the onset and/or manifestations of irritable bowel syndrome (IBS, e.g., [38–41]) and inflammatory bowel diseases (IBD, e.g., [42, 43]). In brief, microbial "dysbiosis" is a hallmark of different manifestations of both IBS and IBD, and is now characterized not just in terms of bacterial taxa, but also in terms of alterations in the relative abundance, prevalence, and/or diversity of archaea, lower eukaryotes, and their respective viromes (e.g., [26, 44–47]). With the expansion (and depth) of the available datasets for comparative analyses, progress has been made in developing faecal microbial-based biomarkers that can discriminate between IBS severity [39], as well as between Crohn's disease (CD) and "non-CD" forms of IBD [42]. Additionally, changes in the gut microbiota have been observed coincident with IBS symptom reduction by the low FODMAP diet (e.g., [48–50]) and with the improved rate of mucosal healing (compared to corticosteroids) associated with exclusive enteral nutrition as induction therapy in new onset pediatric Crohn's disease (e.g., [51–54]).

In relative terms, studies of the mucosa-associated microbiome in IBD and IBS, as well as diseases and disorders of the upper GI tract are relatively scant. Gevers et al. [55] examined the microbial communities resident on ileal and rectal biopsies of treatment-naïve and new onset CD patients. The ileal and rectal mucosal communities were more similar to each other than their matched stool samples, and provided a superior disease classification compered to stool samples. At least part of the improved classification relates to the expansion of bacterial taxa associated with active disease to include members of the Neiserriaceae, Veillonellaceae, and Fusobacteriaceae, perhaps better recognized for their prevalence in the oral cavity; and reductions in bacterial taxa affiliated with the Clostridiales, Erysipelotrichaceae, and Ruminococcaceae, including *Blautia, Coprococcus,* and *Dorea,* compared to non-CD subjects [55]. A small number of microbiome studies have also been undertaken using tissue biopsies collected in longitudinal studies of CD patients at surgery and post-operatively [42, 56, 57] as well as multi-omics methods applied to stool, tissue, and other samples collected at multiple timepoints over a 12 month period, as part of the iHMP project in IBD [58, 59]. Collectively, these findings further established the relevance of the restoration and/or persistence of mucosa-associated populations of *Faecalibacterium prausnitzii* as a microbial biomarker of CD remission, which is also consistent with the conclusions drawn using stool microbiome profiles (e.g., [42, 60]). In that context, Langella, Sokol, and colleagues have been at the forefront establishing the repertoire of peptides and metabolites, in addition to butyrate, which are produced by this bacterium and implicated with promoting anti-inflammatory effects via suppression of NF-κB-modulated pathways, as well as improving gut epithelium integrity (e.g., [57, 61, 62]). These studies augment those by Honda, Atarashi and colleagues that have recovered a collection of bacterial strains affiliated with Clostridiales lineages with the capacity to induce expansion of Treg cell populations, as well as anti-inflammatory cytokines such as IL-10 [63, 64]. In a similar vein, these type of microbiome studies have also revealed key bacterial taxa, such as *Proteus mirabilis* increase the risk of CD recurrence [65–67], which is consistent with the recent establishment that bacterial urease activity can promote the expansion of Proteobacteria and inflammation in animal models of immune-mediated colitis [68].

In summary, despite the advances from microbiome research in understanding the microbial dynamics associated with IBS and IBD, the cause-or-consequence relationships between changes in the microbiome and disease course are not still not clearly established. However, a combination of longitudinal and culture-based studies of the microbiome have revealed candidate bacterial taxa not only involved with inflammation, but also those relevant to the maintenance and/or restoration of gut homeostasis. The key knowledge gaps that remain are not so much the molecular factors at play in immunomodulation from these microbes, but rather, the nature of the ecological characteristics affecting the growth and persistence of key "anti-inflammatory" microbes and restoration of gut homeostasis, as opposed to the commensurate expansion of pro-inflammatory bacteria in IBD "dysbiosis". In that context, the interest in diet both as a potential trigger of disease and as a therapeutic intervention (e.g. [69–73]), as well as modulation of the gut microbiome via faecal microbiota transplantation as outlined below, are of growing interest.

In addition to IBS and IBD, functional dyspepsia (FD) is a disorder of gut-brain interaction (DGBI) which overlaps with IBS and is a common reason for medical consultations throughout the world [74] but there is a paucity of specific diagnostic tests for these conditions, and tests that allow targeted therapy. Despite the development of the international ROME diagnostic criteria for FD [75], both diagnostic and treatment approaches remain symptom-based and do not account for the pathophysiology in a given patient. Indeed, symptom management is arbitrary and subject to frequent treatment failures (e.g. [76]). Mucosal inflammation in the duodenum of FD patients is considered a key trigger of patient symptoms, via modulation of epithelial cells, smooth muscle, and enteric nerves [77–80]. Moderate eosinophilia is observed in some FD patients along with altered duodenal feedback and gastric emptying; as well as the involvement of mast cells in IBS [81, 82]. There is also an increase of small bowel homing CD4+ $\alpha4\beta7$+ CCR9+ lymphocytes, and increased secretions of IL-1β and TNF-α from peripheral blood mononuclear cells (PBMC, [83]). Cytokine activation of the hypothalamic-pituitary-adrenal axis further suggests that immune activation in the gut modulates psychological dysfunction in FD rather than solely vice versa [84].

Collectively, these data strongly suggest that immune dysregulation in the duodenum is a key mechanism that promotes changes in motility, visceral sensitivity, intestinal permeability, and extraintestinal disease manifestations such as anxiety and depression, and consequently may generate symptoms in FD. However, the triggers of this immune activation remain largely unknown and discovery of specific diagnostic biomarkers of FD and its subtypes remain elusive. In comparison to the ileocolonic microbiome, the role and relevance of the duodenal mucosa-associated microbiome (d-MAM) in health and disease is virtually unexplored, and is a key focus of efforts currently underway via the Australian Gastrointestinal Research Alliance (AGIRA, Fig. 1). In particular, our approach employs a patented biopsy device whereby the forceps reside within a sterile sheath and extend only for tissue collection [85]. The loaded forceps are retracted immediately, ensuring the sample is protected from contamination by extraneous materials (e.g. saliva, digesta) and only contains those microbes intimately associated with the tissue. The DNA extracted directly from this biopsy sample can then be used for community of the d-MAM using PCR amplicons produced from the gene encoding 16S rRNA; as well as for measuring bacterial density (load) on the tissue via qPCR, as described by Shanahan et al. [86]. Using this device, our pilot case-control study [87] showed there were significant differences in the d-MAM profiles between FD and asymptomatic, non-FD control subjects. Furthermore, the bacterial load on FD patient duodenal tissue was positively correlated with their symptom response following a standardised nutrient challenge test and negatively correlated with their meal-related quality of life scores using the Nepean Dyspepsia Index. These latter findings have since been expanded by our team [88] to show that d-MAM load is greatest in patients diagnosed with a DGBI compared to asymptomatic control subjects, and both these subject groups possess significantly greater loads in comparison to subjects with IBD. Interestingly, while subjects using proton pump inhibitors (PPI) possessed the greatest d-MAM loads, PPI use was not solely responsible for the differences between DGBI and asymptomatic control subjects. Similar also to the results of the pilot study by Zhong et al. [87] there was a significant negative correlation between d-MAM load and subject symptom response to standardised nutrient challenge test.

In summation, these results show that microbial "dysbiosis" may not simply be explained by alterations in microbial diversity but also in terms of the absolute abundance of the microbiota at various sampling points. Indeed, Raes and colleagues have recently developed methodology that supports "quantitative microbiome profiling" of stool samples, which has advanced our understanding of the "enterotype" concept of the stool microbiome, as well as provided greater differentiation of the absolute as well as compositional changes in the stool microbiota associated with IBD either as the sole disease, or as a co-morbidity with primary sclerosing cholangitis [89, 90].

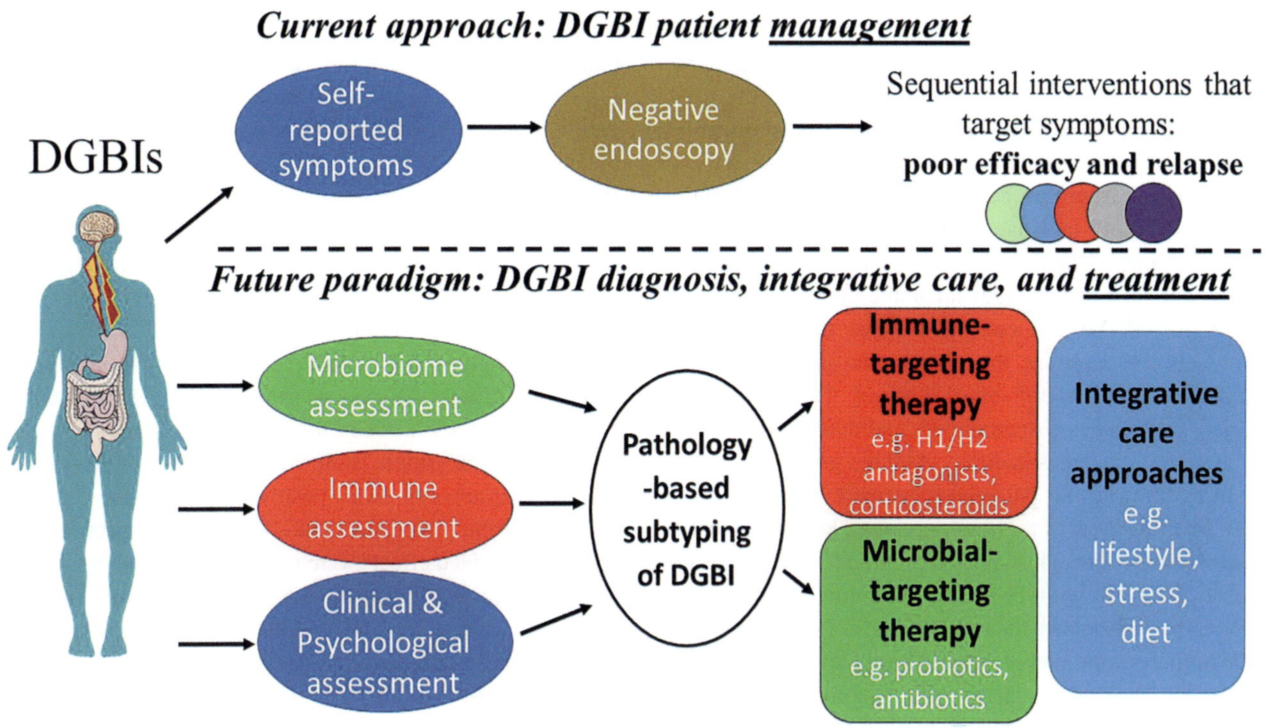

Fig. 1 The Australian Gastrointestinal Research Alliance approach to translating the gut microbiome into medicine for Disorders of Gut-Brain Interaction (DGBI). The microbiome is a key knowledge gap to advance DGBI from symptom-based syndromes into specific diagnostic and treatable entities, and the adoption of a truly multidisciplinary, integrative care approach

4 The Gut Microbiome and Cancer

Gut microbes, either individually or as a community, can be a key modulator of the processes leading to the resilience or susceptibility of host cell transformation into cancer, and thereby of interest in advancing their diagnosis and treatment [91]. Meng et al. note both specific virulence factors restricted to a limited number of bacterial taxa (e.g. *H. pylori* and *Fusobacterium nucleatum*) as well as the microbial "dysbiosis" that have been linked with various gastrointestinal cancers [92]. Recent advances in DNA sequencing methods have now provided greater insights into how the gastric microbiome is altered in subjects with either *H. pylori*-associated gastritis or gastric carcinoma from cohorts in Portugal, China, and Mexico. Interestingly, the relative abundance of *Helicobacter* spp. are reduced in subjects with gastric carcinoma, displaced by "non-*Helicobacter* Proteobacteria" [93]. In similar studies with Chinese subjects, Coker et al. [94] identified key bacterial taxa that were discriminatory of the gastric microbiomes associated with gastric carcinoma from superficial gastritis, and in particular, five bacterial taxa enriched in gastric carcinoma patients that could be used as biomarkers for this condition, which were validated by independent analysis of samples from Mongolian subjects. These datasets have provided important insights into disease progression leading to gastric cancer and their potential diagnostic utility for an improved discrimination of the gastric microbiomes associated with these specific conditions. However, a recent systematic review of select taxonomy-based datasets of the gastric microbiome showed that, with the notable exception of the impacts from *H. pylori* on the gastric microbiota, and the effects of PPI use being associated with an increase in bacterial richness, microbiome profile(s) discriminatory of gastric health and disease could not yet be developed [95] from these taxonomy-based assessments of the microbiome.

In contrast shotgun metagenomic sequencing (MGS) of the stool microbiome, which provides both deep taxonomic and functional information, has shown promise in terms of diagnostic utility for colorectal cancer (CRC). In a case-control study of CRC with Chinese subjects Yu and colleagues [96] identified more than two million genes shared across the MGS datasets from six or more subjects. From this composite dataset, they were able to extract 20 gene markers discriminatory for the case and control groups, and their combined log abundances produced a "CRC index" that clearly separates the CRC and healthy subjects of this cohort. The specificity of this CRC index was confirmed by using other MGS datasets from case-control studies of Type 2 diabetes and IBD including Chinese subjects. More detailed bioinformatics analyses validated that four of these marker genes can differentiate the stool metagenomes from CRC and control subjects in European studies, and qPCR analyses performed with a second cohort of Chinese subjects validated that two marker genes could be used to detect early stage CRC. These studies were augmented by the identification of seven bacterial species universally enriched in the CRC datasets from European, Chinese and North American cohorts [97] and in Asian subjects, these approaches have identified a novel *Lachnoclostridium* genome as a candidate biomarker for pre-cancerous adenoma compared to control subjects [98]. Yu and colleagues have now used similar approaches to examine the mycobiome present in deeply sequenced stool microbiomes [24] of CRC and control subjects, which has also produced candidate fungal biomarkers for CRC diagnosis in different ethnic groups.

In addition to the potential diagnostic utility associated with MGS datasets of the gut microbiota, key ecological and microbiological insights are also provided relating to microbial involvement with the biology of colorectal and perhaps other types of cancer. Importantly, these types of data provide higher resolution (i.e. species/strain level) bacterial and/or fungal networks within the gut microbiomes associated with health and disease, and the keystone species that "hold" these subcommunities together. Such keystone species thereby offer critical control points to either promote protective or thwart deleterious microbial networks linked with the transition from precancerous to cancerous states. For instance, knowledge of these multispecies networks now offer the opportunity to construct defined consortia that can be tested in preclinical animal models, with a view of establishing their protective or deleterious roles in carcinogenesis. In that context, Münch et al. [99] have used a mouse model of Barrett's oesophagus to show that not only does a high fat diet promote changes to the stool microbiota and the development of an inflammatory milieu in oesophageal tissue that accelerates carcinogenesis, but the exposure of their fecal microbiota to animals fed the control diet via animal bedding resulted in tumour development as seen in mice fed a high fat diet. In contrast, germ-free mice were found to possess reductions in the inflammatory milieu and dysregulated cell growth seen in conventionally raised mice, further establishing the links between the gut microbiome, inflammation, and carcinogenesis.

The studies outlined above are an extension of the links now being established between the gut microbiome, cancer immunity, and the efficacy of cancer treatments, particularly immunotherapies [8, 100–102]. This field of research was initially fueled by observations by Zitvogel that in germ-free or antibiotic-treated mice, cyclophosphamide was less effective as a chemotherapy compared to it used in conventionally raised mice [103]. Zitvogel used these observations to postulate the differences were attributable to a loss of microbial-derived stimuli affecting the immune system, and in turn, the susceptibility of cancer cells to the inhibitory effects of immunotherapies with variable efficacy, such as the checkpoint inhibitor anti-CTLA4. Her team later confirmed the limited efficacy of this immunotherapy in germ-free and antibiotic-treated mice, and based on the T cell activation in patients responsive to anti-CTLA4, either inoculated germ-free mice with *Bacteroides fragilis,* or immunized mice with *B. fragilis* polysaccharides, both of which promoted the capacity of this immunotherapy to thwart melanoma tumourigenesis [103, 104]. In parallel, Gajewski and colleagues found there were differences in the degree of spontaneous melanoma anti-tumour activity in genetically similar mice from two different sources and subsequently determined that of the differences between the stool microbiomes of these two groups of mice, only the relative abundance of *Bifidobacterium* spp. was positively associated with the amount of tumour-specific CD8+ T-cells. These presumptive microbial-derived effects were transferrable to other mice by cohousing, fecal transfer, or oral gavage of *Bifidobacterium.* Moreover, treatment of mice with *Bifidobacterium* accentuated the antitumor activity provided via the checkpoint inhibitor anti-PD-1 [103, 105]. These collective findings have prompted more extensive characterizations of the stool microbiome from cancer patients deemed responsive or non-responsive to anti-PD-1 immunotherapy for treatment of epithelial tumours, melanoma, and metastatic melanoma in the three distinct but complementary studies in France and the USA ([106–108], respectively). All three studies showed there are key bacterial taxa that differentiate between the patients deemed responders or non-responders to anti-PD-1 immunotherapy. Each study also found that faecal microbiota transfers from their responder patients to their respective germ-free and/or antibiotic treated mouse models enhanced the antitumour effects of the anti-PD-1 treatment, whereas stool from their non-responder groups did not. However, the studies do show some key differences with respect to those specific bacterial taxa highlighted by each team as being discriminatory of the stool microbiome of the responder groups: being unclassified Gram positive Firmicutes lineages, and in particular *Akkermansia* spp. [106], Ruminococcaceae and in particular *Faecalibacterium* spp. (and Bacteroidales in non responders, [107]), and multiple Gram positive bacteria from the phylum Actinobacteria *(Bifidobacterium longum* and *Colinsella aerofaciens)* as well as *Enteroccoccus faecium* [108]. Interestingly, Honda and colleagues [109] have used a culture-based approach to recover a consortium of 11 bacterial strains, principally comprised of "low abundance" bacterial taxa that collectively promote the relative amount of intestinal IFNγ(CD8+) T-cells. When the 11 strain consortium was used to inoculate germ-free mice engrafted with murine adenocarcinoma cells, anti-tumor immunity was enhanced, and the efficacy of anti-PD-1 treatment was substantially improved relative to germ-free mice, or mice treated with a different 10-strain consortium of microbes with limited capacity to promote intestinal IFNγ(CD8+) T-cells. In summary, while the differences in terms of the key bacterial taxa prognostic for patient response (or non-response) to immune checkpoint inhibitors such as anti-PD-1 remain enigmatic and necessary to understand, the collective findings establish the importance of the gut microbiome in modulating cancer immunity and treatment efficacy, and hold real promise for the translation of microbial augmentation of immunotherapies into more personalised and efficacious treatment for at least some cancer patients.

5 Faecal and Next Generation Microbiota Transplantation

As the preceding sections show, faecal microbiota transplantation (FMT) has emerged as a key approach for establishing proof-of-concept of phenotype transfer by samples of the gut microbiome from human or animal stool samples to naive preclinical animal models. The utilization of FMT in such studies seems to have increased in parallel with the interest and evaluation of FMT in a clinical setting, currently for the treatment of recurrent *Clostridioides difficile* infections (rCDI), and its evaluation for the treatment of IBD, and other idiopathic conditions where the gut microbiome may be implicated (e.g., [110, 111]). A number of systematic reviews have recently been published solely concerning the approaches used for FMT and their impact of primary outcomes for the treatment of rCDI [112, 113]. In general terms FMT efficacy is high and adverse events low, irrespective of the type of stool preparation and/or mode of delivery used (e.g., [114]), although there are contradictory reports in terms of recommending either nasogastric tube or colonoscopy. The growing interest in the use of FMT for the treatment of IBD and other conditions is reflected in the number of systematic reviews recently published on the topic (also discussed in chapter "Fecal Transplant: The Benefits and Harms of Fecal Microbiota Transplantation"). A systematic review and meta-analysis including 53 studies (41 in UC, 11 in CD, 4 in pouchitis) revealed that the pooled proportion of patients achieving clinical remission was 33% for UC and 52% for CD [79]. Narula et al. [115] reported that the four RCTs of FMT for treatment of UC showed FMT overall was superior to placebo in terms of inducing short-term clinical and endoscopic remission. These specific findings are similar to those of Costello et al. [116], who also reported a similar rate of clinical remission to FMT from cohort studies. Encouragingly, a pilot RCT for CD by Sokol et al. [117] suggests that compared to sham treatment, FMT appears to improve rate of steroid-free clinical remission at 10 and 24 weeks, and also results in improved endoscopic scores of CD severity at 6 weeks. In contrast, the systematic reviews of the small number of randomized control trials of FMT for treatment of IBS and obesity have found mixed results in terms of improvement relative to placebo (autologous stool) and less information about longer-term benefits. These systematic reviews as well as others that specifically focus on FMT and adverse events [118, 119] indicate these are uncommon and generally mild, and that the use of FMT for treatment of rCDI, by colonoscopy or capsules, is more cost-effective than other treatment options (e.g., [120]).

Interestingly, while all these systematic reviews emphasize the importance of the gut microbiome cannot be understated with respect to the safety, efficacy, and persistence of benefits associated with FMT, there are still key knowledge gaps in our understanding as it applies to FMT, as well as the development of more defined and perhaps, personalized modalities that eliminate the use of faecal material *per se*. For instance, the systematic review and metanalysis of 168 FMT studies by Lai et al. [114] provides some key insights into the knowledge gaps associated with the global utilization of FMT for rCDI and IBD in particular. First, there appears to be some key differences between the application of FMT in Western and Eastern countries, particularly with respect to the frequency and types of screening criteria used for donor selection. As noted by Lai et al. [114], exclusion criteria should not only address positivity for infectious organisms and diseases, antibiotic use, etc., but also consider how various medications (e.g. PPIs) or historic infections (e.g. *H. pylori* seropositivity in non-Western countries) might impact the donor microbiota. The current era of microbiome research, where deep sequencing of the donor and recipient stool is both time- and cost-effective, and will provide both taxonomic and functional information of the donor samples, offers new and improved opportunities to standardize the selection criteria and safety of donor stools. As the application of FMT continues to expand globally, guidelines for and registries of FMT recipients (e.g., [121, 122]) also need to expand with global input and consensus.

Such data also offers synergies to improve our basic understanding of the bacterial and non-bacterial (i.e. archeal, eukaryote, and virome) components of stool on treatment efficacy (and safety). As such, it is somewhat surprising to note that of 53 cohort and RCT studies reviewed by Paramsothy et al. [123], only 24 provide some details pertaining the microbiome diversity of the donor stool, as well as the recipient post FMT. In addition to the points raised above, the relevance of this knowledge gap is emphasized in part by their sub-group analyses of the four RCTs of FMT for UC, which suggests that remission rate is improved in patients receiving multiple FMTs and lower GI infusions, suggesting engraftment is affected by site and frequency of microbiota administration. Separately Paramsothy et al. [124] provide further insights suggesting that donor microbiome composition as currently measured by diversity metrics is an important factor affecting treatment efficacy for UC. For instance, infusates prepared from multiple donors increased its bacterial diversity and improved treatment efficacy, particularly in those recipients with relatively low bacterial diversity scores at time of treatment. Interestingly, while Moayeddi et al. [125] overall showed no improvement in clinical remission from FMT compared to placebo, their results also reveal infusates prepared from individual donors introduce variability in response. Indeed, those infusates prepared from "Donor B" possessing both greater microbial diversity and efficacy in treatment.

In summation, there are key knowledge gaps of the gut microbiome that need to be addressed to improve treatment efficacy. There is a need to provide a more complete quantitative and compositional assessment of the donor stool, including a more detailed assessment of the ratio of viable to non-viable microbes present in the infusates. How might the recipient microbiome determine treatment efficacy and outcome? An improved understanding of the gut microbiome of the recipient at baseline is also required, not unlike to what is now being employed to stratify cancer patients with respect to their responsiveness to immunotherapies. To what extent is donor microbial engraftment necessary for inducing remission? In that context, Sokol et al. [117] reported that donor microbiome engraftment at 6 weeks following FMT, as measured using 16S rRNA gene amplicon profiling, was not reached in their pilot study of CD. However, their deeper metagenomic sequencing of the microbiome samples collected longitudinally from this cohort was far more informative [126], and has provided some new insights in terms of species/strain levels displacement during FMT, and links of specific taxa with clinical measures of remission. Additionally, the application of other "omics" approaches, and particularly metabolomics in RCTs of FMT for IBD will advance the development of this method to be more safe, efficacious, and longer lasting in their effects.

References

1. Du H, Kuang T-T, Qiu S, Xu T, Gang Huan C-L, Fan G, et al. Fecal medicines used in traditional medical system of China: a systematic review of their names, original species, traditional uses, and modern investigations. Chin Med. 2019;14:31.
2. de Groot PF, Frissen MN, de Clercq NC, Nieuwdorp M. Fecal microbiota transplantation in metabolic syndrome: history, present and future. Gut Microbes. 2017;8(3):253–67.
3. Corliss JO. A salute to Antony van Leeuwenhoek of Delft, most versatile 17th century founding father of protistology. Protist. 2002;153(2):177–90.
4. Kyle RA, Steensma DP, Shampo MA. Barry James Marshall 2014; discovery of *Helicobacter pylori* as a cause of peptic ulcer. Mayo Clin Proc. 2016;91(5):e67–8.
5. Hacker J, Blum-Oehler G. In appreciation of Theodor Escherich. Nat Rev Microbiol. 2007;5(12):902.
6. Savage DC. Microbial biota of the human intestine: a tribute to some pioneering scientists. Curr Issues Intest Microbiol. 2001;2(1):1–15.
7. Grenham S, Clarke G, Cryan JF, Dinan TG. Brain-gut-microbe communication in health and disease. Front Physiol. 2011;2:94.
8. Fessler J, Matson V, Gajewski TF. Exploring the emerging role of the microbiome in cancer immunotherapy. J Immunother Cancer. 2019;7(1):108.
9. Farzi A, Fröhlich EE, Holzer P. Gut microbiota and the neuroendocrine system. Neurotherapeutics. 2018;15(1):5–22.
10. Schroeder BO, Bäckhed F. Signals from the gut microbiota to distant organs in physiology and disease. Nat Med. 2016;22(10):1079–89.
11. Hooper LV, Wong MH, Thelin A, Hansson L, Falk PG, Gordon JI. Molecular analysis of commensal host-microbial relationships in the intestine. Science. 2001;291(5505):881–4.
12. Stappenbeck TS, Hooper LV, Gordon JI. Developmental regulation of intestinal angiogenesis by indigenous microbes via Paneth cells. PNAS. 2002;99(24):15451–5.
13. Turnbaugh PJ, Ley RE, Mahowald MA, Magrini V, Mardis ER, Gordon JI. An obesity-associated gut microbiome with increased capacity for energy harvest. Nature. 2006;444(7122):1027–31.
14. Turnbaugh PJ, Ridaura VK, Faith JJ, Rey FE, Knight R, Gordon JI. The effect of diet on the human gut microbiome: a metagenomic analysis in humanized gnotobiotic mice. Sci Transl Med. 2009;1(6):6ra14.
15. Ridaura VK, Faith JJ, Rey FE, Cheng J, Duncan AE, Kau AL, et al. Gut microbiota from twins discordant for obesity modulate metabolism in mice. Science. 2013;341(6150):1241214.
16. Ducarmon QR, Zwittink RD, Hornung BVH, van Schaik W, Young VB, Kuijper EJ. Gut microbiota and colonization resistance against bacterial enteric infection. Microbiol Mol Biol Rev. 2019;83(3):e00007-19.
17. Becattini S, Littmann ER, Carter RA, Kim SG, Morjaria SM, Ling L, et al. Commensal microbes provide first line defense against Listeria monocytogenes infection. J Exp Med. 2017;214(7):1973–89.
18. Bhattarai Y, Kashyap PC. Germ-free mice model for studying host-microbial interactions. Methods Mol Biol. 2016;1438:123–35.
19. Walter J, Armet AM, Finlay BB, Shanahan F. Establishing or exaggerating causality for the gut microbiome: lessons from human microbiota-associated rodents. Cell. 2020;180(2):221–32.
20. Quigley EM. Basic definitions and concepts: organization of the gut microbiome. Gastroenterol Clin North Am. 2017;46(1):1–8.
21. Morgan XC, Huttenhower C. Chapter 12: Human microbiome analysis. PLoS Comput Biol. 2012;8(12):e1002808.
22. Malla MA, Dubey A, Kumar A, Yadav S, Hashem A, Abd_Allah EF. Exploring the human microbiome: the potential future role of next-generation sequencing in disease diagnosis and treatment. Front Immunol. 2019;9:2868.
23. Garmaeva S, Sinha T, Kurilshikov A, Fu J, Wijmenga C, Zhernakova A. Studying the gut virome in the metagenomic era: challenges and perspectives. BMC Biol. 2019;17(1):84.
24. Coker OO, Nakatsu G, Dai RZ, Wu WKK, Wong SH, Ng SC, et al. Enteric fungal microbiota dysbiosis and ecological alterations in colorectal cancer. Gut. 2019;68(4):654–62.
25. Di Paola M, Rizzetto L, Stefanini I, Vitali F, Massi-Benedetti C, Tocci N, et al. Comparative immunophenotyping of Saccharomyces cerevisiae and Candida spp. strains from Crohn's disease patients and their interactions with the gut microbiome. J Transl Autoimmun. 2020;3:100036.
26. Tito RY, Chaffron S, Caenepeel C, Lima-Mendez G, Wang J, Vieira-Silva S, et al. Population-level analysis of Blastocystis subtype prevalence and variation in the human gut microbiota. Gut. 2019;68(7):1180–9.
27. Berg G, Rybakova D, Fischer D, Cernava T, Vergès M-CC, Charles T, et al. Microbiome definition re-visited: old concepts and new challenges. Microbiome. 2020;8(1):103.

28. Proctor L, LoTempio J, Marquitz A, Daschner P, Xi D, Flores R, et al. A review of 10 years of human microbiome research activities at the US National Institutes of Health, Fiscal Years 2007-2016. Microbiome. 2019;7(1):31.
29. Robles-Alonso V, Guarner F. From basic to applied research: lessons from the human microbiome projects. J Clin Gastroenterol. 2014;48(Suppl 1):S3–4.
30. Proctor LM, Creasy HH, Fettweis JM, Lloyd-Price J, Mahurkar A, Zhou W, et al. The integrative human microbiome project. Nature. 2019;569(7758):641–8.
31. Hadrich D. Microbiome research is becoming the key to better understanding health and nutrition. Front Genet. 2018;9:212.
32. Aagaard K, Ma J, Antony KM, Ganu R, Petrosino J, Versalovic J. The placenta harbors a unique microbiome. Sci Transl Med. 2014;6(237):237ra65.
33. Theis KR, Romero R, Winters AD, Greenberg JM, Gomez-Lopez N, Alhousseini A, et al. Does the human placenta delivered at term have a microbiota? Results of cultivation, quantitative real-time PCR, 16S rRNA gene sequencing, and metagenomics. Am J Obstet Gynecol. 2019;220(3):267.e1–39.
34. Neuman H, Koren O. The pregnancy microbiome. Nestle Nutr Inst Workshop Ser. 2017;88:1–9.
35. Neu J, Rushing J. Cesarean versus vaginal delivery: long-term infant outcomes and the hygiene hypothesis. Clin Perinatol. 2011;38(2):321–31.
36. Shao Y, Forster SC, Tsaliki E, Vervier K, Strang A, Simpson N, et al. Stunted microbiota and opportunistic pathogen colonization in caesarean-section birth. Nature. 2019;574(7776):117–21.
37. Korpela K, Helve O, Kolho KL, Saisto T, Skogberg K, Dikareva E, et al. Maternal fecal microbiota transplantation in cesarean-born infants rapidly restores normal gut microbial development: a proof-of-concept study. Cell. 2020;183(2):324–34.e5.
38. Chong PP, Chin VK, Looi CY, Wong WF, Madhavan P, Yong VC. The microbiome and irritable bowel syndrome—a review on the pathophysiology, current research and future therapy. Front Microbiol. 2019;10:1136.
39. Tap J, Derrien M, Tornblom H, Brazeilles R, Cools-Portier S, Dore J, et al. Identification of an intestinal microbiota signature associated with severity of irritable bowel syndrome. Gastroenterology. 2017;152(1):111–23.e8.
40. Rajilić-Stojanović M, Jonkers DM, Salonen A, Hanevik K, Raes J, Jalanka J, et al. Intestinal microbiota and diet in IBS: causes, consequences, or epiphenomena? Am J Gastroenterol. 2015;110(2):278–87.
41. Pimentel M, Lembo A. Microbiome and its role in irritable bowel syndrome. Dig Dis Sci. 2020;65(3):829–39.
42. Pascal V, Pozuelo M, Borruel N, Casellas F, Campos D, Santiago A, et al. A microbial signature for Crohn's disease. Gut. 2017;66(5):813–22.
43. Mondot S, Kang S, Furet JP, Aguirre de Carcer D, McSweeney C, Morrison M, et al. Highlighting new phylogenetic specificities of Crohn's disease microbiota. Inflamm Bowel Dis. 2011;17(1):185–92.
44. Matijašić M, Meštrović T, Paljetak HČ, Perić M, Barešić A, Verbanac D. Gut microbiota beyond bacteria-mycobiome, virome, archaeome, and eukaryotic parasites in IBD. Int J Mol Sci. 2020;21(8):2668.
45. Ungaro F, Massimino L, D'Alessio S, Danese S. The gut virome in inflammatory bowel disease pathogenesis: from metagenomics to novel therapeutic approaches. United European Gastroenterol J. 2019;7(8):999–1007.
46. Lam S, Zuo T, Ho M, Chan FKL, Chan PKS, Ng SC. Review article: fungal alterations in inflammatory bowel diseases. Aliment Pharmacol Ther. 2019;50(11–12):1159–71.
47. Zuo T, Lu XJ, Zhang Y, Cheung CP, Lam S, Zhang F, et al. Gut mucosal virome alterations in ulcerative colitis. Gut. 2019;68(7):1169–79.
48. Halmos EP, Christophersen CT, Bird AR, Shepherd SJ, Gibson PR, Muir JG. Diets that differ in their FODMAP content alter the colonic luminal microenvironment. Gut. 2015;64(1):93–100.
49. Staudacher HM, Whelan K. Altered gastrointestinal microbiota in irritable bowel syndrome and its modification by diet: probiotics, prebiotics and the low FODMAP diet. Proc Nutr Soc. 2016;75(3):306–18.
50. Staudacher HM, Whelan K. The low FODMAP diet: recent advances in understanding its mechanisms and efficacy in IBS. Gut. 2017;66(8):1517–27.
51. Pigneur B, Lepage P, Mondot S, Schmitz J, Goulet O, Doré J, et al. Mucosal healing and bacterial composition in response to enteral nutrition vs steroid-based induction therapy—a randomised prospective clinical trial in children with Crohn's disease. J Crohns Colitis. 2019;13(7):846–55.
52. Murtaza N, Cuív PÓ, Morrison M. Diet and the microbiome. Gastroenterol Clin North Am. 2017;46(1):49–60.
53. MacLellan A, Moore-Connors J, Grant S, Cahill L, Langille MGI, Van Limbergen J. The impact of exclusive enteral nutrition (EEN) on the gut microbiome in Crohn's disease: a review. Nutrients. 2017;9(5):447.
54. Kaakoush NO, Day AS, Leach ST, Lemberg DA, Nielsen S, Mitchell HM. Effect of exclusive enteral nutrition on the microbiota of children with newly diagnosed Crohn's disease. Clin Transl Gastroenterol. 2015;6(1):e71.
55. Gevers D, Kugathasan S, Denson LA, Vazquez-Baeza Y, Van Treuren W, Ren B, et al. The treatment-naive microbiome in new-onset Crohn's disease. Cell Host Microbe. 2014;15(3):382–92.
56. De Cruz P, Kang S, Wagner J, Buckley M, Sim WH, Prideaux L, et al. Association between specific mucosa-associated microbiota in Crohn's disease at the time of resection and subsequent disease recurrence: a pilot study. J Gastroenterol Hepatol. 2015;30(2):268–78.
57. Sokol H, Pigneur B, Watterlot L, Lakhdari O, Bermudez-Humaran LG, Gratadoux JJ, et al. Faecalibacterium prausnitzii is an anti-inflammatory commensal bacterium identified by gut microbiota analysis of Crohn disease patients. Proc Natl Acad Sci U S A. 2008;105(43):16731–6.
58. Lloyd-Price J, Arze C, Ananthakrishnan AN, Schirmer M, Avila-Pacheco J, Poon TW, et al. Multi-omics of the gut microbial ecosystem in inflammatory bowel diseases. Nature. 2019;569(7758):655–62.
59. Schirmer M, Franzosa EA, Lloyd-Price J, McIver LJ, Schwager R, Poon TW, et al. Dynamics of metatranscription in the inflammatory bowel disease gut microbiome. Nat Microbiol. 2018;3(3):337–46.
60. Varela E, Manichanh C, Gallart M, Torrejon A, Borruel N, Casellas F, et al. Colonisation by Faecalibacterium prausnitzii and maintenance of clinical remission in patients with ulcerative colitis. Aliment Pharmacol Ther. 2013;38(2):151–61.
61. Miquel S, Leclerc M, Martin R, Chain F, Lenoir M, Raguideau S, et al. Identification of metabolic signatures linked to anti-inflammatory effects of Faecalibacterium prausnitzii. mBio. 2015;6(2):e00300-15.
62. Quevrain E, Maubert MA, Michon C, Chain F, Marquant R, Tailhades J, et al. Identification of an anti-inflammatory protein from Faecalibacterium prausnitzii, a commensal bacterium deficient in Crohn's disease. Gut. 2016;65(3):415–25.
63. Atarashi K, Tanoue T, Oshima K, Suda W, Nagano Y, Nishikawa H, et al. Treg induction by a rationally selected mixture of Clostridia strains from the human microbiota. Nature. 2013;500(7461):232–6.
64. Narushima S, Sugiura Y, Oshima K, Atarashi K, Hattori M, Suematsu M, et al. Characterization of the 17 strains of regulatory T cell-inducing human-derived Clostridia. Gut Microbes. 2014;5(3):333–9.

65. Wright EK, Kamm MA, Wagner J, Teo SM, Cruz P, Hamilton AL, et al. Microbial factors associated with postoperative Crohn's disease recurrence. J Crohns Colitis. 2017;11(2):191–203.
66. Zhang J, Hoedt EC, Liu Q, Berendsen E, Teh JJ, Hamilton A, et al. Elucidation of Proteus mirabilis as a key bacterium in Crohn's disease inflammation. Gastroenterology. 2021;160(1):317–330.e11.
67. Mondot S, Lepage P, Seksik P, Allez M, Treton X, Bouhnik Y, et al. Structural robustness of the gut mucosal microbiota is associated with Crohn's disease remission after surgery. Gut. 2016;65(6):954–62.
68. Ni J, Shen TD, Chen EZ, Bittinger K, Bailey A, Roggiani M, et al. A role for bacterial urease in gut dysbiosis and Crohn's disease. Sci Transl Med. 2017;9(416):eaah6888.
69. Lee D, Albenberg L, Compher C, Baldassano R, Piccoli D, Lewis JD, et al. Diet in the pathogenesis and treatment of inflammatory bowel diseases. Gastroenterology. 2015;148(6):1087–106.
70. Charlebois A, Rosenfeld G, Bressler B. The impact of dietary interventions on the symptoms of inflammatory bowel disease: a systematic review. Crit Rev Food Sci Nutr. 2016;56(8):1370–8.
71. Halmos EP, Gibson PR. Dietary management of IBD—insights and advice. Nat Rev Gastroenterol Hepatol. 2015;12(3):133–46.
72. Levine A, Sigall Boneh R, Wine E. Evolving role of diet in the pathogenesis and treatment of inflammatory bowel diseases. Gut. 2018;67(9):1726–38.
73. Weng YJ, Gan HY, Li X, Huang Y, Li ZC, Deng HM, et al. Correlation of diet, microbiota and metabolite networks in inflammatory bowel disease. J Dig Dis. 2019;20(9):447–59.
74. Ford AC, Lacy BE, Talley NJ. Irritable bowel syndrome. N Engl J Med. 2017;376(26):2566–78.
75. Drossman DA. Functional gastrointestinal disorders: what's new for Rome IV? Lancet Gastroenterol Hepatol. 2016;1(1):6–8.
76. Talley NJ. Functional gastrointestinal disorders as a public health problem. Neurogastroenterol Motil. 2008;20(Suppl 1):121–9.
77. Keely S, Walker MM, Marks E, Talley NJ. Immune dysregulation in the functional gastrointestinal disorders. Eur J Clin Investig. 2015;45(12):1350–9.
78. Talley NJ, Walker MM, Holtmann G. Functional dyspepsia. Curr Opin Gastroenterol. 2016;32(6):467–73.
79. Powell N, Walker MM, Talley NJ. The mucosal immune system: master regulator of bidirectional gut-brain communications. Nat Rev Gastroenterol Hepatol. 2017;14(3):143–59.
80. Burns G, Carroll G, Mathe A, Horvat J, Foster P, Walker MM, et al. Evidence for local and systemic immune activation in functional dyspepsia and the irritable bowel syndrome: a systematic review. Am J Gastroenterol. 2019;114(3):429–36.
81. Walker MM, Aggarwal KR, Shim LS, Bassan M, Kalantar JS, Weltman MD, et al. Duodenal eosinophilia and early satiety in functional dyspepsia: confirmation of a positive association in an Australian cohort. J Gastroenterol Hepatol. 2014;29(3):474–9.
82. Walker MM, Talley NJ, Prabhakar M, Pennaneac'h CJ, Aro P, Ronkainen J, et al. Duodenal mastocytosis, eosinophilia and intraepithelial lymphocytosis as possible disease markers in the irritable bowel syndrome and functional dyspepsia. Aliment Pharmacol Ther. 2009;29(7):765–73.
83. Liebregts T, Adam B, Bredack C, Roth A, Heinzel S, Lester S, et al. Immune activation in patients with irritable bowel syndrome. Gastroenterology. 2007;132(3):913–20.
84. Spinedi E, Hadid R, Daneva T, Gaillard RC. Cytokines stimulate the CRH but not the vasopressin neuronal system: evidence for a median eminence site of interleukin-6 action. Neuroendocrinology. 1992;56(1):46–53.
85. Shanahan ER, Zhong L, Talley NJ, Morrison M, Holtmann G. Characterisation of the gastrointestinal mucosa-associated microbiota: a novel technique to prevent cross-contamination during endoscopic procedures. Aliment Pharmacol Ther. 2016;43(11):1186–96.
86. Shanahan ER, Shah A, Koloski N, Walker MM, Talley NJ, Morrison M, et al. Influence of cigarette smoking on the human duodenal mucosa-associated microbiota. Microbiome. 2018;6(1):150.
87. Zhong L, Shanahan ER, Raj A, Koloski NA, Fletcher L, Morrison M, et al. Dyspepsia and the microbiome: time to focus on the small intestine. Gut. 2017;66(6):1168–9.
88. Shah A, Talley NJ, Koloski N, Macdonald GA, Kendall BJ, Shanahan ER, et al. Duodenal bacterial load as determined by quantitative polymerase chain reaction in asymptomatic controls, functional gastrointestinal disorders and inflammatory bowel disease. Aliment Pharmacol Ther. 2020;52:155–67.
89. Vandeputte D, Kathagen G, D'Hoe K, Vieira-Silva S, Valles-Colomer M, Sabino J, et al. Quantitative microbiome profiling links gut community variation to microbial load. Nature. 2017;551(7681):507–11.
90. Vieira-Silva S, Sabino J, Valles-Colomer M, Falony G, Kathagen G, Caenepeel C, et al. Quantitative microbiome profiling disentangles inflammation- and bile duct obstruction-associated microbiota alterations across PSC/IBD diagnoses. Nat Microbiol. 2019;4(11):1826–31.
91. Garrett WS. Cancer and the microbiota. Science. 2015;348(6230):80–6.
92. Meng C, Bai C, Brown TD, Hood LE, Tian Q. Human gut microbiota and gastrointestinal cancer. Genomics Proteomics Bioinformatics. 2018;16(1):33–49.
93. Ferreira RM, Pereira-Marques J, Pinto-Ribeiro I, Costa JL, Carneiro F, Machado JC, et al. Gastric microbial community profiling reveals a dysbiotic cancer-associated microbiota. Gut. 2018;67(2):226–36.
94. Coker OO, Dai Z, Nie Y, Zhao G, Cao L, Nakatsu G, et al. Mucosal microbiome dysbiosis in gastric carcinogenesis. Gut. 2018;67(6):1024–32.
95. Rajilic-Stojanovic M, Figueiredo C, Smet A, Hansen R, Kupcinskas J, Rokkas T, et al. Systematic review: gastric microbiota in health and disease. Aliment Pharmacol Ther. 2020;51(6):582–602.
96. Yu J, Feng Q, Wong SH, Zhang D, Liang QY, Qin Y, et al. Metagenomic analysis of faecal microbiome as a tool towards targeted non-invasive biomarkers for colorectal cancer. Gut. 2017;66(1):70–8.
97. Dai Z, Coker OO, Nakatsu G, Wu WKK, Zhao L, Chen Z, et al. Multi-cohort analysis of colorectal cancer metagenome identified altered bacteria across populations and universal bacterial markers. Microbiome. 2018;6(1):70.
98. Liang JQ, Li T, Nakatsu G, Chen YX, Yau TO, Chu E, et al. A novel faecal Lachnoclostridium marker for the non-invasive diagnosis of colorectal adenoma and cancer. Gut. 2020;69(7):1248–57.
99. Münch NS, Fang HY, Ingermann J, Maurer HC, Anand A, Kellner V, et al. High-fat diet accelerates carcinogenesis in a mouse model of Barrett's esophagus via interleukin 8 and alterations to the gut microbiome. Gastroenterology. 2019;157(2):492–506.e2.
100. Gopalakrishnan V, Helmink BA, Spencer CN, Reuben A, Wargo JA. The influence of the gut microbiome on cancer, immunity, and cancer immunotherapy. Cancer Cell. 2018;33(4):570–80.
101. Helmink BA, Khan MAW, Hermann A, Gopalakrishnan V, Wargo JA. The microbiome, cancer, and cancer therapy. Nat Med. 2019;25(3):377–88.
102. Zitvogel L, Ma Y, Raoult D, Kroemer G, Gajewski TF. The microbiome in cancer immunotherapy: diagnostic tools and therapeutic strategies. Science (New York, NY). 2018;359(6382):1366–70.
103. Guglielmi G. How gut microbes are joining the fight against cancer. Nature. 2018;557(7706):482–4.
104. Vetizou M, Pitt JM, Daillere R, Lepage P, Waldschmitt N, Flament C, et al. Anticancer immunotherapy by CTLA-4 blockade relies on the gut microbiota. Science. 2015;350(6264):1079–84.
105. Sivan A, Corrales L, Hubert N, Williams JB, Aquino-Michaels K, Earley ZM, et al. Commensal Bifidobacterium promotes anti-

tumor immunity and facilitates anti-PD-L1 efficacy. Science. 2015;350(6264):1084–9.
106. Routy B, Le Chatelier E, Derosa L, Duong CPM, Alou MT, Daillere R, et al. Gut microbiome influences efficacy of PD-1-based immunotherapy against epithelial tumors. Science. 2018;359(6371):91–7.
107. Gopalakrishnan V, Spencer CN, Nezi L, Reuben A, Andrews MC, Karpinets TV, et al. Gut microbiome modulates response to anti-PD-1 immunotherapy in melanoma patients. Science. 2018;359(6371):97–103.
108. Matson V, Fessler J, Bao R, Chongsuwat T, Zha Y, Alegre ML, et al. The commensal microbiome is associated with anti-PD-1 efficacy in metastatic melanoma patients. Science. 2018;359(6371):104–8.
109. Tanoue T, Morita S, Plichta DR, Skelly AN, Suda W, Sugiura Y, et al. A defined commensal consortium elicits CD8 T cells and anti-cancer immunity. Nature. 2019;565(7741):600–5.
110. Borody T, Fischer M, Mitchell S, Campbell J. Fecal microbiota transplantation in gastrointestinal disease: 2015 update and the road ahead. Expert Rev Gastroenterol Hepatol. 2015;9(11):1379–91.
111. Paramsothy S, Walsh AJ, Borody T, Samuel D, van den Bogaerde J, Leong RW, et al. Gastroenterologist perceptions of faecal microbiota transplantation. World J Gastroenterol. 2015;21(38):10907–14.
112. Gough E, Shaikh H, Manges AR. Systematic review of intestinal microbiota transplantation (fecal bacteriotherapy) for recurrent Clostridium difficile infection. Clin Infect Dis. 2011;53(10):994–1002.
113. Quraishi MN, Widlak M, Bhala N, Moore D, Price M, Sharma N, et al. Systematic review with meta-analysis: the efficacy of faecal microbiota transplantation for the treatment of recurrent and refractory Clostridium difficile infection. Aliment Pharmacol Ther. 2017;46(5):479–93.
114. Lai CY, Sung J, Cheng F, Tang W, Wong SH, Chan PKS, et al. Systematic review with meta-analysis: review of donor features, procedures and outcomes in 168 clinical studies of faecal microbiota transplantation. Aliment Pharmacol Ther. 2019;49(4):354–63.
115. Narula N, Kassam Z, Yuan Y, Colombel JF, Ponsioen C, Reinisch W, et al. Systematic review and meta-analysis: fecal microbiota transplantation for treatment of active ulcerative colitis. Inflamm Bowel Dis. 2017;23(10):1702–9.
116. Costello SP, Soo W, Bryant RV, Jairath V, Hart AL, Andrews JM. Systematic review with meta-analysis: faecal microbiota transplantation for the induction of remission for active ulcerative colitis. Aliment Pharmacol Ther. 2017;46(3):213–24.
117. Sokol H, Landman C, Seksik P, Berard L, Montil M, Nion-Larmurier I, et al. Fecal microbiota transplantation to maintain remission in Crohn's disease: a pilot randomized controlled study. Microbiome. 2020;8(1):12.
118. Baxter M, Colville A. Adverse events in faecal microbiota transplant: a review of the literature. J Hosp Infect. 2016;92(2):117–27.
119. Wang S, Xu M, Wang W, Cao X, Piao M, Khan S, et al. Systematic review: adverse events of fecal microbiota transplantation. PLoS One. 2016;11(8):e0161174.
120. Luo Y, Lucas AL, Grinspan AM. Fecal transplants by colonoscopy and capsules are cost-effective strategies for treating recurrent Clostridioides difficile infection. Dig Dis Sci. 2020;65(4):1125–33.
121. Mullish BH, Quraishi MN, Segal JP, McCune VL, Baxter M, Marsden GL, et al. The use of faecal microbiota transplant as treatment for recurrent or refractory Clostridium difficile infection and other potential indications: joint British Society of Gastroenterology (BSG) and Healthcare Infection Society (HIS) guidelines. Gut. 2018;67(11):1920–41.
122. Kelly CR, Yen EF, Grinspan AM, Kahn SA, Atreja A, Lewis JD, et al. Fecal microbiota transplantation is highly effective in real-world practice: initial results from the FMT National Registry. Gastroenterology. 2020;158:S-14–5.
123. Paramsothy S, Paramsothy R, Rubin DT, Kamm MA, Kaakoush NO, Mitchell HM, et al. Faecal microbiota transplantation for inflammatory bowel disease: a systematic review and meta-analysis. J Crohns Colitis. 2017;11(10):1180–99.
124. Paramsothy S, Nielsen S, Kamm MA, Deshpande NP, Faith JJ, Clemente JC, et al. Specific bacteria and metabolites associated with response to fecal microbiota transplantation in patients with ulcerative colitis. Gastroenterology. 2019;156(5):1440–54.e2.
125. Moayyedi P, Surette MG, Kim PT, Libertucci J, Wolfe M, Onischi C, et al. Fecal microbiota transplantation induces remission in patients with active ulcerative colitis in a randomized controlled trial. Gastroenterology. 2015;149(1):102–9.e6.
126. Kong L, Lloyd-Price J, Vatanen T, Seksik P, Beaugerie L, Simon T, et al. Linking strain engraftment in fecal microbiota transplantation with maintenance of remission in Crohn's disease. Gastroenterology. 2020;159(6):2193–202.e5.

The Players Within the Intestinal Microbiome (Bacteria, Fungi, Parasites, and Viruses)

Jintong Liu, Marie Fogarty, Soo Chan Lee, Michael Dougherty, Renay Ngobeni-Nyambi, Yi Ge, and Ian M. Carroll

Highlights

- The gut microbiota include multiple, diverse, symbiotic and commensal microorganisms that provide benefits to their host, including bacteria, archaea, fungi, parasites, and viruses.
- Antifungal drug-mediated changes in the gut mycobiome can result in increased inflammation in the gut and lead to colitis. The severity of colitis is associated with a gene encoding the dectin-1 receptor, which is a pattern-recognition receptor that specifically recognizes the key fungal cell component β-1,3-glucan.
- Eukaryotic gut parasites can be non-pathogenic, e.g. helminths promote a T-regulatory cell response characterized by the anti-inflammatory cytokines IL-10 and TGF-β, and human trials of *helminth therapy* have been carried out in autoimmune diseases with the roundworms *Trichuria suis* and *Necator americanus.*
- Enteric viruses can also be beneficial to the host—monocolonization with murine norovirus (MNV) can reverse the abnormal intestinal morphology and immune system typically found in germ-free mice.

J. Liu · Y. Ge · I. M. Carroll (✉)
Department of Nutrition, University of North Carolina at Chapel Hill, Chapel Hill, NC, USA
e-mail: jintong.liu@cuanschutz.edu; gloriagy@live.unc.edu; ian_carroll@med.unc.edu

M. Fogarty
Department of Biological Sciences, Durham Technical Community College, Durham, NC, USA
e-mail: fogartym@durhamtech.edu

S. C. Lee
South Texas Center for Emerging Infectious Diseases (STCEID), Department of Molecular Microbiology and Immunology, University of Texas, San Antonio, San Antonio, TX, USA
e-mail: soochan.lee@utsa.edu

M. Dougherty
Division of Gastroenterology and Hepatology, Department of Medicine, University of North Carolina at Chapel Hill, Chapel Hill, NC, USA
e-mail: michael.dougherty@unchealth.unc.edu

R. Ngobeni-Nyambi
Department of Microbiology, Stellenbosch University, Matieland, Cape Town, South Africa
e-mail: renay@sun.ac.za

1 Introduction

The gut microbiota is a complex and dynamic community of microorganisms that reside within the gastrointestinal (GI) tract of animals [1]—communities are associated with the GI mucosa or the luminal contents [2]. This complex microbial community encompasses multiple, diverse microorganisms including bacteria, archaea (discussed in chapter "Changes in the Gut Microbiome as Seen in Diabetes and Obesity"), fungi, parasites, and viruses (including eukaryotic viruses and bacteriophages) (Fig. 1). In contrast to pathogenic microorganisms that cause disease, the gut microbiota largely contains symbiotic and commensal microbes, which provide various benefits to their host. Specifically, gut microbes help break down nutrients in the intestine, maintain the integrity of the intestinal epithelium, and help the proper development of the host immune system [3]. Emerging data suggest that the gut microbiota also influences the host's behavior by exerting its effect via the gut-brain axis [4]. While a healthy gut microbiota promotes the health of the host, an abnormal (often termed *dysbiotic*) gut microbiota has been shown to contribute to GI dysfunction and factors related to metabolic syndrome (e.g. obesity), inflammatory bowel diseases (IBD), and cardiovascular diseases [5–8].

To date the most widely used method to investigate the functional influence of an abnormal gut microbiota on the host is to transplant all microbial elements of a *dysbiotic* gut microbiota into a recipient animal devoid of (germ-free, GF) or depleted in (antibiotic treatment) enteric microbes, and

M. Pimentel et al. (eds.), *Clinical Understanding of the Human Gut Microbiome*, https://doi.org/10.1007/978-3-031-46712-7_2

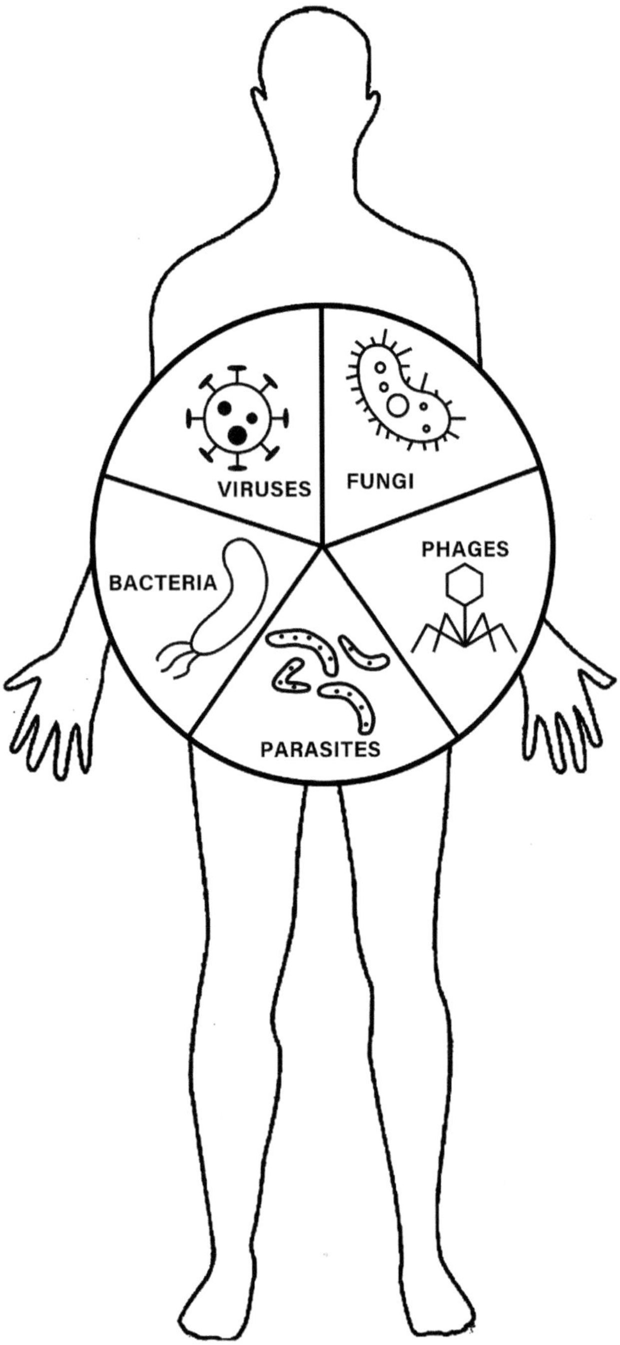

Fig. 1 Microbial composition of the intestinal microbiota

observe changes in specific host phenotypes. This approach has been used in landmark studies to determine the effect of human fecal microbiota on obesity [9], diseases associated with starvation [10], IBD [11], and autism spectrum disorders [12]—it is important to note this approach is also influenced by mouse diet and is donor-dependent. These studies comprehensively characterized the bacterial component of the intestinal microbiota and highlight associations between specific bacterial taxa and abnormal host phenotypes; however, the role of non-bacterial members of the intestinal microbiota are largely overlooked in these studies. Indeed, investigation of supplemental material accompanying the fecal microbiota-Kwashiorkor study [10] reveals that specific intestinal parasites (*Giardia*) were transferred to GF mice and may influence weight loss in mice fed a typical Malawian diet. Therefore, ignoring the contribution of non-bacterial members of the gut microbiota in the initiation and maintenance of abnormal host phenotypes may lead to missing microbes that are truly associated with disease.

As there is strong evidence for a causal relationship between an abnormal gut microbiota and particular diseases, it is not surprising that attempts have been made to restore host health via reestablishing a normal balance in gut microbial communities. Indeed, repopulating the GI tract with a normal, healthy microbiota is an active field of research. Probiotics, prebiotics, and fecal microbiota transplantation (FMT) rank among the most popular approaches to restore health to the host via modulating indigenous gut microbes. Probiotics are living microorganisms that when consumed in adequate amounts are beneficial to the host, while a prebiotic is a substrate that is selectively utilized by host microorganisms conferring a health benefit [13]. FMT (discussed in chapter "Fecal Transplant: The Benefits and Harms of Fecal Microbiota Transplantation") is a procedure where the fecal microbiota from a healthy donor is transferred into a recipient host believed to harbor a *dysbiotic* gut microbiota. FMT is an effective treatment for *Clostridioides difficile* infection (CDI), with resolution rates usually higher than 85% [14, 15]. In theory, both probiotics and prebiotics promote restoration of a healthy gut microbiota; however, the impact of probiotics and prebiotics on the non-bacterial members within the GI ecosystem remains generally ignored. Moreover, FMT infuses all microbial elements of the fecal microbiota into a donor, yet the engraftment and activity of transplanted fecal bacteria are typically the focus of investigation. Ultimately, understanding the influence of non-bacterial elements of gut microbial communities on the host will likely produce more effective microbiota-based treatments for GI diseases. Indeed, the future of gut microbiome research needs to consider the human *holobiont* (inclusive unit of biological organization including host and all microbes) [16] as we attempt to develop gut microbiota-based therapies.

2 The Gut Bacteriome

With 3.8×10^{13} cells and over 1900 species residing in the human intestine, bacteria are the most diverse group of microorganisms within the gut microbiota [3]. It is therefore not surprising that while all microbial members contribute to the makeup and function of this community, enteric bacteria are the most commonly studied members of this community.

2.1 Characterization of the Gut Bacteriome

Initially, investigators were restricted to studying the bacterial component of the gut microbiota using traditional culture-based methods; however, this approach provides limited information as it is estimated that only 10–50% of intestinal bacteria are culturable [17]. Additionally, the costly and time-consuming nature of this approach limits the number of samples investigators can analyze. With the reduced cost of high throughput sequencing and advances in *omics* technologies (e.g., metabolomics, metagenomics, and transcriptomics), investigators are now able to study the gut microbiome (cumulative genetic material of the gut microbiota) in a more comprehensive manner [6]. These technological approaches also allow investigators to identify the taxonomic composition of most bacteria within a microbial community, including non-culturable microbes. Present in all bacteria but not in eukaryotes, the 16S rRNA gene is a valuable genetic locus for characterizing the bacterial component of the gut microbiota [7, 8]. With conserved regions that are universal across bacteria and variable regions that are shared within bacterial groups, the 16S rRNA gene offers sufficient information for researchers to identify taxonomic groups within microbial communities. These technological advances have led to seminal discoveries regarding the intricate relationships between enteric bacterial microbes and human health.

2.2 Contribution of Enteric Bacteria to Host Health

Since the onset of molecular characterization of gut microbial communities, the gut bacteriome has been associated with multiple notable diseases, such as obesity, IBD, irritable bowel syndrome (IBS), and cancer [9, 18, 19]. As previously mentioned, by ignoring the non-bacterial components of the gut microbial community, the associations these studies report may not be solely attributed to enteric bacteria. To confidently attribute bacterial members of the gut microbiota with a disease of interest, experimental designs that investigate specific bacterial strains in isolation (or in the absence of non-bacterial microorganisms) and attempt to reveal the mechanism by which they cause disease have been applied to gut microbiota studies. For example, initial studies investigating the role of enteric bacteria in the development of IBD used a GF mouse model of intestinal inflammation (Interleukin (IL) 10-deficient mouse) mono-colonized with different intestinal bacterial strains that are harbored within the gut microbiota [20]. It was discovered that different members of the bacterial gut microbiota result in varying degrees of intestinal inflammation, with an *Escherichia coli* strain eliciting a more rapid onset of intestinal inflammation. More recently, colonization of formerly GF IL 10-deficient mice with *Akkermansia muciniphila* found that this organism does not lead to intestinal inflammation [21], highlighting the need for further investigation as to whether this microorganism may be beneficial for GI diseases. Indeed, *A. muciniphila* is a known beneficial participant in host metabolism and is a potential candidate for treatment of metabolic abnormalities that accompany obesity.

Emerging evidence has firmly placed the gut microbiota as a regulator of the gut-brain axis both in mice and humans [22]. Recently, this complex microbial community was shown to influence the set points of the brain and GI serotonergic systems (one of the oldest transmitter systems in the brain) and affected their response to acute stress [23]. As this study used GF and conventionally-raised mice (i.e. harboring a normal gut microbiota) the influence of enteric microbes on mouse behavior must be attributed to all types of microorganisms within this community. Recently, mono-colonization of formerly GF mice with pure cultures of common enteric bacterial strains, identified via high throughput sequencing of 16S rRNA genes, found that members of the *Lactobacillus* genus (*Lactobacillus reuteri*, *Lactobacillus plantarum*, or *Lactobacillus brevis*) significantly improved memory compared to controls (GF or *E. coli* mono-colonized mice) [24]. Together these studies show that the intestinal microbiota has a significant effect on host behavior and that specific bacterial strains within this complex microbial community can impact the brain.

3 The Gut Mycobiome

Fungi are eukaryotic organisms that are, in most cases, microscopic. They interact with other microbes within complex microbial communities located in and on many different parts of the human body, including the intestine, skin, and mouth. It has been established that a diverse community of fungi reside and are abundant in the human gut. Indeed, it is estimated that fungi compose 0.01–0.1% of the gut microbiota [25, 26]. Although not as numerous as bacteria in the gut microbiota, their significantly larger size compared to bacteria make enteric fungi an important element of this community. For example, the average diameter of the *Candida albicans* yeast cell is approximately 10–12 μm, whereas that of *E. coli* is as small as 0.25 μm.

3.1 Characterization of the Gut Mycobiome

The cumulative genomes of the fungal component of the gut microbiota (the gut mycobiome) has been characterized by analyzing genetic loci that are unique to fungi—the internal transcribed spacers (ITS) 1 and 2 regions situated between the small- and large-subunit rRNA genes [27, 28]. This approach is considered a highly sensitive and effective method of characterizing the gut mycobiome [29]. The UNITE [30], and targeted host-associated fungi (THF) ITS [27] databases are commonly used to identify fungal operational taxonomic units (OTUs). Metagenomic and fungal-specific 18S rRNA gene analyses have also been used; however, the data obtained from the ITS1 and ITS2 loci are considered more accurate [29]. Using ITS1 and ITS2 can impose some limitations when calculating the abundance of fungal species. Specifically, unlike bacteria, fungi can present as haploid or diploid [31]. To add to this complexity, within the same species, ploidy status may vary from strain to strain. In addition, some species have multinucleate spores or yeast, and even have multicellular-like stages, such as hyphae. These limitations therefore have to be considered when measuring the abundance of fungal species in the gut microbiota.

Taking these limitations into account, in a recent study the gut mycobiomes of 147 individuals' stool samples have been analyzed. One striking finding was that intestinal fungal diversity is generally low compared to gut bacterial diversity [29]. Yeast-forming fungi were found to dominate, and the more abundant genera, including *Saccharomyces*, *Malassezia*, and *Candida*, were found in 80–97% of the samples, providing evidence of a core gut mycobiome. Tremellomycetes such as *Cryptococcus* and *Trichosporon* were common. Mold-forming fungi such as *Aspergillus*, *Penicillium*, *Cladosporium*, *Phoma*, *Alternaria*, *Mucor*, and *Sclerotinia* were also detected [32, 33].

3.2 Contribution of the Gut Mycobiome to Host Health

Due to their lower abundance in the gut microbiota, or a lack of robust analysis methods, fungi have been given less attention than bacterial microbiomes in most gut microbiota studies [34]. This can be considered shortsighted given the evidence for the role of enteric fungi in GI tract diseases. For example, antibiotic treatment that induce alterations in the gut bacterial population often cause the overgrowth of fungi in the intestine, resulting in GI discomfort [35]. Interestingly, the anti-*Saccharomyces cerevisiae* antibody (ASCA), which recognizes fungal cell wall components, has been used as a serological marker for IBD for decades [36]. In addition, two other antibodies that recognize β-1,3-glucan and chitin in the fungal cell wall have also been used as markers for GI tract inflammation [36]. Moreover, antifungal drug-mediated changes in the gut mycobiome can result in increased inflammation in the gut and lead to colitis [37, 38]. Together these findings demonstrate that enteric fungi play an important role as a component of the gut microbiome's contribution to the health of the host.

Alterations in the taxonomic composition of the gut mycobiome in specific GI diseases have been reported (e.g. IBD) [39–41]. Indeed, the role of intestinal fungi in GI inflammation has begun to be illustrated in great detail. Iliev et al. [42] found that the severity of colitis is associated with a gene encoding the dectin-1 receptor. The dectin-1 receptor is a pattern-recognition receptor that specifically recognizes a key fungal cell component—β-1,3-glucan [43]. Mice deficient in dectin-1 exhibit exacerbated colitis induced by dextran sulphate sodium (DSS) compared to mice not lacking dectin-1 [42]. Moreover, the abundance of *Candida tropicalis* (a common enteric fungus) is significantly elevated in the gut of dectin-1 deficient mice. Interestingly, single nucleotide polymorphisms (SNPs) in the dectin-1 gene are linked to the severity of ulcerative colitis (UC) in humans [42]. Ultimately, the authors argue that improper recognition of fungi in the gut therefore results in worsening inflammation, demonstrating a role for enteric fungi in GI tract inflammation.

As fungal communities can potentially play detrimental roles in the GI tract, it is not surprising that the mammalian immune system exhibits an innate immunity to gut fungi. It has been reported that specific immune cells (macrophages) expressing fracktalkine receptor 1 (CX3CR1) provide the vast majority of antifungal immunity. Additionally, SNPs in the gene encoding CX3CR1 are also associated with the response against fungi in Crohn's disease (CD), suggesting a clinical relevance for the mycobiome in GI disease [37]. Furthermore, two independent research groups have demonstrated inter-kingdom correlations between bacterial and fungal microbiomes [44], one group reported synergistic interactions between the fungus *C. tropicalis* and the bacteria *E. coli* and *Serratia marcescens* in the formation of biofilms [41].

The role of some bacterial species harbored within the gut microbiome has been associated with cancer development. Specifically, *Fusobacterium nucleatum* is clearly associated with colorectal cancer (CRC) [reviewed in [45]], and a recent study reported that *E. coli* is linked to ~5% of CRC [46]. Additionally, the causal role of *Helicobacter pylori* (an inhabitant of the stomach) in stomach cancer has been established [47]; however, the mechanism(s) by which the mycobiome potentially aids in cancer development is poorly understood. A seminal study; however, demonstrated a link between the gut mycobiome and carcinogenesis, where the dandruff fungus *Malassezia* was linked to pancreatic cancer as this fungus was capable of infiltrating the GI barrier to reach the pancreas [48]. Together these studies highlight the contribution of the gut mycobiome to human health and highlight the importance of this element of the intestinal microbiota to the health of the host.

4 Gut Parasites

In addition to bacteria and fungi the gut microbiota also includes a rich community of eukaryotes. Among these are organisms traditionally referred to as *parasites*, due to the adult form's reliance on a host for nutrients. *Endo*-parasites comprise the helminths and protozoa. The former are multicellular, often macroscopically visible members of the kingdom Animalia, while protozoa are unicellular organisms within the more nebulous, polyphyletic kingdom Protista. Protists differ from fungi in that they do not use chitin in their cell walls, though there are notable examples (*Microsporidium*, recently reclassified from protozoa to fungus, and *Blastocystis*, a stramenopile previously thought to be a protozoan and prior to that a yeast) with features blurring traditional taxonomic distinctions. Within the gut, parasites are usually studied as the miscellaneous conglomerate of non-fungal eukaryotic microbiota [49].

The species richness and diversity of the gut parasitic microbiota is significantly lower than those of the smaller sized, bacterial inhabitants [50]. However, the diversity of the gut eukaryome follows similar tendencies to the bacterial component of the gut microbiota, with inhabitants of low-resource, poor sanitation settings having a much larger microbial repertoire [51]. Parasites have traditionally been studied for their potential pathogenicity, but it is increasingly appreciated that, particularly in low-resource settings, asymptomatic colonization and occasionally even symbiosis are the most common scenarios [49, 51–56].

The abundance of gut parasites depends on a variety of factors, and is generally much greater in low-resource settings (including non-pathogenic parasites) and in symptomatic disease rather than colonization. However, there is also great variation according to the type of parasite and particularly the type of life cycle. Helminths generally lay eggs that require passage outside of the human host, and often mature in soil (*Ascaris*, hookworm, whipworm) or a zoonotic host (*Schistosoma*) before acquiring the ability to re-infect a human host. Intraluminal helminths therefore generally depend on reinfection from repetitive exposures in endemic regions for long-term carriage. Egg burden can still be significant, with thousands (whipworm) to hundreds of thousands (tapeworms) of eggs per organism per day. A notable anomaly among the helminths is *Strongyloides*, capable of completing its entire lifecycle within the human host, and therefore has an indefinite period of infection, as well as potentially fatal *hyperinfection syndrome* during immune suppression.

While some protozoa (e.g. *Plasmodium* and *Leishmania)* have complex life cycles requiring multiple hosts, the majority of gut protists can replicate and auto-infect the human host, with clearance depending on the host immune system and likely other members of the gut microbiota. Multiplication occurs via binary or multiple fission (schizogony) of the trophozoite, the latter being particularly prominent in the Apicomplexans (e.g. *Cryptosporidium*) and allowing for rapid amplification of parasite burden. Most also have an oocyst stage with protective walls to exist in the environment and promote transmission.

Traditionally, parasites are considered detrimental to the host and are often eradicated with antimicrobials if effective therapies are available—which is appropriate in cases of clear symptomatic disease. Interestingly, an emerging concept explores the immune-modulating potential of human-parasite symbiosis. While inverse associations with protozoal infection and autoimmune disorders have been found—particularly with less- or non-pathogenic species such as *Blastocystis* [51, 53]—most therapeutic interventions to date have involved helminths. Helminths are known to promote a T-regulatory cell response characterized by the anti-inflammatory cytokines IL-10 and transforming growth factor (TGF)-β, among others, ostensibly for the evolutionary benefit of reducing parasite clearance [56]. The clinical downstream effects of this immune modulation are varied, from increased severity in other infections such as tuberculosis [57, 58], to an inverse association with allergic and autoimmune diseases such as asthma, multiple sclerosis, celiac disease, and IBD [56, 59]. On this basis, multiple human trials of *helminth therapy* have been carried out in autoimmune diseases with the roundworms *Trichuria suis* and *Necator americanus,* though results to date have been disappointing [60–62].

4.1 Characterization of the Gut Eukaryome

While protozoa may be the first element of the gut microbiota to be described—van Leeuwenhoek's observation in 1681 of *Giardia* trophozoites in his diarrheal stools [63]—their impact on inter-kingdom gut ecology and mucosal health is only beginning to be explored. Prior microscopy-based studies were extremely insensitive, relying on detection of intermittently-shed eggs or larvae. While other largely qualitative tests such as serology and stool antigen tests have proven useful clinically, as with most enteric microbes, the scientific understanding of the human eukaryome has seen the greatest recent advancement from next-generation sequencing technologies. However, sequencing-based diagnostics for eukaryotes have lagged behind those for bacteria.

Targeted PCR-based tests, such as multiplex stool pathogen assays, have become the standard of care for clinical detection of protozoal infections. Quantitative PCR of stool samples has proved a valuable research tool. For example, re-analysis of the largest epidemiologic studies of childhood diarrhea in low-resource countries increased the yield of causative pathogens and confirmed that parasites (largely *Cryptosporidia*) were responsible for 1–17% of cases [64, 65]. Amplicon sequencing using the 18S rRNA gene subunit has also permitted studying the eukaryotic contribution to the whole gut microbiome [66]. However, the more recent emergence of whole genome shotgun metagenomics has done more to enhance the inclusion of eukaryotic signatures to gut microbiota studies than any other technology [67]. Challenges unique to parasitology remain. For instance, the high intra-species diversity due to sexual recombination poses obstacles for creation of taxonomic databases [68], and DNA-based techniques struggle to quantify abundance for multicellular organisms that vary in size (e.g. helminths).

4.2 Contribution of Gut Parasites to Host Health

Parasites are a major cause of infectious disease worldwide, though the incidence of asymptomatic colonization is much higher still. Parasitic diseases of the gut consist primarily of diarrhea from protozoa and nutrient deficiency in helminthic infections [69]. The *Apicomplexa* are perhaps the most likely parasite to cause symptomatic GI disease, particularly in children and the immunocompromised. *Cryptosporidium* alone was responsible for almost 50,000 deaths, 44 million episodes of acute diarrhea, and 13 million disability-adjusted life years (DALYs) in children under 5 in 2016 [70]. Less universally virulent protozoa still capable of causing severe disease include *Giardia* causing malabsorptive diarrhea, and *Entamoeba histolytica* causing dysentery and liver abscess. Despite the usually subtle symptoms, the soil-transmitted helminths still constitute a major burden of disease due to prevalence, affecting 1.5 billion individuals (one quarter of the world's population), and near-universal infection of children in low-resource settings [71]. Lastly, parasites that only transiently infect the GI tract but persist in extra-luminal spaces are responsible for even greater global burden of disease (e.g. schistosomiasis, neurocysticercosis, toxoplasmosis and *Trypansoma cruzi*).

Apart from the classic clinical syndromes, multiple parasitic infections have been associated with malnutrition, stunting, and impaired cognitive development in resource-poor settings. In the case of helminths much of these effects are likely mediated by parasite consumption of resources [71], while protozoa may play a role with pathogenic intestinal bacteria to drive an *environmental enteropathy* with resulting malabsorption. In fact, this syndrome is likely one of the best examples of the clinical impact of the complex interaction between parasites and gut microbiota [72, 73].

One of the most fascinating findings of the growing sequencing-based understanding of the gut eukaryome is the interaction with the rest of the gut microbiome, including possible co-modulation of the host immune system. Parasites (e.g. *Blastocystis hominis* and non-histolytica *Entamoeba* species) have been associated with more favorable intestinal bacterial profiles—including greater overall richness, lower *Bacteroidales* and *Enterobacteriaceae*, and more *Faecalibacterium* and Firmicutes. These microbiota profiles in turn are associated with lower incidences of autoimmune and other industrialized diseases, such as IBD and IBS [54, 68, 74, 75]. These parasite-bacterial effects could occur via secretion of bactericidal or trophic factors (e.g. chitin in helminth eggs), changing the epithelial mucus barrier, direct consumption of certain bacterial species, or interaction with the host immune system [50, 53, 76]. However, bacterial profiles also likely modulate parasitic disease, perhaps even playing a principal role in determining whether a colonization or overt clinical disease will develop for parasites such as *Blastocystis, Giardia,* or *Entamoeba histolytica* [59, 77, 78]. Specifically, diarrheal stools with *Entamoeba histolytica* coincide with increased levels of the bacterial pathobiont *Prevotella copri*, while non-diarrheal stools with *E. histolytica* or non-pathogenic *Entamoeba* species do not [79, 80]. Other pathogenic species (*Shigella, Staphylococcus aureus*) can also induce virulence in *E. histolytica*, but commensal *E. coli* do not. More seminal discoveries from mouse models include the gut microbiota's preservation of host T-helper 1 response to intestinal invasion by *Toxoplasma* (an Apicomplexan like *Cryptosporidium*) in specific microbe recognition receptor-deficient mice [77], and *Trichuria's* prevention of CD via promotion of an anti-inflammatory gut bacterial *enterotype* [75].

5 The Intestinal Virome

Viruses are infectious agents that live and replicate only in living cells, including animals, plants, bacteria, and archaea and are estimated to have 1×10^{31} particles on Earth [81, 82]. Though some viruses, such as influenza and the human immunodeficiency virus (HIV) result in disease, a significant proportion of viruses naturally reside in the human host without detrimental consequences. For example, the cumulative viruses that reside within the GI tract form a significant component of the gut microbiota, termed the intestinal virome. Approximately 1.2×10^9 to 5.5×10^9 virus-like particles (VLPs) can be found in 1 g of feces; however, this estimate increases when accounting for viruses that reside in gut bacterial cells (bacteriophages—discussed later) [83]. The mammalian intestinal virome consists of bacteriophages, eukaryotic viruses, viruses that infect archaea, as well as endogenous retroviruses (viral elements embedded in the host genome) [84, 85]. Although the majority of enteric viruses are bacteriophages, a recent discovery revealed that eukaryotic viruses also play an important role in the gut microbiota, and they are linked to several human diseases [86].

5.1 Characterization of Intestinal Eukaryotic Viruses

Similarly to other microbial components of the intestinal microbiota, the intestinal virome has not been as intensively studied as the intestinal bacteriome—mainly due to limited available technologies. Traditionally, characterization of the intestinal virome involved viral culture and microscopy. These approaches provided limited information about the intestinal virome as the majority of viruses are not easily culturable [87, 88]. While 16S rRNA gene sequencing provides numerous insights into the gut bacteriome, due to its culture-independent nature, viruses do not harbor universally conserved genes such as the 16S rRNA gene and exhibit high genetic diversity [87]. Fortunately, directly sequencing the entire genomes of organisms present in a common environment (e.g. feces) using shotgun metagenomics has enabled investigators to characterize the intestinal virome [67, 89]. Though originally only DNA viruses could be sequenced, next-generation high-throughput sequencing techniques can now be applied to single-stranded DNA and RNA viruses [90]. However, the lack of analysis pipelines, such as QIIME [91] or Mothur [92] (commonly used for bacterial microbiota), for investigators to characterize the intestinal virome holistically and easily, hinders intensive exploration of the gut virome [93].

5.2 Contribution of Intestinal Eukaryotic Viruses to Host Health

Although the popular conception is that viruses are detrimental to the host, it has been demonstrated that enteric viruses can be just as beneficial to the host as gut bacteria. Specifically, a common gut virus—murine norovirus (MNV)—was found to be sufficient to replace the beneficial functions (e.g. immune system development) of commensal intestinal bacteria in mice [86]. Additionally, GF mice have been commonly used to investigate the functional influence of gut bacteria on the host; however, this model for host-microbe interactions is also an ideal platform to study the influence of enteric viruses on the mammalian host. For example, mono-colonization of GF mice with MNV was capable of reversing the abnormal intestinal morphology and immune system typically found in mice devoid of microbes. Interestingly the intestinal morphology of MNV mono-colonized mice resembled that of conventionally raised mice with a normal gut microbiota, suggesting that this virus alone can transmit some of the beneficial effects provided by a complex microbial community. Moreover, it was found that MNV prevented intestinal damage resulting from infectious and non-infectious secondary challenges to the GI tract in antibiotic-treated mice. These findings suggest that MNV, and other hitherto uncovered eukaryotic viruses, might be beneficial during developmental or recovery stages after antibiotic administration when the immune system and the intestinal microbiota are vulnerable.

As previously mentioned IBD, presenting as UC or CD, is characterized by chronic intestinal inflammation accompanied by gut epithelial barrier disruption and bacterial translocation to the mucosal tissue [94]. Though a vast majority of research has focused on the effects of intestinal bacteria on IBD, some notable studies have investigated the role enteroviruses play in IBD, although the association between commensal eukaryotic viruses and IBD remained relatively unexplored [95]. The gut virome was characterized in the rectal mucosal samples obtained from 91 patients with UC and 76 non-IBD controls, and revealed that *Pneumoviridae* were more abundant in UC patients than in controls, while the abundance of *Anelloviridae* was lower in patients with UC compared to controls [96]. Another recent study involved young patients (less than 20 years old) with CD (N = 216) or UC (N = 61), as well as 42 healthy controls. The mucosa virome characterization revealed that UC patients exhibited higher levels of eukaryotic *Hepadnaviridae* transcripts than other groups, while CD patients had a higher abundance of *Hepeviridae* than other patients [97]. Furthermore, this study reported that patients with UC had a lower level of *Polydnaviridae* and *Tymoviridae* and patients with CD showed a lower level of *Virgaviridae* in their gut mucosa, when compared to healthy controls. These association studies suggest that eukaryotic viruses might play a role in IBD, either by triggering intestinal inflammation and contribute to IBD pathogenesis, or by providing protection against inflammation. More research is therefore warranted to investigate their potential mechanisms of actions and pave the way for novel diagnostic biomarkers and treatment development for IBD.

The enteric virome has also been associated with CRC. The development of colorectal cancer (CRC) is multifaceted; including host, environmental, and gut microbial factors, but the structural variations in the nonbacterial component of gut microbiota has not been explored until recently. A study involving 74 patients with CRC and 92 healthy controls investigated the viral component of the CRC-associated gut microbiome [98]. The study revealed that the abundance of *Orthobunyavirus* in colonic contents was significantly higher in patients with CRC than in controls. However, the role of *Orthobunyavirus* in the context of the gut microbiota has not yet been explored [99]. This study suggests that enteric eukaryotic viruses might contribute to the development of CRC, but further research is needed to uncover the mechanism.

5.3 Gut Bacteriophages

Bacteriophages (or phages—derived from Greek for *bacteria eaters*) are viruses that exclusively infect bacteria and require this microbial host for replication. Phages were discovered independently by Frederick W. Twort in Great Britain and Félix d'Hérelle in France between 1915 and 1918 [100]. These ubiquitous particles are the most numerous entities in the biosphere, outnumbering all other organisms put together [101]. In nature, the ratio of phage to bacteria is estimated to be between 10:1, suggesting that bacteria are constantly subjected to phage infection in the natural environment. Phages interact dynamically with each other and their host—acquiring genes from, and contributing genes to, phage genomes and bacterial genomes by horizontal gene transfer. Out of necessity for survival, phages are highly diverse and play vital roles in the evolution, physiology, and pathogenicity of their hosts through cycles of predation and co-evolution [102]. The vast majority of phages found in nature fall into the *Caudovirales* order, primarily having tails and double stranded (ds) DNA genomes [103]. The three main families of *Caudovirales* are named based on tail type—*Myoviridae* have long contractile tails, *Siphoviridae* have long non-contractile and flexible tails, and *Podoviridae* have short, stubby, and non-contractile tails [104]. Phages are often referred to as the *dark matter* of the biological world due to the large amount of unclassified genetic information they carry in their genomes [102]. This is more evident in the complex microbial gut environment, where most phages cannot be taxonomically classified or linked to their particular hosts [105].

Phage replication can be lytic or lysogenic [106]. In both life cycles, a phage attaches to a bacterium using a receptor-binding protein and injects its genome into the host. In the lytic cycle, the phage replicates using the host's machinery, assembles new virions and then disintegrates (lyses) the cell. These phages are termed virulent phages. In the lysogenic life cycle, the phage genome is commonly integrated into the host chromosome as a prophage and replicates along with the host genome when the cell divides. In a mutually beneficial interaction, bacteria harboring a prophage are often rendered immune to superinfection by closely related phages. Less commonly the phage genome may be propagated independently (termed an episome) within the host cell but outside the host chromosome [106]. A phage capable of forming lysogens is known as a temperate phage. Temperate phages can be induced into the lytic cycle, often in response to environmental stressors or DNA damage.

5.4 Traditional and Molecular Characterization

The plaque assay first developed by Felix d'Hérelle remains the standard method for detecting and quantifying phage particles in culture. This involves making dilutions of phage samples, culturing them in the presence of a bacterial host on solid nutrient media, and counting the individual clear areas in a lawn of bacteria (plaques) where microbes have been killed. Since a large portion of the bacteria in the distal gut are difficult to culture, the plaque assay is not ideal for large-scale characterization of phage component of the gut virome. Instead, next generation and metagenomic sequencing is more useful for assessing the diversity and abundance of phages in the gut. This involves the physical separation of DNA and RNA of virus-like particles (VLPs) from gut bacteria, followed by recreation and characterization of genomes using assembly software [105]. Although challenges still remain with this approach, including the limited availability of reference viral genomes, these studies have begun to illuminate the phage component of the gut microbiota—the human gut phageome.

Phages (prophages and viral particles) dominate the gut virome, with an estimated 10^8–10^{11} VLPs per gram of feces [83, 107]. The majority of gut phages are thought to be temperate, residing as prophages within their bacterial host [107, 108]. In addition, both lysogenic and lytic phage can adhere to the GI mucus layer, where they have been proposed to prevent microbial colonization and protect underlying epithelial cells from invading bacteria [109]. A comparison of 21 adult monozygotic twin pairs found that phage diversity correlates with microbiome diversity and the richer the gut microbial community the richer the phageome [110]. Gut phageomes are predicted to be highly individual-specific, with more than 50% being predicted to be unique to each individual [107, 108, 111]. On the other hand, within individuals the gut virome displays high temporal stability for up to 2.5 years [108, 112]. There is evidence that some phages are globally distributed, suggesting the existence of a *core* healthy gut phageome [111]. Transmission Electron Microscopy and metagenomic sequencing studies show that ds DNA phages of the *Caudovirales* order and single stranded (ss) DNA containing phages of the *Microviradae* family dominate the gut [83, 112]. The host range of the *Caudovirales* include the most prevalent bacterial phyla found in the gut: Firmicutes, Bacteroidetes, Proteobacteria, and Actinobacteria [105]. In contrast to other environments where phages outnumber their bacterial hosts by a factor of 10, evidence suggests a closer to 1:1 ratio of phage to host in the human gut

[83, 113]. In 2014, computational analysis of publicly available metagenomic data identified what is now accepted as the most abundant phage in the gut microbiota, a unique podovirus-like phage, named crAssphage (named after the cross Assembly software used) [114]. CrAssphage is likely part of the *core* healthy gut virome and present in ~50% of individuals. Specific gut phage-host interactions remain elusive and the first phage-host pair was only recently published when one CrAssphage (crAss001) was successfully cultured from *Bacteroides intestinalis* using a phage-enriched filtrate from human fecal samples [115].

5.5 Contribution of Gut Phages to Host Health

As antibiotic-resistant bacterial infections rapidly increase around the world, there has been a resurgence of interest in phage therapy. Exploitation of the natural ability of lytic phage to selectively target a particular bacterial host have received the most interest. There have been some exciting cases of effective clinical use of phage therapy, including a dramatic race by infectious disease epidemiologist, Dr. Steffanie Strathdee, to find a novel phage therapy for her husband's multidrug resistant *Acinetobacter baumannii* infection [116]. This resulted in the first U.S. patient to be successfully treated by intravenous phage therapy for a systemic *superbug* infection and the subsequent launching of the first dedicated phage therapy center in North America—the center for Innovative Phage Applications and Therapeutics (IPATH). In 2019, the first safe and effective use of an engineered phage to treat a drug-resistant infection in a 15-year old Cystic Fibrosis lung transplant patient was reported [117]. The potential of phage therapy to positively modulate the gut microbiota in human GI diseases such as IBD and CRC is tantalizing; however, a lot remains to be understood about the complex role of phage in shaping the human gut microbiome before the true usefulness of gut phage therapy can be assessed. Indeed, studies have been contradictory, likely due to the ongoing refinement and analysis of publicly-available gut metagenomic data.

It has been established that differences in gut phage composition exist between healthy and diseased cohorts. Specifically, phage composition from the VLP fraction is different in individuals with IBD compared to healthy controls [118], with an increased abundance of *Caudovirales* phages observed in both UC and CD [119, 120]. Of concern is the potential for phage to interact directly with the mammalian host immune system, causing expansion of immune cells in the gut, increased intestinal inflammation and exacerbated colitis [121]. On a positive note, the same study showed that phage therapy with human GI tract-derived phages could protect against an adherent invasive *E. coli*-exacerbated CRC, reducing tumor growth and mortality in mice [121]. Additionally, in FMT studies to treat reoccurring CDI, phages from the donor fecal material were found in recipient feces 6 weeks following FMT treatment with stable engraftments of the donor phage population detectable for up to 12 months [14, 122]. In addition, there is evidence that phages may play a role in the success of FMT for relapsing CDI, as FMT treatment from a sterile filtrate effectively reduced CD infection [15]. Moreover, in UC, a positive response to FMT is reported to be inversely correlated with *Caurovirales* abundance [121].

6 Clinical Perspectives

The future of gut microbiota research must take account of the entire repertoire of microorganisms harbored within this complex microbial community (including archaea, bacteria, fungi, parasites, and viruses), and indeed the *holobiont* as a unit, with inter-kingdom interactions in a specific setting of nutrient and host immune factors. Indeed, neglecting to account for these interactions, the conclusions of gut microbiota studies will be biased and incomplete. In order to accomplish this, investigators in the field need to: (1) acknowledge the significance of the non-bacterial components of the gut microbiota, (2) develop relevant sequence databases and platforms to adequately characterize all members of the gut microbiota, and (3) develop animal models that take the *holobiont* into account. Presently there is room for optimism, through data collaborations such as the parasite microbiome project [123] and IPATH, coupled with the development of new animal models, and increasing affordability of multi-omic platforms to elucidate the mechanisms of microbial effects [124]. When paired with longitudinal, controlled, and interventional study designs, these elements should lead to a more comprehensive understanding of the human intestinal microbiota. Manipulation of the bacterial as well as associated protist, helminth, fungal, and viral microbiota harbors enormous potential for therapy for infectious as well as autoimmune, allergic, and metabolic disease.

References

1. Faith JJ, Guruge JL, Charbonneau M, et al. The long-term stability of the human gut microbiota. Science. 2013;341:1237439.
2. Ringel Y, Maharshak N, Ringel-Kulka T, et al. High throughput sequencing reveals distinct microbial populations within the mucosal and luminal niches in healthy individuals. Gut Microbes. 2015;6:173–81.

3. Thursby E, Juge N. Introduction to the human gut microbiota. Biochem J. 2017;474:1823–36.
4. Carabotti M, Scirocco A, Maselli MA, et al. The gut-brain axis: interactions between enteric microbiota, central and enteric nervous systems. Ann Gastroenterol. 2015;28:203–9.
5. Ussar S, Griffin NW, Bezy O, et al. Interactions between gut microbiota, host genetics and diet modulate the predisposition to obesity and metabolic syndrome. Cell Metab. 2015;22:516–30.
6. Nishino K, Nishida A, Inoue R, et al. Analysis of endoscopic brush samples identified mucosa-associated dysbiosis in inflammatory bowel disease. J Gastroenterol. 2018;53:95–106.
7. Trøseid M. Gut microbiota and acute coronary syndromes: ready for use in the emergency room? Eur Heart J. 2017;38:825–7.
8. Hasan N, Yang H. Factors affecting the composition of the gut microbiota, and its modulation. PeerJ. 2019;7:e7502.
9. Ridaura VK, Faith JJ, Rey FE, et al. Gut microbiota from twins discordant for obesity modulate metabolism in mice. Science. 2013;341:1241214.
10. Smith MI, Yatsunenko T, Manary MJ, et al. Gut microbiomes of Malawian twin pairs discordant for kwashiorkor. Science. 2013;339:548–54.
11. Britton GJ, Contijoch EJ, Mogno I, et al. Microbiotas from humans with inflammatory bowel disease alter the balance of gut Th17 and RORγt. Immunity. 2019;50:212–224.e4.
12. Sharon G, Cruz NJ, Kang DW, et al. Human gut microbiota from autism spectrum disorder promote behavioral symptoms in mice. Cell. 2019;177:1600–1618.e17.
13. Gibson GR, Hutkins R, Sanders ME, et al. Expert consensus document: the International Scientific Association for Probiotics and Prebiotics (ISAPP) consensus statement on the definition and scope of prebiotics. Nat Rev Gastroenterol Hepatol. 2017;14:491–502.
14. Jalanka J, Hillamaa A, Satokari R, et al. The long-term effects of faecal microbiota transplantation for gastrointestinal symptoms and general health in patients with recurrent Clostridium difficile infection. Aliment Pharmacol Ther. 2018;47:371–9.
15. Ott SJ, Waetzig GH, Rehman A, et al. Efficacy of sterile fecal filtrate transfer for treating patients with Clostridium difficile infection. Gastroenterology. 2017;152:799–811.e7.
16. Theis KR, Dheilly NM, Klassen JL, Brucker RM, Baines JF, Bosch TCG, et al. Getting the hologenome concept right: an eco-evolutionary framework for hosts and their microbiomes. mSystems. 2016;1:e00028-16. PMID:27822520.
17. Browne HP, Forster SC, Anonye BO, Kumar N, Neville BA, Stares MD, Goulding D, Lawley TD. Culturing of 'unculturable' human microbiota reveals novel taxa and extensive sporulation. Nature. 2016;533:543–6.
18. Carroll IM, Ringel-Kulka T, Siddle JP, et al. Alterations in composition and diversity of the intestinal microbiota in patients with diarrhea-predominant irritable bowel syndrome. Neurogastroenterol Motil. 2012;24(6):521–30, e248.
19. Arthur JC, Perez-Chanona E, Mühlbauer M, et al. Intestinal inflammation targets cancer-inducing activity of the microbiota. Science. 2012;338:120–3.
20. Kim SC, Tonkonogy SL, Albright CA, et al. Variable phenotypes of enterocolitis in interleukin 10-deficient mice monoassociated with two different commensal bacteria. Gastroenterology. 2005;128:891–906.
21. Ring C, Klopfleisch R, Dahlke K, et al. Akkermansia muciniphila strain ATCC BAA-835 does not promote short-term intestinal inflammation in gnotobiotic interleukin-10-deficient mice. Gut Microbes. 2019;10:188–203.
22. Cryan JF, Dinan TG. Mind-altering microorganisms: the impact of the gut microbiota on brain and behaviour. Nat Rev Neurosci. 2012;13:701–12.
23. Lyte JM, Gheorghe CE, Goodson MS, et al. Gut-brain axis serotonergic responses to acute stress exposure are microbiome-dependent. Neurogastroenterol Motil. 2020;32:e13881.
24. Mao JH, Kim YM, Zhou YX, et al. Genetic and metabolic links between the murine microbiome and memory. Microbiome. 2020;8:53.
25. Qin J, Li R, Raes J, et al. A human gut microbial gene catalogue established by metagenomic sequencing. Nature. 2010;464:59–65.
26. Oh J, Byrd AL, Park M, et al. Temporal stability of the human skin microbiome. Cell. 2016;165:854–66.
27. Tang J, Iliev ID, Brown J, et al. Mycobiome: approaches to analysis of intestinal fungi. J Immunol Methods. 2015;421:112–21.
28. Lafontaine DL, Tollervey D. The function and synthesis of ribosomes. Nat Rev Mol Cell Biol. 2001;2:514–20.
29. Nash AK, Auchtung TA, Wong MC, et al. The gut mycobiome of the Human Microbiome Project healthy cohort. Microbiome. 2017;5:153.
30. Kõljalg U, Larsson KH, Abarenkov K, et al. UNITE: a database providing web-based methods for the molecular identification of ectomycorrhizal fungi. New Phytol. 2005;166:1063–8.
31. Todd RT, Forche A, Selmecki A. Ploidy variation in fungi: polyploidy, aneuploidy, and genome evolution. Microbiol Spectr. 2017;5:10.1128.
32. Paterson MJ, Oh S, Underhill DM. Host-microbe interactions: commensal fungi in the gut. Curr Opin Microbiol. 2017;40:131–7.
33. Mar Rodríguez M, Pérez D, Javier Chaves F, et al. Obesity changes the human gut mycobiome. Sci Rep. 2015;5:14600.
34. Huseyin CE, O'Toole PW, Cotter PD, et al. Forgotten fungi-the gut mycobiome in human health and disease. FEMS Microbiol Rev. 2017;41:479–511.
35. Erdogan A, Rao SS. Small intestinal fungal overgrowth. Curr Gastroenterol Rep. 2015;17:16.
36. Paul S, Boschetti G, Rinaudo-Gaujous M, et al. Association of anti-glycan antibodies and inflammatory bowel disease course. J Crohns Colitis. 2015;9:445–51.
37. Leonardi I, Li X, Semon A, et al. CX3CR1+ mononuclear phagocytes control immunity to intestinal fungi. Science. 2018;359:232–6.
38. Mueller KD, Zhang H, Serrano CR, et al. Gastrointestinal microbiota alteration induced by Mucor circinelloides in a murine model. J Microbiol. 2019;57:509–20.
39. Sohail MA, Adnan Sohail M, Ikram U. Flucytosine-induced colitis. BMJ Case Rep. 2014;2014:bcr2013203381.
40. Sokol H, Leducq V, Aschard H, et al. Fungal microbiota dysbiosis in IBD. Gut. 2017;66:1039–48.
41. Hoarau G, Mukherjee PK, Gower-Rousseau C, et al. Bacteriome and mycobiome interactions underscore microbial dysbiosis in familial Crohn's disease. mBio. 2016;7:e01250-16.
42. Iliev ID, Funari VA, Taylor KD, et al. Interactions between commensal fungi and the C-type lectin receptor Dectin-1 influence colitis. Science. 2012;336:1314–7.
43. Ferwerda B, Ferwerda G, Plantinga TS, et al. Human dectin-1 deficiency and mucocutaneous fungal infections. N Engl J Med. 2009;361:1760–7.
44. Garsin DA, Lorenz MC. Candida albicans and Enterococcus faecalis in the gut: synergy in commensalism? Gut Microbes. 2013;4:409–15.
45. Shang FM, Liu HL. Fusobacterium nucleatum and colorectal cancer: a review. World J Gastrointest Oncol. 2018;10:71–81.
46. Pleguezuelos-Manzano C, Puschhof J, Rosendahl Huber A, et al. Mutational signature in colorectal cancer caused by genotoxic pks. Nature. 2020;580:269–73.
47. Sokic-Milutinovic A, Alempijevic T, Milosavljevic T. Role of Helicobacter pylori infection in gastric carcinogenesis: cur-

rent knowledge and future directions. World J Gastroenterol. 2015;21:11654–72.
48. Aykut B, Pushalkar S, Chen R, et al. The fungal mycobiome promotes pancreatic oncogenesis via activation of MBL. Nature. 2019;574:264–7.
49. Hamad I, Raoult D, Bittar F. Repertory of eukaryotes (eukaryome) in the human gastrointestinal tract: taxonomy and detection methods. Parasite Immunol. 2016;38:12–36.
50. Laforest-Lapointe I, Arrieta MC. Microbial eukaryotes: a missing link in gut microbiome studies. mSystems. 2018;3:e00201-17.
51. Lukeš J, Stensvold CR, Jirků-Pomajbíková K, et al. Are human intestinal eukaryotes beneficial or commensals? PLoS Pathog. 2015;11:e1005039.
52. Muhsen K, Levine MM. A systematic review and meta-analysis of the association between Giardia lamblia and endemic pediatric diarrhea in developing countries. Clin Infect Dis. 2012;55:S271–93.
53. Chabé M, Lokmer A, Ségurel L. Gut protozoa: friends or foes of the human gut microbiota? Trends Parasitol. 2017;33:925–34.
54. Scanlan PD, Stensvold CR, Rajilić-Stojanović M, et al. The microbial eukaryote Blastocystis is a prevalent and diverse member of the healthy human gut microbiota. FEMS Microbiol Ecol. 2014;90:326–30.
55. Herwaldt BL, de Arroyave KR, Wahlquist SP, et al. Multiyear prospective study of intestinal parasitism in a cohort of Peace Corps volunteers in Guatemala. J Clin Microbiol. 2001;39:34–42.
56. Wiria AE, Djuardi Y, Supali T, et al. Helminth infection in populations undergoing epidemiological transition: a friend or foe? Semin Immunopathol. 2012;34:889–901.
57. Mabbott NA. The influence of parasite infections on host immunity to co-infection with other pathogens. Front Immunol. 2018;9:2579.
58. Monin L, Griffiths KL, Lam WY, et al. Helminth-induced arginase-1 exacerbates lung inflammation and disease severity in tuberculosis. J Clin Invest. 2015;125:4699–713.
59. Rowan-Nash AD, Korry BJ, Mylonakis E, et al. Cross-domain and viral interactions in the microbiome. Microbiol Mol Biol Rev. 2019;83:e00044-18.
60. Tanasescu R, Tench CR, Constantinescu CS, et al. Hookworm treatment for relapsing multiple sclerosis: a randomized double-blinded placebo-controlled trial. JAMA Neurol. 2020;77:1089–98.
61. Huang X, Zeng LR, Chen FS, et al. Trichuris suis ova therapy in inflammatory bowel disease: a meta-analysis. Medicine (Baltimore). 2018;97:e12087.
62. Helmby H. Human helminth therapy to treat inflammatory disorders—where do we stand? BMC Immunol. 2015;16:12.
63. Baron S. Medical microbiology. Galveston, TX: University of Texas Medical Branch at Galveston; 1996.
64. Liu J, Platts-Mills JA, Juma J, et al. Use of quantitative molecular diagnostic methods to identify causes of diarrhoea in children: a reanalysis of the GEMS case-control study. Lancet. 2016;388:1291–301.
65. Platts-Mills JA, Liu J, Rogawski ET, et al. Use of quantitative molecular diagnostic methods to assess the aetiology, burden, and clinical characteristics of diarrhoea in children in low-resource settings: a reanalysis of the MAL-ED cohort study. Lancet Glob Health. 2018;6:e1309–18.
66. Maritz JM, Rogers KH, Rock TM, et al. An 18S rRNA workflow for characterizing protists in sewage, with a focus on zoonotic Trichomonads. Microb Ecol. 2017;74:923–36.
67. Quince C, Walker AW, Simpson JT, et al. Shotgun metagenomics, from sampling to analysis. Nat Biotechnol. 2017;35:833–44.
68. Stensvold CR, van der Giezen M. Associations between gut microbiota and common luminal intestinal parasites. Trends Parasitol. 2018;34:369–77.
69. Mohapatra S, Singh DP, Alcid D, et al. Beyond O&P times three. Am J Gastroenterol. 2018;113:805–18.
70. Khalil IA, Troeger C, Rao PC, et al. Morbidity, mortality, and long-term consequences associated with diarrhoea from Cryptosporidium infection in children younger than 5 years: a meta-analyses study. Lancet Glob Health. 2018;6:e758–68.
71. Bethony J, Brooker S, Albonico M, et al. Soil-transmitted helminth infections: Ascariasis, Trichuriasis, and hookworm. Lancet. 2006;367:1521–32.
72. Ahmed T, Auble D, Berkley JA, et al. An evolving perspective about the origins of childhood undernutrition and nutritional interventions that includes the gut microbiome. Ann N Y Acad Sci. 2014;1332:22–38.
73. Bartelt LA, Bolick DT, Guerrant RL. Disentangling microbial mediators of malnutrition: modeling environmental enteric dysfunction. Cell Mol Gastroenterol Hepatol. 2019;7:692–707.
74. Petersen AM, Stensvold CR, Mirsepasi H, et al. Active ulcerative colitis associated with low prevalence of Blastocystis and Dientamoeba fragilis infection. Scand J Gastroenterol. 2013;48:638–9.
75. Ramanan D, Bowcutt R, Lee SC, et al. Helminth infection promotes colonization resistance via type 2 immunity. Science. 2016;352:608–12.
76. Rapin A, Harris NL. Helminth-bacterial interactions: cause and consequence. Trends Immunol. 2018;39:724–33.
77. Bär AK, Phukan N, Pinheiro J, et al. The interplay of host microbiota and parasitic protozoans at mucosal interfaces: implications for the outcomes of infections and diseases. PLoS Negl Trop Dis. 2015;9:e0004176.
78. Partida-Rodríguez O, Serrano-Vázquez A, Nieves-Ramírez ME, et al. Human intestinal microbiota: interaction between parasites and the host immune response. Arch Med Res. 2017;48:690–700.
79. Ngobeni R, Samie A, Moonah S, et al. Entamoeba species in South Africa: correlations with the host microbiome, parasite burdens, and first description of Entamoeba bangladeshi outside of Asia. J Infect Dis. 2017;216:1592–600.
80. Gilchrist CA, Petri SE, Schneider BN, et al. Role of the gut microbiota of children in diarrhea due to the protozoan parasite Entamoeba histolytica. J Infect Dis. 2016;213:1579–85.
81. Koonin EV, Senkevich TG, Dolja VV. The ancient Virus World and evolution of cells. Biol Direct. 2006;1:29.
82. Breitbart M, Rohwer F. Here a virus, there a virus, everywhere the same virus? Trends Microbiol. 2005;13:278–84.
83. Hoyles L, McCartney AL, Neve H, et al. Characterization of virus-like particles associated with the human faecal and caecal microbiota. Res Microbiol. 2014;165:803–12.
84. Virgin HW. The virome in mammalian physiology and disease. Cell. 2014;157:142–50.
85. Virgin HW, Wherry EJ, Ahmed R. Redefining chronic viral infection. Cell. 2009;138:30–50.
86. Kernbauer E, Ding Y, Cadwell K. An enteric virus can replace the beneficial function of commensal bacteria. Nature. 2014;516:94–8.
87. Carding SR, Davis N, Hoyles L. Review article: the human intestinal virome in health and disease. Aliment Pharmacol Ther. 2017;46:800–15.
88. Mokili JL, Rohwer F, Dutilh BE. Metagenomics and future perspectives in virus discovery. Curr Opin Virol. 2012;2:63–77.
89. Hugenholtz P, Tyson GW. Microbiology: metagenomics. Nature. 2008;455:481–3.
90. Ansorge WJ. Next-generation DNA sequencing techniques. New Biotechnol. 2009;25:195–203.
91. Bolyen E, Rideout JR, Dillon MR, et al. Reproducible, interactive, scalable and extensible microbiome data science using QIIME 2. Nat Biotechnol. 2019;37:852–7.

92. Schloss PD, Westcott SL, Ryabin T, et al. Introducing mothur: open-source, platform-independent, community-supported software for describing and comparing microbial communities. Appl Environ Microbiol. 2009;75:7537–41.
93. Abeles SR, Ly M, Santiago-Rodriguez TM, et al. Effects of long term antibiotic therapy on human oral and fecal viromes. PLoS One. 2015;10:e0134941.
94. Ungaro RC, Colombel JF. Editorial: biologics in inflammatory bowel disease-time for direct comparisons. Aliment Pharmacol Ther. 2017;46:68–9.
95. Lopetuso LR, Ianiro G, Scaldaferri F, et al. Gut virome and inflammatory bowel disease. Inflamm Bowel Dis. 2016;22:1708–12.
96. Zuo T, Lu XJ, Zhang Y, et al. Gut mucosal virome alterations in ulcerative colitis. Gut. 2019;68:1169–79.
97. Ungaro F, Massimino L, Furfaro F, et al. Metagenomic analysis of intestinal mucosa revealed a specific eukaryotic gut virome signature in early-diagnosed inflammatory bowel disease. Gut Microbes. 2019;10:149–58.
98. Nakatsu G, Zhou H, Wu WKK, et al. Alterations in enteric virome are associated with colorectal cancer and survival outcomes. Gastroenterology. 2018;155:529–541.e5.
99. Elliott RM. Orthobunyaviruses: recent genetic and structural insights. Nat Rev Microbiol. 2014;12:673–85.
100. Duckworth D. History and basic properties of bacterial viruses. In: Goyal SM, Gerba CP, Bitton G, editors. Phage ecology. New York: John Wiley & Sons; 1987. p. 1–44.
101. Hendrix RW, Smith MC, Burns RN, et al. Evolutionary relationships among diverse bacteriophages and prophages: all the world's a phage. Proc Natl Acad Sci U S A. 1999;96:2192–7.
102. Pedulla ML, Ford ME, Houtz JM, et al. Origins of highly mosaic mycobacteriophage genomes. Cell. 2003;113:171–82.
103. Ackermann HW. Bacteriophage taxonomy in 1987. Microbiol Sci. 1987;4:214–8.
104. Ackermann HW. Bacteriophage observations and evolution. Res Microbiol. 2003;154:245–51.
105. Sutton TDS, Hill C. Gut bacteriophage: current understanding and challenges. Front Endocrinol (Lausanne). 2019;10:784.
106. Weinbauer MG. Ecology of prokaryotic viruses. FEMS Microbiol Rev. 2004;28:127–81.
107. Reyes A, Haynes M, Hanson N, et al. Viruses in the faecal microbiota of monozygotic twins and their mothers. Nature. 2010;466:334–8.
108. Minot S, Sinha R, Chen J, et al. The human gut virome: inter-individual variation and dynamic response to diet. Genome Res. 2011;21:1616–25.
109. Barr JJ, Auro R, Furlan M, et al. Bacteriophage adhering to mucus provide a non-host-derived immunity. Proc Natl Acad Sci U S A. 2013;110:10771–6.
110. Moreno-Gallego JL, Chou SP, Di Rienzi SC, et al. Virome diversity correlates with intestinal microbiome diversity in adult monozygotic twins. Cell Host Microbe. 2019;25:261–272.e5.
111. Manrique P, Bolduc B, Walk ST, et al. Healthy human gut phageome. Proc Natl Acad Sci U S A. 2016;113:10400–5.
112. Shkoporov AN, Clooney AG, Sutton TDS, et al. The human gut virome is highly diverse, stable, and individual specific. Cell Host Microbe. 2019;26:527–541.e5.
113. Sender R, Fuchs S, Milo R. Revised estimates for the number of human and bacteria cells in the body. PLoS Biol. 2016;14:e1002533.
114. Dutilh BE, Cassman N, McNair K, et al. A highly abundant bacteriophage discovered in the unknown sequences of human faecal metagenomes. Nat Commun. 2014;5:4498.
115. Shkoporov AN, Khokhlova EV, Fitzgerald CB, et al. ΦCrAss001 represents the most abundant bacteriophage family in the human gut and infects Bacteroides intestinalis. Nat Commun. 2018;9:4781.
116. Schooley RT, Biswas B, Gill JJ, et al. Development and use of personalized bacteriophage-based therapeutic cocktails to treat a patient with a disseminated resistant Acinetobacter baumannii infection. Antimicrob Agents Chemother. 2017;61:e00954-17.
117. Dedrick RM, Guerrero-Bustamante CA, Garlena RA, et al. Engineered bacteriophages for treatment of a patient with a disseminated drug-resistant Mycobacterium abscessus. Nat Med. 2019;25:730–3.
118. Clooney AG, Sutton TDS, Shkoporov AN, et al. Whole-virome analysis sheds light on viral dark matter in inflammatory bowel disease. Cell Host Microbe. 2019;26:764–778.e5.
119. Lepage P, Colombet J, Marteau P, et al. Dysbiosis in inflammatory bowel disease: a role for bacteriophages? Gut. 2008;57:424–5.
120. Norman JM, Handley SA, Baldridge MT, et al. Disease-specific alterations in the enteric virome in inflammatory bowel disease. Cell. 2015;160:447–60.
121. Gogokhia L, Buhrke K, Bell R, et al. Expansion of bacteriophages is linked to aggravated intestinal inflammation and colitis. Cell Host Microbe. 2019;25:285–299.e8.
122. Draper LA, Ryan FJ, Smith MK, et al. Long-term colonisation with donor bacteriophages following successful faecal microbial transplantation. Microbiome. 2018;6:220.
123. Dheilly NM, Martínez Martínez J, Rosario K, et al. Parasite microbiome project: grand challenges. PLoS Pathog. 2019;15:e1008028.
124. Leung JM, Graham AL, Knowles SCL. Parasite-microbiota interactions with the vertebrate gut: synthesis through an ecological lens. Front Microbiol. 2018;9:843.

The Role of Enteric Infection and the Microbiome in Human Health and Disease

Mark S. Riddle

Highlights

- The gut microbiota's capability to prevent and limit pathogen colonization, termed 'colonization resistance', includes direct effects such as production of bacteriocins, and indirect effects such as induction of immunomodulatory molecules, and stimulation of immune cell expansion.
- Efforts to promote disease resistance include the use of nonpathogenic bacteria and yeast (probiotics), or nutrients which enhance the growth of favorable microbes (prebiotics). Studies using *Saccharomyces boulardii*, *Lactobacillus* GG, *Enterococcus* LAB SF68, and the galacto-oligosaccharide mixture B-GOS, have shown limited efficacy in the prevention and treatment of travelers' diarrhea.
- GI tract infections can alter host microbiota composition and have long-term effects on GI tract function, as in the case of post-infectious irritable bowel syndrome (PI-IBS). Alternative mechanisms such as that observed with molecular mimicry between cytolethal distending toxin B (produced by *Campylobacter jejuni* and other pathogens) and the host protein vinculin have also been observed.
- Co-evolution between humans and infectious exposures are important in development of functional innate and adaptive immune development. Immune-mediated inflammatory conditions that have being increasing over the past decades may be related in part to recent secular changes of human populations to the diversity of microbial antigenic stimuli.

1 Introduction

Given the recent interest in the human gut microbiome in health and disease, a logical focus of this attention is on the role of the microbiome in relationship to the host and its interface with the external world of infection, particularly enteric infections. Much work has historically focused on the notion of colonization resistance and the possibility that a unique microbiome may put one at a lesser or greater risk for an enteric infection. However, the relationship between the host microbiome and infection can also be seen from the lens of how infection could impact the host microbiome and result in disease or possibly health (Fig. 1). While much has been learned, there is much more that we have yet to understand (summarized in Table 1).

M. S. Riddle (✉)
University of Nevada, Reno School of Medicine, Reno, NV, USA

M. Pimentel et al. (eds.), *Clinical Understanding of the Human Gut Microbiome*, https://doi.org/10.1007/978-3-031-46712-7_3

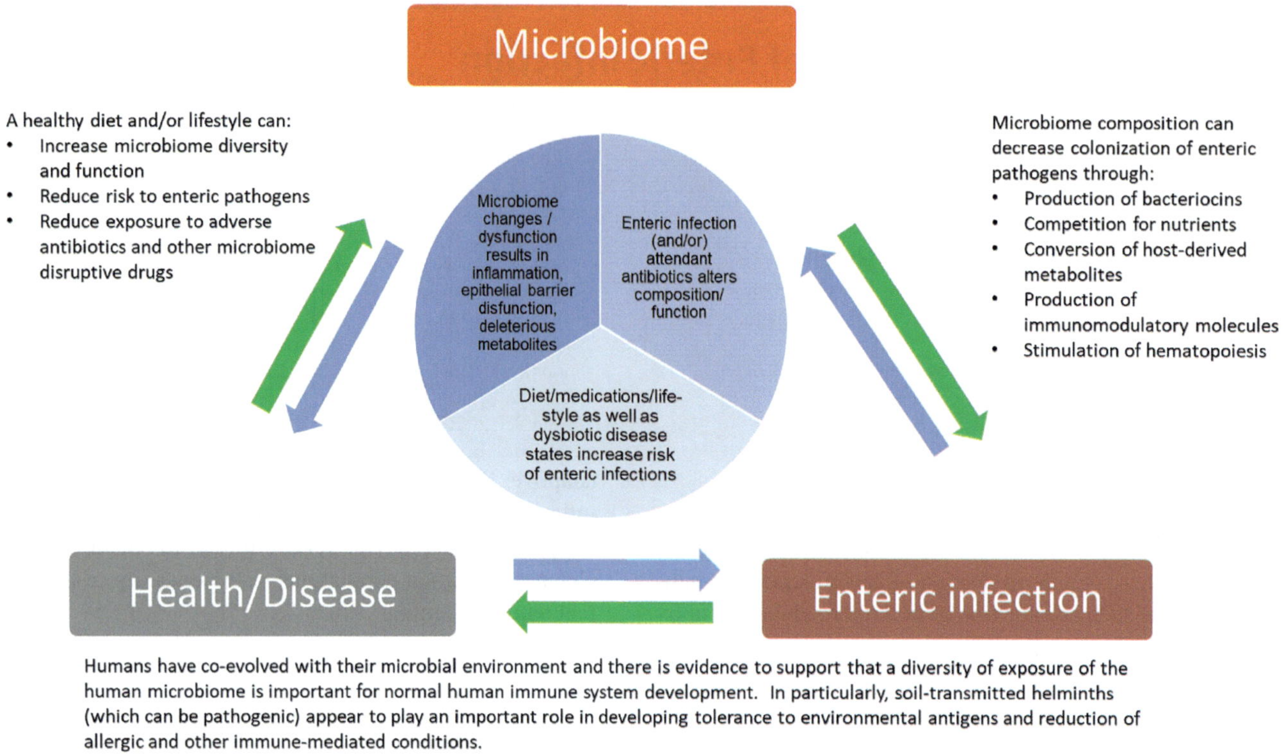

Fig. 1 Conceptual framework of the relationship between gastrointestinal infection, the microbiome, and human health and disease

Table 1 Knowledge gaps in the domains of the role of the microbiome and enteric infections in human health and disease

Knowledge area	Gap
Colonization resistance	1. Identification of a microbiome profile that resists the common infections of enteric infectious agents
	2. Functional aspects of the microbiome that lead to mechanisms of pathogen diverse resistance to colonization and infection
	3. Trials of probiotics, prebiotics and fecal microbiome supplementation on resistance to enteric infectious agents in geographically diverse settings
Chronic health consequences	1. Establishment of infection-induced dysbiosis and association with onset and persistent of chronic gastrointestinal problems, particularly IBS, dyspepsia and constipation
	2. Understanding of the underlying mechanisms of infection induction and development of chronic gastrointestinal consequences
	3. Treatment of post-enteric infection chronic gastrointestinal disorders (dysbiotic) with trials of probiotic, prebiotic and or human fecal microbiome augmentation
Other novel areas of study related to the microbiome, acute enteric infection and health	1. Impact of soil-transmitted helminth antigenic exposure on allergic and immune-mediated disease
	2. Impact of host-microbiome on response to parenteral and oral vaccines
	3. Mechanisms and interventions to reduce risk of acquisition of anti-microbial resistant pathogen acquisition or clearance after colonization

2 The Problem

Infections of the gastrointestinal tract are extremely common world-wide. Most recent estimates from the World Health Organization provide the staggering enormity of the problem [1, 2]. In 2010, the latest date available for comprehensive assessment, 33 different viral, protozoal, bacterial and parasitic infections was estimated to cause over 2 billion (95% uncertainty interval [UI] 1.5–3.0 billion) infections, over 1 million (95% UI 0.9–1.5 million) deaths, and 87.5 million (95% UI 72.6–110.2 million) disability adjusted life-years. A large proportion of this burden is borne by children in the developing world and these numbers do not consider the second order detrimental effects on nutrition, susceptibility to other infections, and apparent declines in cognitive function and physical stature and fitness. While fortunately these trends are falling, they still represent a huge burden of disease that largely could be prevented, and thus remain a focus of the global community.

From a less macroscopic perspective, food-borne illness also is a major problem for high and middle-income countries, as well as those who travel from these countries to the developing world. For example, in the United States, a report from the Centers for Disease Control and Prevention from 2011 (the latest available) estimated that approximately 1 in 6 Americans develop a food-borne (gastrointestinal) infection each year and were linked to 179 million episodes, 486,777 hospitalizations, and 6186 deaths. Particular to the latter category, travel continues to grow and expand at a little over 3% year with an estimated 1.8 billion annual trips to be reached by 2030 [3], and with an estimated 10–40% per trip attack rate, results in a significant health problem for international travelers alike.

3 The Microbiome as a Barrier to Infection

All these infections, which exist naturally in the environment and make their way to the potential human host through contaminated food and water, must first make their way through the gut epithelial barrier. The human gut has the largest surface area to exposure ratio with 300 sq. m of surface (or about the size of a tennis court or a studio apartment) [4]. This large surface area is critical for the host's survival as it is needed for the digestion of food and absorption of nutrients. However, the human gut surface also serves to shield the host from infectious organisms. The external layer of this shield (extrinsic barrier) contains an increasing density and complexity gradient of microbial flora (including bacteria, protozoa and viruses) that are found in the lumen of the gastrointestinal tract in the oral to aboral (relating to or denoting the side or end that is furthest from the mouth) direction. There is also a microbial gradient highest at the lumen and decreasing as you get closer to the surface of the gastrointestinal epithelial cell lining [4]. These microbial species are in a diverse matrix of mucous forming a complex biofilm in which bacterial species are found to be in various motile and non-motile forms [5].

Underlying this extrinsic barrier is the intrinsic barrier—a layer of a variety of host cells including stem cells, absorptive enterocytes, mucus-secreting goblet cells, enteroendocrine cells, and Paneth cells [6]. This highly dynamic layer undergoes complete turnover every few days and relies on a fragile balance between cell proliferation and cell death. This cell layer is multi-functional and produces antibacterial peptides, samples the luminal contents, and absorbs important nutrients and fluid into the host circulatory system. This layer is at an interface between the luminal extrinsic barrier and the host innate and adaptive immune system. Finally, it is important to note that the intrinsic and extrinsic barriers have bidirectional effects on the structure, composition, and function of each.

Within this context, much research has gone into understanding how exactly this barrier functions in host resistance to enteric pathogens. This concept of host resistance to enteric infection, commonly referred to as colonization resistance, describes the microbiota's capability to prevent and limit pathogen colonization and growth. It is not a new concept, and was first described over 60 years ago where the use of antibiotics was shown to immediately increase susceptibility to *Salmonella* infection in mice [7]. Over the past few decades our understanding has matured and understanding of direct effects such as production of bacteriocins (proteins produced by bacteria of one strain and active against those of a closely related strain), competition for nutrients, cross-feeding, and conversion of host-derived metabolites, and indirect effects such as induction of immunomodulatory molecules, and stimulation of immune cell expansion have been exquisitely characterized [8, 9].

However exquisite the mechanisms that have evolved to shield us from would be invaders, the bacterial, virus, and parasitic organisms have similarly evolved with creative mechanisms [10]. Examples include the antibacteriocin resistance of *Clostridoides difficile*, the alteration of microbiota composition and nutrient availability through local inflammatory effects of *Salmonella typhimurium*, and mucosal layer penetrating mechanisms of enterohemmorhagic *Escherichia coli* and *Shigella flexneri*, to name a few. Less is known about gut barrier defeating mechanisms of enteric viral and protozoal pathogens.

4 Modification of the Microbiome to Enhance Colonization Resistance: The Case of Travelers' Diarrhea

Given what is known about the importance of the microbiome as part of our extrinsic barrier against pathogen colonization and infection, efforts have been put forth towards using nonpathogenic bacteria and yeast (probiotics), as well as nutrients which enhance the growth of favorable microbes (prebiotics) as an intervention to promote disease resistance [11]. For a microorganism to be defined as a probiotic it must possess non-pathogenic properties, be viable in delivery vehicles, be stable in acid and bile, adhere to target epithelial tissue, persist within the gastrointestinal tract, produce antimicrobial substances, modulate the immune system, and influence metabolic activities [12]. Similarly, prebiotics which are usually non-digestible food ingredients that are fermentable in the colon and intended to stimulate health promoting bacteria, should be documented to confer a beneficial shift in the microbial equilibrium of the host gut flora [13]. However, the attempts to impact the risk and severity of the variety of pathogens that cause travelers' diarrhea have not met with much success. Two meta-analyses suggest a marginal benefit of probiotics in prevention of TD, though both suggest there is insufficient evidence for extrapolation to global recommendations for their use [14, 15]. Individual trials with two different probiotic classes demonstrate that location of travel (and perhaps attributable pathogen etiology by extension) may be an important factor. In a study using the yeast *Saccharomyces boulardii*, 3000 Austrian travelers were randomized in a two-dose level versus placebo effectiveness randomized controlled trial (RCT) (~1000 meeting efficacy evaluable criteria), where a clinically modest dose-dependent benefit in TD prevention was found, with greater effectiveness in travelers to North Africa and Turkey compared to other regions of travel [16]. This finding was consistent with a prior study which demonstrated better overall protection, but similar relative difference based on travel destination [17].

Evaluation of a different probiotic bacteria, *Lactobacillus* GG, among 756 Finnish travelers in 2 separate destinations in Turkey yielded a very modest efficacy overall (Treatment: 41% attack rate (AR); Placebo: 46.5% AR), the rate of protection was 39.5% at one location and almost nil at the other location [18]. In a U.S. study of 245 (out of 400 initially enrolled) travelers to various destinations using a similar *Lactobacillus* GG product, a protective efficacy of 47% was found, though overall attack rates were low and regional stratification was not performed [19]. None of these studies evaluated specific etiology of infections so it is difficult to know whether there are unique pathogen-specific effects or other host variability in travelers to different destinations.

Additional attempts have been explored with prebiotic formulations. A double blind placebo-controlled trial of a novel galacto-oligosaccharide mixture (B-GOS) in 159 healthy travelers to countries of low and high risk for TD showed a significant reduction in diarrhea in the prebiotic group compared to those who consumed placebo (maltodextrin) [20]. This particular formulation was studied because of earlier work showing this agent increased *Bifidobacferium* numbers in the colonic content of piglets and inhibited the attachment of entero-hepatic *E. coli* and *Salmonella enterica* Typhimurium to HT29 cells in vitro [21]. However, a large follow-on randomized controlled trial of this compound did not provide convincing evidence for clinically meaningful protective efficacy [22]. Lastly, a small TD prophylaxis study of 42 adult patients traveling to tropical countries (22 in the study group, 20 in the placebo group) evaluated the use of sodium butyrate and short chain fatty acids as a prebiotic and found an impressive reduction in the occurrence of TD in the prebiotic group (4.5%) versus 40% in the placebo group [23]. However, no further study with this formulation has been performed.

Beyond prevention, attempts have also tried microbiome modifying interventions to treat these enteric infections, through various mechanisms including production of acids, hydrogen peroxide, or other antimicrobial substances, increasing mucus production and gut barrier defense, as well as competition for nutrients or adhesion receptors, antitoxin action, and stimulation of the immune system [24]. A 2010 Cochrane Collaboration systematic review on probiotics and treatment of intestinal infection identified 63 randomized and quasi-randomized controlled trials comparing specific probiotic agent(s) versus placebo or no-treatment for acute diarrhea of presumed infectious etiology [25]. Of these, six trials were among adults, none of which were in a travelers' diarrhea setting although the pathogen-attributable infections described were somewhat comparable to common TD etiologies [26–31]. These six studies evaluated two different probiotics (five studies with the single product *Enterococcus* LAB SF68 and one study with *Sacchromyces boulardii*). While heterogeneous in study designs, settings, outcomes and populations, there were four studies on *Enterococcus* LAB SF68 that had an endpoint of diarrhea lasting greater than 4 days, and were aggregated and reported in the Cochrane review. In the probiotic arms of these studies, 21/168 (12.5%) compared to 102/165 (62%) in the placebo arms, had diarrhea lasting greater than 4 days after treatment initiation [26–29]. When combined, 79% efficacy was achieved (Relative risk: 0.21, 95% Confidence Interval 0.08–0.52) for this outcome though with significant heterogeneity ($Tau^2 = 0.56$; $Chi^2 = 10.47$, df = 3 ($P = 0.01$); $I^2 = 71\%$). Unfortunately, theoretical safety concerns raised about this product limits further recommendation [32]. One study evaluating *S. boulardii* did not appear to confer any advantage in the primary or secondary outcomes evaluated [31]. With respect to prebiotics, no studies have been done in travelers, and only a single poorly described study of a polyphenol-based prebiotic has been described in the treatment of acute diarrhea in Nicaraguan children and adults seeking treatment at community health centers in Managua, with a favorable effect [33].

In summary, available studies on probiotics and prebiotics for prevention and treatment of TD suffer from variability in age groups, setting, causes of acute diarrhea, formulations, dose and regimens. As such, their use cannot be recommended at this time; additional research is needed. Strategies which include complex microbiome augmentation which may be needed to create a stable and colonization resistant ecology, or with prebiotic formulations given their lesser cold-chain requirements, are warranted in future research.

5 Changes to the Microbiome Following Enteric Infection

While the microbiome can exert an effect on changing the susceptibility to enteric pathogens, the corollary of how enteric pathogens can exert an effect on changing the host microbiome is a topic of interest. While this may be intuitive, there are few studies in humans which explore this aspect. One such study which characterized patients in Bangladesh with *Vibrio cholerae* or enterotoxigenic *E. coli* infection longitudinally [34]. A model is described where infection and/or antibiotics clear both obligate and oxygen-consuming facultative anaerobes from the gut, which allows oxygen and dietary/host substrates to accumulate, then facultative anaerobes recolonize the gut. Next, aerobic respiration by the early recolonizers lower oxygen tensions to the point where obligate anaerobes can begin recolonizing the gut. It is also described that bacteriophages targeting the early colonizers may have an effect on reducing their numbers. Finally, competition for dietary and host substrates promotes the return to the original state. This entire process can take up to 30 days post infection.

Other studies which have tried to explore the infection-associated changes in the microbiome include one from Sweden, where the research team compared microbiota structure with and without enteric infection among returning travelers [35]. Among those who did not receive any antibiotics during travel (12 *Campylobacter* positive, 1 *Salmonella* positive and 36 uninfected), no significant changes in the composition of fecal microbiota were reported between pre- and post-travel samples, independent of infected or non-infected status. In contrast, a study among poultry abattoir workers with *Campylobacter* infection found long-term alterations to fecal microbiota compositions [36]. Lastly, a report on 38 subjects with confirmed acute norovirus infection (sampled ~1 day after acute infection onset) was compared to 22 unmatched healthy controls and found that approximately 20% of individuals with norovirus infection had a significant change in microbiota structure characterized by a loss in diversity and increased Proteobacteria species [37], though long-term follow-up sampling was not performed.

The disentanglement of independent effects of infection and the antibiotics that are often used to treat these infections is a challenge. However, these observations suggest that an infection of the intestinal tract may alter the milieu and certainly has persistent effects on both the innate and adaptive immune responses in the gastrointestinal tract. The potential impacts of these infections on long term changes in the gut that may affect health of diseases is the next topic.

6 Microbiome-Associated Disturbances in Chronic GI Disorders After Acute Enteric Infections

In most cases, the gastrointestinal effects of intestinal infection are self-limited, however in a subset of individuals, persistent changes in GI function may occur following infection. While temporary conditions such as post-illness lactose intolerance or other dietary intolerance has been described, and other more serious, but rarer, conditions of celiac and inflammatory bowel disease may be unmasked by an acute enteric infection [38], the most frequently reported (and studied) gastrointestinal sequelae following a gastrointestinal infection is irritable bowel syndrome (IBS). Perhaps the first account of this disorder was described by Sir Arthur Hurst in his 1918 Second Edition of Medical Diseases of the War where he writes [39],

> *Patients who have recovered from an acute attack of dysentery frequently remain unfit for a considerable period, which may even extend to years. The symptoms are due to the chronic colitis, which may follow either amoebic or bacillary dysentery after the specific infection has died out...In most cases the patient suffers from alternating attacks of constipation and diarrhoea, the latter often being brought on by aperients taken for the relief of the former, or it may follow an indiscretion in diet or exposure to cold...The diarrhoea may only last for a few hours, or it may continue for two or three days, the attacks being separated by intervals of several weeks or months...Sometimes the attacks of diarrhoea cease to occur, but intractable constipation remains and the general symptoms persist, though in a lessened degree.*

Similar reports were subsequently identified among British soldiers returning from WWII where multiple endoscopy and fecal examinations revealed an absence of inflammation or infection [40, 41], and most recently a systematic review was published encompassing six well-designed observational studies, specifically studying this phenomenon in the traveler setting [42], and are further supported by the dozens of other studies evaluating post-infectious IBS (PI-IBS) risk after acute enteric infection in the non-traveler setting [43].

While epidemiologic association is strong and consistent, a full understanding of disease mechanisms (causality) has yet to emerge [44, 45]. Multiple mechanisms at the root of these disorders including brain-gut-immune dysregulation, loss of epithelial barrier integrity, innate immunity defects, and dysbiosis have been studied [46–51]. Elaboration of these mechanisms has been the focus of a number of recent reviews [52, 53]. From a microbiome perspective, animal and human data demonstrate how changes in these microbial communities become evident in pathologic states including IBS [54, 55]. For example, one such study utilizing current microbiota sequencing techniques as well as intestinal biopsy was performed on 57 subjects from 5 different study groups: patients with diagnosed PI-IBS, patients who 6 months after gastroenteritis had either persisting bowel dysfunction or no IBS symptoms, benchmarked against patients with IBS-D (diarrhea predominant IBS) and healthy controls [56]. In this study, 27 discriminant bacterial groups distinguished the fecal microbiota profile of post-infectious IBS patients from that of healthy controls. Furthermore, pain quality and multiple somatic symptoms, but not anxiety or depression, were partially explained by the associated microbiota profile and significant associations between the host mucosal gene expression and histology and fecal microbiota were recognized. These findings indicate a situation of increased inflammatory response and compromised intestinal barrier in PI-IBS patients associated with a change in microbiome structure. Thus, the emerging evidence suggests that in some subjects, particularly those with PI-IBS, there is an associated dysbiosis which can be attributed to the associated clinical symptoms.

In addition to the link between dysbiosis and aberrant inflammation with PI-IBS, the triggering mechanisms of infection have also been studied. For example, *Campylobacter* and *Salmonella* infections are known to facilitate translocation of commensal flora across the intestinal epithelium, which may induce an aberrant host immune response to commensal bacteria, or host or dietary antigens [57, 58]. This was perhaps most elegantly shown by Hand et al., who demonstrated induction of microbiota-specific T-cell responses in mice infected with *Toxoplasma gondii* [58]. In a set of experiments it was found that following infection clearance, non-specific interferon-gamma producing CD4+ T cells were identified, pointing towards a response to commensal bacteria. Additionally, these T cells had the potential for proliferation following a second infectious event, defining a pathway whereby re-infection or exposure to similar inducing epitopes could trigger reactivation (and persistence of disease). Another mechanistic example is provided by Pimentel and colleagues, who have provided unique insight into the mechanism of post-*Campylobacter* small intestinal bacterial overgrowth (SIBO) through possible mimicry between the cytolethal distending toxin of *Campylobacter* and the protein vinculin found in high density in host cells [59]. SIBO, which is common in individuals with IBS, was demonstrated in a rat model in which *C. jejuni* infection induced changes in stool consistency, rectal inflammation, and lower density of important enteric nervous system cells, the interstitial cells of Cajal; an effect which could be inhibited by antimicrobial chemoprophylaxis [60–63]. A small study utilizing serum from the Department of Defense repository identified analogous anti-vinculin immune responses in post-*Campylobacter* IBS, but not in other pathogen-associated IBS [64]. Of note, this study also found unique antimicrobial markers in post-infectious IBS compared to "idiopathic" IBS, suggesting a role for host-immune interaction with the microbiome in chronic gastrointestinal disease.

In summary, while more study is needed, infection by enteric pathogens has been shown to have a disruptive effect on the balance within the gut epithelial barrier and microbiome. It is likely that, in susceptible individuals (though genetic susceptibility has not been fully worked out), these infections can result in abnormal immune activation which may further alter the balance of the microbiome in the gut (dysbiosis). This dysbiosis may have direct effects on the neuroendocrine system with central effects on the brain (perception/pain), as well as local effects on the enteric system (e.g. motility), leading to outcomes of chronic GI disorders which may take various forms, including functional and organic disorders.

7 Can Enteric Infection Be Good for You?

A confluence of observations decades ago including studies by Strachan demonstrating increase risk of seasonal allergic rhinitis and atopy related to increased family size and position in the household [65, 66], as well as Bach who described an ecological inverse relationship between population-based rates of infections and the rise of allergic and auto-immune diseases [67], brought about the notion of the potential importance of crowding-associated infectious exposures during childhood in the shaping of human health and disease. This framework was originally considered under a "hygiene hypothesis" whereby improvements in sanitation and cleanliness were thought to somehow leave individuals vulnerable to immune-mediated diseases. This hypothesis has subsequently been debunked as evolutionarily crowd infections were not part of the small hunter-gatherer experience and subsequent epidemiological studies have confirmed that childhood infections do not protect against atopic diseases [68].

However, what has emerged from the efforts to understand the observed epiphenomenon of allergic and immune-mediated diseases is a more complex and interconnected understanding of how early (and ongoing) life exposures may modulate the microbiome and one's innate and adaptive immune development to make individuals susceptible to these disorders of the immune system. For example, Rook coined the term the "Old Friends Hypothesis" where it is postulated that microbes present during hunter-gatherer times co-evolved with their human counterparts and were necessary components for humans to develop a "normal" immune system [69]. A rich body of literature has resulted specifically documenting the impact of soil-transmitted helminths on the immune system regulation and the important role that antigenic stimulation from these organisms plays in immune tolerance and disease attribution of allergic and other immune-mediated diseases [70]. The rationale and mechanistic support of the role of helminthic antigens has led to a large body of research that has shown promise in both animal and humans.

In summary, gastrointestinal exposure of microorganisms beyond helminth exposures are likely important for human health and to assure a diversity of microbiome exposures from both food and environmental microbial ingestions. The exact role of pathogenic species of human gastrointestinal infection are not fully known and a rich area of investigation. How we can garner the benefits of the diversity of enteric exposure to "healthy" infections without the negative consequences of such infections is important to consider.

References

1. Torgerson PR, Devleesschauwer B, Praet N, Speybroeck N, Willingham AL, Kasuga F, et al. World Health Organization estimates of the global and regional disease burden of 11 foodborne parasitic diseases, 2010: a data synthesis. PLoS Med. 2015;12(12):e1001920.
2. Kirk MD, Pires SM, Black RE, Caipo M, Crump JA, Devleesschauwer B, et al. World Health Organization estimates of the global and regional disease burden of 22 foodborne bacterial, protozoal, and viral diseases, 2010: a data synthesis. PLoS Med. 2015;12(12):e1001921.
3. Glaesser D, Kester J, Paulose H, Alizadeh A, Valentin B. Global travel patterns: an overview. J Travel Med. 2017;24(4).
4. Helander HF, Fandriks L. Surface area of the digestive tract—revisited. Scand J Gastroenterol. 2014;49(6):681–9.
5. Macfarlane S, Bahrami B, Macfarlane GT. Mucosal biofilm communities in the human intestinal tract. Adv Appl Microbiol. 2011;75:111–43.
6. Hooper LV, Littman DR, Macpherson AJ. Interactions between the microbiota and the immune system. Science. 2012;336(6086):1268–73.
7. Bohnhoff M, Miller CP. Enhanced susceptibility to Salmonella infection in streptomycin-treated mice. J Infect Dis. 1962;111:117–27.
8. Kamada N, Chen GY, Inohara N, Nunez G. Control of pathogens and pathobionts by the gut microbiota. Nat Immunol. 2013;14(7):685–90.
9. Leslie JL, Young VB. The rest of the story: the microbiome and gastrointestinal infections. Curr Opin Microbiol. 2015;23:121–5.
10. Ducarmon QR, Zwittink RD, Hornung BVH, van Schaik W, Young VB, Kuijper EJ. Gut microbiota and colonization resistance against bacterial enteric infection. Microbiol Mol Biol Rev. 2019;83(3):e00007-19.
11. Roberfroid MB. Prebiotics and probiotics: are they functional foods? Am J Clin Nutr. 2000;71(6 Suppl):1682S–7S; discussion 8S–90S.
12. Dunne C, O'Mahoney L, Murphy L. In vitro selection criteria for probiotic bacteria of human origin: correlation with in vivo findings. Am J Clin Nutr. 2001;73:3865–925.
13. Gibson GR, Probert HM, Loo JV, Rastall RA, Roberfroid MB. Dietary modulation of the human colonic microbiota: updating the concept of prebiotics. Nutr Res Rev. 2004;17(2):259–75.
14. Sazawal S, Hiremath G, Dhingra U, Malik P, Deb S, Black RE. Efficacy of probiotics in prevention of acute diarrhoea: a meta-analysis of masked, randomised, placebo-controlled trials. Lancet Infect Dis. 2006;6(6):374–82.
15. McFarland LV. Meta-analysis of probiotics for the prevention of traveler's diarrhea. Travel Med Infect Dis. 2007;5(2):97–105.
16. Kollaritsch H, Holst H, Grobara P, Wiedermann G. Prevention of travelers' diarrhea with Saccharomyces boulardii: results of a placebo controlled double-blind study. Fortschr Med. 1993;111(9):152–6.
17. Kollaritsch HH, Wiedermann G. Prevention of traveler's diarrhea: a double-blind randomized trial with Saccharomyces cerevisiae Hansen CBS 5926. Travel medicine. Berlin: Springer Science + Business Media; 1989. p. 328–32.
18. Oksanen PJ, Salminen S, Saxelin M, Hamalainen P, Ihantola-Vormisto A, Muurasniemi-Isoviita L, et al. Prevention of travellers' diarrhoea by Lactobacillus GG. Ann Med. 1990;22(1):53–6.
19. Hilton E, Kolakowski P, Singer C, Smith M. Efficacy of Lactobacillus GG as a diarrheal preventive in travelers. J Travel Med. 1997;4(1):41–3.
20. Drakoularakou A, Tzortzis G, Rastall RA, Gibson GR. A double-blind, placebo-controlled, randomized human study assessing the capacity of a novel galacto-oligosaccharide mixture in reducing travellers' diarrhoea. Eur J Clin Nutr. 2010;64(2):146–52.
21. Tzortzis G. Development and functional properties of Bimuno®: a second-generation prebiotic mixture. Food Sci Technol Bull Funct Foods. 2010;6(7):81–9.
22. Hasle G, Raastad R, Bjune G, Jenum PA, Heier L. Can a galacto-oligosaccharide reduce the risk of traveller's diarrhoea? A placebo-controlled, randomized, double-blind study. J Travel Med. 2017;24(5).
23. Krokowicz L, Kaczmarek BF, Krokowicz P, Stojcev Z, Mackiewicz J, Walkowiak J, et al. Sodium butyrate and short chain fatty acids in prevention of travellers' diarrhoea: a randomized prospective study. Travel Med Infect Dis. 2014;12(2):183–8.
24. Power SE, O'Toole PW, Stanton C, Ross RP, Fitzgerald GF. Intestinal microbiota, diet and health. Br J Nutr. 2014;111(3):387–402.
25. Allen SJ, Martinez EG, Gregorio GV, Dans LF. Probiotics for treating acute infectious diarrhoea. Cochrane Database Syst Rev. 2010;2010(11):CD003048.
26. Bruno F, Frigerio G. [A new therapeutic alternative for the treatment of enteritis—controlled double-blind tests with the strain SF 68]. Schweiz Rundsch Med Prax. 1981;70(39):1717–20.
27. Bruno F, Nastasi A, Bruno M. [Double-blind controlled study of the effect of the lactogenic enterococcus SF68 strain on various enterocolitis associated manifestations and on Salmonella infections]. Clin Ter. 1983;105(3):203–7.
28. Buydens P, Debeuckelaere S. Efficacy of SF 68 in the treatment of acute diarrhea. A placebo-controlled trial. Scand J Gastroenterol. 1996;31(9):887–91.
29. Wunderlich PF, Braun L, Fumagalli I, D'Apuzzo V, Heim F, Karly M, et al. Double-blind report on the efficacy of lactic acid-producing Enterococcus SF68 in the prevention of antibiotic-associated diarrhoea and in the treatment of acute diarrhoea. J Int Med Res. 1989;17(4):333–8.
30. Mitra AK, Rabbani GH. A double-blind, controlled trial of bioflorin (Streptococcus faecium SF68) in adults with acute diarrhea due to Vibrio cholerae and enterotoxigenic Escherichia coli. Gastroenterology. 1990;99(4):1149–52.
31. Höchter W, Chase D, Hagenhoff G. Saccharomyces boulardii in the treatment of acute adult diarrhoea. [Saccharomyces boulardii bei acuter Erwachsenendiarrhoea]. Münchener medizinische Wochenschrift. 1990;132(12):188–92.
32. Lund B, Edlund C. Probiotic Enterococcus faecium strain is a possible recipient of the vanA gene cluster. Clin Infect Dis. 2001;32(9):1384–5.
33. Noguera T, Wotring R, Melville CR, Hargraves K, Kumm J, Morton JM. Resolution of acute gastroenteritis symptoms in children and

adults treated with a novel polyphenol-based prebiotic. World J Gastroenterol. 2014;20(34):12301–7.
34. David LA, Weil A, Ryan ET, Calderwood SB, Harris JB, Chowdhury F, et al. Gut microbial succession follows acute secretory diarrhea in humans. mBio. 2015;6(3):e00381-15.
35. Kampmann C, Dicksved J, Engstrand L, Rautelin H. Composition of human faecal microbiota in resistance to Campylobacter infection. Clin Microbiol Infect. 2016;22(1):61.e1–8.
36. Dicksved J, Ellstrom P, Engstrand L, Rautelin H. Susceptibility to Campylobacter infection is associated with the species composition of the human fecal microbiota. mBio. 2014;5(5):e01212-14.
37. Nelson AM, Walk ST, Taube S, Taniuchi M, Houpt ER, Wobus CE, et al. Disruption of the human gut microbiota following Norovirus infection. PLoS One. 2012;7(10):e48224.
38. Connor BA, Riddle MS. Post-infectious sequelae of travelers' diarrhea. J Travel Med. 2013;20(5):303–12.
39. Hurst AF, Barber HW, Knott FA, Ross TA. Medical diseases of war. Arnold; 1944.
40. Chaudhary NA, Truelove SC. The irritable colon syndrome. A study of the clinical features, predisposing causes, and prognosis in 130 cases. Q J Med. 1962;31:307–22.
41. Stewart GT. Post-dysenteric colitis. Br Med J. 1950;1(4650):405–9.
42. Schwille-Kiuntke J, Mazurak N, Enck P. Systematic review with meta-analysis: post-infectious irritable bowel syndrome after travellers' diarrhoea. Aliment Pharmacol Ther. 2015;41(11):1029–37.
43. Halvorson HA, Schlett CW, Riddle MS. Post-infectious irritable bowel syndrome—a meta-analysis. Am J Gastroenterol. 2006;101:1894–9.
44. Deising A, Gutierrez RL, Porter CK, Riddle MS. Postinfectious functional gastrointestinal disorders: a focus on epidemiology and research agendas. Gastroenterol Hepatol (N Y). 2013;9(3): 145–57.
45. Morabia A. Until the lab takes it away from epidemiology. Prev Med. 2011;53(4–5):217–20.
46. Collins SM, Denou E, Verdu EF, Bercik P. The putative role of the intestinal microbiota in the irritable bowel syndrome. Dig Liver Dis. 2009;41(12):850–3.
47. Rutella S, Locatelli F. Intestinal dendritic cells in the pathogenesis of inflammatory bowel disease. World J Gastroenterol. 2011;17(33):3761–75.
48. Wells JM, Rossi O, Meijerink M, van Baarlen P. Epithelial crosstalk at the microbiota-mucosal interface. Proc Natl Acad Sci U S A. 2011;108(Suppl 1):4607–14.
49. Fukudo S, Kanazawa M. Gene, environment, and brain-gut interactions in irritable bowel syndrome. J Gastroenterol Hepatol. 2011;26(Suppl 3):110–5.
50. Mearin F, Perello A, Balboa A, Perona M, Sans M, Salas A, et al. Pathogenic mechanisms of postinfectious functional gastrointestinal disorders: results 3 years after gastroenteritis. Scand J Gastroenterol. 2009;44(10):1173–85.
51. Jung IS, Kim HS, Park H, Lee SL. The clinical course of postinfectious irritable bowel syndrome: a five-year follow-up study. J Clin Gastroenterol. 2009;43:534–40.
52. Verdu EF, Riddle MS. Chronic gastrointestinal consequences of acute infectious diarrhea: evolving concepts in epidemiology and pathogenesis. Am J Gastroenterol. 2012;107(7):981–9.
53. Bashashati M, Rezaei N, Andrews CN, Chen CQ, Daryani NE, Sharkey KA, et al. Cytokines and irritable bowel syndrome: where do we stand? Cytokine. 2012;57(2):201–9.
54. Rajilic-Stojanovic M, Biagi E, Heilig HG, Kajander K, Kekkonen RA, Tims S, et al. Global and deep molecular analysis of microbiota signatures in fecal samples from patients with irritable bowel syndrome. Gastroenterology. 2011;141(5):1792–801.
55. Talley NJ, Fodor AA. Bugs, stool, and the irritable bowel syndrome: too much is as bad as too little? Gastroenterology. 2011;141(5):1555–9.
56. Jalanka-Tuovinen J, Salojarvi J, Salonen A, Immonen O, Garsed K, Kelly FM, et al. Faecal microbiota composition and host-microbe cross-talk following gastroenteritis and in postinfectious irritable bowel syndrome. Gut. 2013;63:1737–45.
57. Kalischuk LD, Leggett F, Inglis GD. Campylobacter jejuni induces transcytosis of commensal bacteria across the intestinal epithelium through M-like cells. Gut Pathog. 2010;2:14.
58. Hand TW, Dos Santos LM, Bouladoux N, Molloy MJ, Pagan AJ, Pepper M, et al. Acute gastrointestinal infection induces long-lived microbiota-specific T cell responses. Science. 2012;337(6101):1553–6.
59. Pimentel M, Morales W, Pokkunuri V, Brikos C, Kim SM, Kim SE, et al. Autoimmunity links vinculin to the pathophysiology of chronic functional bowel changes following Campylobacter jejuni infection in a rat model. Dig Dis Sci. 2015;60(5):1195–205.
60. Pokkunuri V, Pimentel M, Morales W, Jee SR, Alpern J, Weitsman S, et al. Role of cytolethal distending toxin in altered stool form and bowel phenotypes in a rat model of post-infectious irritable bowel syndrome. J Neurogastroenterol Motil. 2012;18(4):434–42.
61. Morales W, Pimentel M, Hwang L, Kunkel D, Pokkunuri V, Basseri B, et al. Acute and chronic histological changes of the small bowel secondary to C. jejuni infection in a rat model for post-infectious IBS. Dig Dis Sci. 2011;56(9):2575–84.
62. Pimentel M, Morales W, Jee SR, Low K, Hwang L, Pokkunuri V, et al. Antibiotic prophylaxis prevents the development of a post-infectious phenotype in a new rat model of post-infectious IBS. Dig Dis Sci. 2011;56(7):1962–6.
63. Pimentel M, Chatterjee S, Chang C, Low K, Song Y, Liu C, et al. A new rat model links two contemporary theories in irritable bowel syndrome. Dig Dis Sci. 2008;53(4):982–9.
64. Pike BL, Paden KA, Alcala AN, Jaep KM, Gormley RP, Maue AC, et al. Immunological biomarkers in postinfectious irritable bowel syndrome. J Travel Med. 2015;22(4):242–50.
65. Strachan DP. Hay fever, hygiene, and household size. BMJ. 1989;299(6710):1259–60.
66. Strachan DP. Family size, infection and atopy: the first decade of the "hygiene hypothesis". Thorax. 2000;55 Suppl 1(Suppl 1):S2–10.
67. Bach JF. The effect of infections on susceptibility to autoimmune and allergic diseases. N Engl J Med. 2002;347(12):911–20.
68. Bloomfield SF, Rook GA, Scott EA, Shanahan F, Stanwell-Smith R, Turner P. Time to abandon the hygiene hypothesis: new perspectives on allergic disease, the human microbiome, infectious disease prevention and the role of targeted hygiene. Perspect Public Health. 2016;136(4):213–24.
69. Rook GA. 99th Dahlem conference on infection, inflammation and chronic inflammatory disorders: Darwinian medicine and the 'hygiene' or 'old friends' hypothesis. Clin Exp Immunol. 2010;160(1):70–9.
70. Alexandre-Silva GM, Brito-Souza PA, Oliveira ACS, Cerni FA, Zottich U, Pucca MB. The hygiene hypothesis at a glance: early exposures, immune mechanism and novel therapies. Acta Trop. 2018;188:16–26.

Gut Microbes: The Gut Brain Connection

Caoimhe M. K. Lynch, Kenneth J. O'Riordan, Gerard Clarke, and John F. Cryan

Highlights

- The microbiota-gut-brain axis plays a central role in regulating brain function and behavior and has been implicated in a myriad of brain diseases and disorders.
- The gut microbiota is dynamic and undergoes constant change throughout the lifespan, developing from birth to a mature, stable, and dynamic ecosystem later in life, followed by gradual decline in aging.
- The gut microbiota is susceptible to environmental signals throughout the lifespan, which modifies the risk of neurodevelopmental, neuropsychiatric disorders, and neurodegenerative diseases.
- The microbiota-gut-brain axis facilitates bidirectional communication via numerous direct and indirect routes including endocrine, metabolic, immune, and neural pathways.
- While cross-sectional clinical studies have enabled the implication of the gut-brain axis in the pathophysiology of disease, longitudinal and randomized controlled trials in humans are merited to fully understand the molecular underpinnings of the gut microbiota, and enable the development of novel microbiota-based therapeutic interventions.

C. M. K. Lynch · J. F. Cryan (✉)
APC Microbiome Ireland, University College Cork, Cork, Ireland

Department of Anatomy and Neuroscience, University College Cork, Cork, Ireland
e-mail: caoimhe_lynch@umail.ucc.ie; J.Cryan@ucc.ie

K. J. O'Riordan
APC Microbiome Ireland, University College Cork, Cork, Ireland
e-mail: ken.oriordan@ucc.ie

G. Clarke
APC Microbiome Ireland, University College Cork, Cork, Ireland

Department of Psychiatry and Neurobehavioural Science, University College Cork, Cork, Ireland
e-mail: G.Clarke@ucc.ie

1 Introduction

The human gastrointestinal tract is inhabited by an estimated 10^{13} to 10^{14} microorganisms [1]. Humans consist of more than 99% microbial genes and contain more microbial cells than human cells [2]. A relationship between the gut microbiota and the brain has long been surmised. For instance, over 2000 years ago the Greek physician Hippocrates, also known as the father of modern medicine, reportedly proclaimed that "*All disease begins in the gut*" (Hippokrátes ho Kṓos: c. 460–c. 370 BCE). Indeed, over recent decades, research into the gut-brain axis has grown exponentially as microbiologists, neuroscientists and nutritionists combine to further investigate the scope of influence and mechanistic basis for the relationship between our gut microbiota and our brain. More recently, a research revolution has occurred with the key player being the gut microbiome and the trillions of microorganisms that reside within, collectively known as the gut microbiota. Studies that have fueled this interest not only highlight the role of the gut microbiome in maintaining homeostasis and in regulating brain function, but also its role in key aspects of neurodevelopment, neuroinflammation and behavior [3–5]. Indeed, these studies arose following the emergence of the technology necessary for large scale surveys of the gut microbiome and associated metagenomic projects such as the *Human Microbiome Project,* placing microbiome research at the forefront of health and disease [6–8].

Models used to interrogate the microbiota-gut-brain axis have been instrumental in demonstrating the fundamental role of the gut microbiota in regulating a variety of physiological processes [9–11]. Studies involve the use of a range of experimental strategies including germ-free (GF) animals (born and reared devoid of all microbes), fecal microbiota transplant (FMT), antibiotic-induced depletion, prebiotic/probiotic supplementation, and animals exposed to specific pathogens. A wealth of research has focused on defining and elucidating the bidirectional communication pathways

M. Pimentel et al. (eds.), *Clinical Understanding of the Human Gut Microbiome*, https://doi.org/10.1007/978-3-031-46712-7_4

between the gut microbiota and the brain: defined as the microbiota-gut-brain axis [4, 12–14]. Determining the mechanistic pathways that underly gut microbiota to brain communication is an exciting multidisciplinary area of research that will provide novel insights into the effects that the gut microbiota have on human health. Harnessing this knowledge will enable the development of state-of-the-art therapeutic interventions targeting the promotion of good mental health and the aberrant aspects of brain function that manifest across a range of neuropsychiatric, neurodevelopmental, and neurodegenerative disorders. This chapter aims to provide a comprehensive overview of advances in the field of microbiota-gut-brain axis research with an emphasis on modes of communication, mechanisms of action, its role in behavior and disease, and the effects of influential factors, thus providing the reader with an insight into how this field has developed and what yet remains to be determined.

2 Life Cycle of the Microbiota-Gut-Brain-Axis

Throughout time, the microbiota has co-evolved with its host organism in a symbiotic relationship, benefiting the survival of both the host and its microbial colonizers [15–17]. It is important to note that mammals have never existed without their microbial counterparts, except in highly restrictive laboratory experiments. Consequently, the human brain has not experienced a time where it does not receive signals from host microbial inhabitants.

The gut microbiota undergoes constant change throughout the human life cycle [18] (Fig. 1). Alterations in microbial composition can be observed on a small scale due to day-to-day variations; however, these changes demonstrate dramatic effects when observed across the life cycle [19]. At both extremes of life, the gut microbiota exhibits different levels of diversity and composition of specific microbial taxa [20–22]. It is thought that these periods of change in the microbiota are a result of "sensitive periods" or "critical windows". Critical windows are defined as specific developmental windows during which a system exhibits heightened plasticity and is particularly sensitive to environmental stimuli [23]. A cascade of intricately timed sensitive periods have also been described for higher brain function and behavior [24, 25]. Thus, it is hypothesized that sensitive periods in the gut microbiota coincide not only with key neurodevelopmental windows but also the deterioration of bodily functions such as general brain function, immune function, and hypothalamic-pituitary-adrenal (HPA) signaling [26–29].

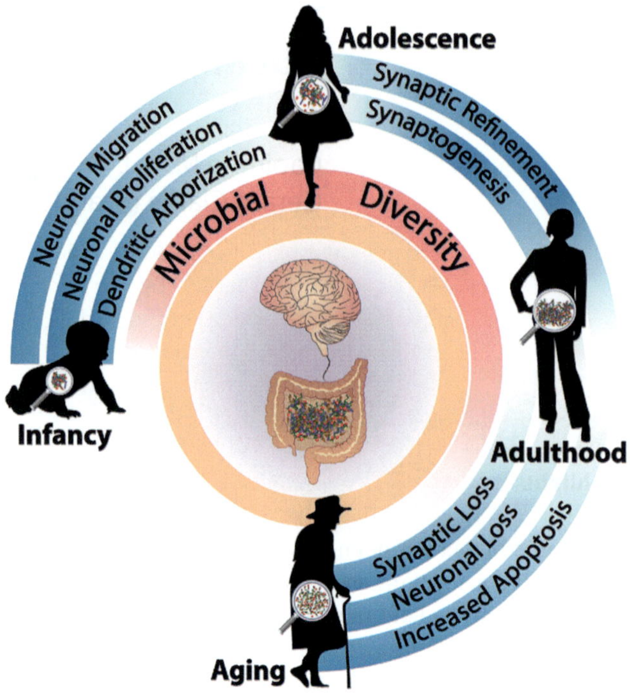

Fig. 1 Lifecycle of the microbiota-gut-brain axis throughout our lifespan. The composition of the gut microbiota dynamically changes throughout life, establishing a relationship with the host at critical windows during infancy, adolescence and aging. These changes in the microbiota are accompanied by concomitant changes in neuronal processes during specific stages of life. During these time windows there is increased vulnerability to environmental insults, which may predispose an individual to specific brain disorders. Intensity of bar colour signifies time period in which indicated processes are greatest

2.1 The Microbiota and Brain Development

Although it is thought that brain development and function are modulated by the microbiota-gut-brain axis, the extent of its influence during critical windows in early life remain elusive. Moreover, whether microbial colonization begins *in utero* is of major debate [30–36]. However, it is clear that colonization of the gut microbiota rapidly increases postnatally after initial exposure to the maternal gut and vaginal microbiome [37, 38]. Studies utilizing germ-free (GF) mice have been instrumental in highlighting the crucial role of the gut microbiota in key neurodevelopmental processes [39–43]. In these models, a variety of fundamental developmental processes such as neurogenesis [44], myelination [45, 46], neuroplasticity [47], axonogenesis [43], blood-brain barrier (BBB) permeability [48, 49], blood-cerebrospinal fluid barrier [50], neurotransmitters [51], microglia activation and neuroinflammation [52, 53], have all been shown to be regulated by the gut microbiota.

The absence of a gut microbiota during pregnancy has dramatic effects on offspring neurodevelopment. For instance,

GF mouse embryos exhibit increased BBB permeability and decreased expression of tight junction proteins [48]. Additional prenatal maternal factors which influence offspring brain development, behavior and microbial composition include stress, diet, obesity, and immune activation have been found to interact with the gut microbiome and elicit long-term changes in offspring development [54–60]. Shortly after birth the microbiota is initially dominated by high abundances of *Enterobacteriaceae*, *Bifidobacteriacea*, and *Clostridiaceae*. As the infant develops, the microbiota begins to diversify and shift from mostly facultative taxa to more anaerobic taxa. This additional surge in development aligns with weaning of the infant [61]. Some studies suggest that weaning is followed by a period of microbiota stability [62], whilst others suggest further plasticity across childhood and even into adolescence [63]. This early microbiota is enriched in nutrient-rich pathways essential for driving gradual development [64]. Further, numerous factors influence initial colonization of the microbiota including birth mode (vaginal vs. Cesarean section) [65], prematurity [66], birth place [67], diet (breast vs. formula fed) [68], stress [69, 70], antibiotic exposure [71, 72], pets [73], and illness [74]. Although some of these factors may elicit only transient effects on the gut microbiota, and may initially be clinically silent, it has been hypothesized that these early life influences may have long-lasting effects on an individual's physical and mental health and/or cognitive performance [69, 75, 76].

Indeed, the use of GF mice has been imperative for advancing our knowledge regarding the role of the gut microbiota in specific neurodevelopmental processes; however, it is an extreme approach to which the microbiota is absent across the life cycle with modest translatability to humans [40, 42]. Consequently, emerging research is focused on determining the impact of altered microbiota composition in early life and the enduring effects that specific microbes may have on brain health and behavior. Thus, preclinical studies utilizing antibiotic treatments have offered a useful alternative [77–79]. These studies aim to identify critical windows in which the developmental trajectory of the gut microbiota coincides with periods of neurodevelopmental plasticity [27, 77, 80, 81].

Currently, there have only been a few studies in humans that have investigated the effects of early-life microbiota alterations on host physiology and brain health. One study directly measured both microbiota composition and cognitive function. In a cohort of 89 infants, fecal microbiota composition at 1 year of age predicted cognitive performance at 2 years of age (assessed with the Mullen Scales of Early Learning) [82]. Lower microbiota alpha (α) diversity (a statistic used to describe variability within a sample) at 1 year of age was associated with better cognitive performance, language development, and visual reception a year later [83]. Further studies have also observed correlations between early childhood microbiota composition and behavioral temperament [69, 84, 85]. For instance, the Barwon Infant Study found a clear association between decreased abundance of *Prevotella* in fecal samples at 12 months of age and increased behavioral problems at 2 years [86]. Another study investigated the effects of perinatal antibiotic administration given either to the mother during gestation, or the infant whilst in the Neonatal Intensive Care Unit. This study found that perinatal antibiotic exposure was linked to increased attentional dysfunction at 4–5 years of age [87]. Correlations have also been found between early-life microbiota composition and later metabolic and immune function [88, 89]. Furthermore, numerous preclinical trials have demonstrated the ability to reduce an infant's risk of developing GI problems, autism spectrum disorder, and attention deficit hyperactivity disorder (ADHD) using probiotic interventions [90, 91]. Moreover, a large area of preclinical research has focused on early-life interventions which diminish the effects of early-life disruptions, e.g. antibiotics, C-section, diet, and early-life stress, on infant development [54, 92–98]. Overall, it is becoming increasingly clear that early-life disruption of the gut microbiota affects a wide range of developmental processes that can have long term effects on behavior and brain neurocircuitry.

2.2 The Microbiota and Aging: Causation or Correlation?

Aging is a highly complex process affecting an array of physiological, metabolic, genomic and immunological systems [99–101]. While gut microbes are incapable of aging *per se,* the prevalence of diseases associated with the gut microbiota increases as the host ages. Yet, it remains unclear whether alterations in the gut microbiota are a direct result or cause of aging. Early studies in GF mice showed that mice deficient of microbiota exhibit increased longevity compared to conventional controls, highlighting that the microbiota plays a role in aging [102, 103]. Recent preclinical studies have offered mechanistic insight into age-associated changes of microbiota composition and its implications for host health and behavior. For instance, age-related enrichment of the *Ruminococcacea* metabolite isoamylamine (IAA) was found to promote cognitive dysfunction by promoting microglial cell death [104]. Thus, in recent years aging research has increasingly focused on understanding how changes in the aging gut microbiota influences host health, and whether aging patterns within the gut microbiota reflect or contribute to healthy aging and survival outcomes.

Compositional analysis of the adult human gut microbiota shows that it is generally stable throughout life; however, this stability deteriorates as an individual ages. These changes may be attributed to alterations in lifestyle, diet, mobility, a weakened immune system, or an individual's overall health status [105, 106]. Of these, diet and exercise are two factors which can dramatically affect an aging individual by affecting cognition and microbial composition [107]. Notably, characterization of age-related changes in the microbiota is challenging due to individual heterogeneity among human populations, perhaps due to geographical, cultural, or methodological differences. In particular, a human study of the Irish population, the ELDERMET consortium, reported that age-related shifts in microbial composition correlated with global health measures including frailty, cognitive performance, depression and immune function [20, 108]. Specifically, the microbiota composition of the elderly was strongly influenced by diet, where an increase in microbial diversity was associated with improved health outcomes. A further study demonstrated that aged individuals (>60 years) demonstrated higher variability in microbiota composition when compared to younger individuals (20–60 years), attributed to increased abundance of pathobionts and reduced beneficial bacteria [109]. More recently, it was found that as an individual ages their gut microbiome becomes increasingly unique and this gut signature highly correlated with changes in plasma metabolites and the steady decrease of core bacterial genera including *Bacteroides* [110].

Indeed, microbiota based therapeutic interventions have demonstrated positive effects on microbial composition and associated host health [111, 112]. For instance, the NU-AGE project profiled 612 non-frail or pre-frail subjects across 5 European countries and found that a 1-year Mediterranean diet had the potential to promote healthier aging. Diet-modulated microbial alterations correlated with decreased frailty, increased cognition, and an increase in SCFA production [113]. However, the most promising therapeutic strategy to date has been dietary restriction, with a range of species exhibiting an increase in lifespan after caloric restriction [114–116].

Research has also focused on identifying the role of the gut microbiota in "inflammaging", a long-term proinflammation associated with physiological aging and the pathogenesis of age-related disorders [117–120]. The inflammaging process is thought to be accelerated by aging, age-related disorders and stress, all of which have been shown to negatively affect gastrointestinal barrier function and BBB permeability via the gut microbiota [48, 121–123]. Additionally, gut microbiota have been shown to regulate activation of microglia, the brain's resident immune cells. Interestingly, SCFAs are key bacterial metabolites required for normal microglia maturation and function [52]. However, during aging microglia transform into a highly reactive and unstable state, promoting cognitive deficits, altered brain plasticity, and the onset of chronic neurodegenerative diseases [124]. Recently, a study showed that gut microbiota transplanted from young mice into old mice reversed specific age-associated differences in peripheral and brain immunity, hippocampal metabolome, and attenuated age associated cognitive decline [117, 125]. Collectively, these studies suggest that targeting the gut microbiota could offer a novel non-invasive therapeutic intervention for promoting healthy aging.

3 Mechanisms of Gut-Brain Communication

There are several known pathways involved in the bidirectional communication between the gut microbiota and the brain (Fig. 2). The mechanisms of gut-brain signaling are complex and involve numerous direct and indirect routes including endocrine, metabolic, immune, or neural pathways. However, further research is merited to fully elucidate the molecular mechanisms in which the gut microbiota can influence brain function and behavior.

The vagus nerve, the tenth cranial nerve, is the fastest and most direct route of communication between the gut microbiota and the brain [126]. It is the longest cranial nerve and consists of an extensive branched network of afferent and efferent neurons, hence it's derived from the Latin word for *wandering*. It tonically transmits information from visceral organs (gastrointestinal tract, cardiovascular, and respiratory systems) to the brain, and *vice versa*. To date, research has primarily focused on the role of the vagus nerve in stress, hunger, inflammation, and satiety. However, more recent studies have highlighted the importance of the vagus nerve as a central route of communication between the gut microbes and the brain for normal brain function and behavior. In humans, ablation of the vagus nerve, a consequence of gastrectomy used for the treatment of peptic ulcers, resulted in individuals exhibiting increased incidences of psychiatric-related disorders [127]. Epidemiological studies from Denmark and Sweden have shown that individuals who underwent selective vagotomy in early life have a decreased risk of developing specific neurologic disorders [128, 129]. Conversely, vagal nerve stimulation exerts anti-inflammatory effects and offers a novel intervention for treatment-resistant major depressive disorder (MDD) [130, 131]. Moreover, the beneficial effects of specific bacterial strains were abrogated in vagotomized animals [132, 133]. Indirectly the vagus nerve has been shown to be activated by gut microbiota-associated metabolites such as serotonin [134], SCFAs [135], and indole [136]. Using optogenetic approaches in animal models, the vagus nerve has also been implicated in aspects of gut-induced CNS reward [137]. While the vagus nerve has been implicated as a key conduit for microbiota-brain signaling, research is now focused on mapping the underlying neural networks and molecular mechanisms that mediate its effects [138].

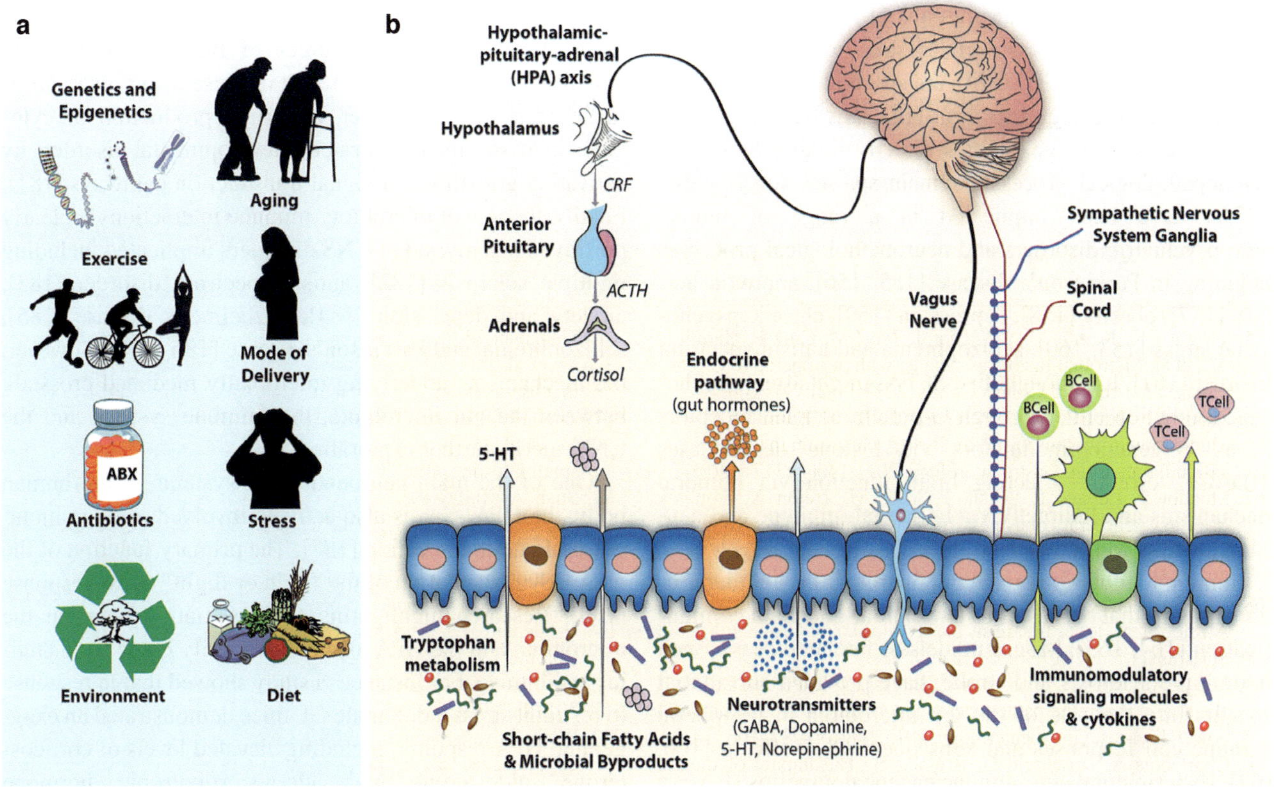

Fig. 2 Bidirectional communication between the gut microbiota and the brain. (**a**) Common factors which are known to affect microbiota-gut-brain axis communication. (**b**) Multiple potential direct (e.g., vagus nerve) and indirect (e.g., short-chain fatty acids, cytokines and key dietary amino acids such as tryptophan) pathways of communication between the gut microbiota and the brain exist. These include the immune pathway (signaling molecules and cytokines); the neuroactive pathway (neurotransmitters and neuroactive metabolites); short-chain fatty acids and microbial metabolites; the neural pathway (spinal nerves, the vagus nerve, enteric nervous system); the HPA axis and the endocrine pathway. A putative mechanism in which gut microbes can influence brain behavior and function is by the synthesis of neurotransmitters (GABA, dopamine, serotonin, and norepinephrine). Neuroactive bacterial metabolites, as well as dietary metabolites, can influence gut-brain communication by modulating gut-barrier function, stimulating enteroendocrine cells to release hormones, and regulating dendritic cell modulation of immune (B cells, T cells, and cytokines) and microglial function. Similarly, stress and emotions can influence microbial composition of the gut through the HPA axis, which regulates stress hormones (e.g., cortisol) and sympathetic neurotransmitter release. These consequently affect gut physiology by altering gut-barrier function and permeability, and by modulating immune cells in the gut. However, the precise molecular signaling pathways involved in gut-brain communication have yet to be fully elucidated. *ACTH* adrenocorticotropic hormone, *CRF* corticotropin-releasing factor, *5-HT* serotonin, *GABA* γ-aminobutyric acid, *HPA* hypothalamic-pituitary-adrenal axis

The microbiota also play a crucial role in modulating essential neurotransmitters by altering the availability of precursors [139, 140]. For instance, studies have demonstrated that *B. infantis* can influence central serotonin transmission by increasing plasma levels of tryptophan [141]. Microbially derived metabolites also influence the biosynthesis of serotonin by promoting the expression of its rate limiting enzyme tryptophan hydroxylase 1 (TPH1) in gut enterochromaffin cells [142]. Similarly, neuropod cells located in the gut epithelium can synthesize and release neurotransmitters such as glutamate which can transmit sensory signals to the brain directly via the vagus nerve [143]. Moreover, gut bacteria have been found to synthesize a range of neurotransmitters *de novo*; the inhibitory neurotransmitter gamma-aminobutyric acid (GABA) is produced by *Lactobacillus* and *Bifidobacterium* species. Additionally, *Escherichia*, *Saccharomyces* and *Bacillus* have been shown to produce noradrenaline and *Bacillus* can produce dopamine, whereas *Lactobacillus* produces acetylcholine [144–146]. While microbially synthesised neurotransmitters can cross the mucosal barrier of the intestine, it is doubtful that they can permeate across the BBB. Thus, an abnormal microbial composition may lead to abnormal neurotransmitter release, which may indirectly promote the development of neuropsychiatric disorders. For example, a case-control study hypothesized that compositional differences in the gut microbiota of ADHD children may contribute to changes in the gut-brain axis, affecting neurotransmitter levels, which may contribute to ADHD symptoms [147].

Furthermore, SCFAs including acetate, butyrate and propionate have been implicated in a plethora of physiological processes including GI function [148], neuroimmune modulation [149, 150], circadian rhythms [151], energy balance and metabolism [152] as well as stress regulation [153]. However, more recently SCFAs are speculated to play a key role in modulating microbiota-gut-brain axis signaling [154]. Despite our knowledge, studies directly linking SCFAs to psychopathological processes remain sparse. Nevertheless, SCFAs have been implicated in a range of human neuropsychiatric disorders and neuropathological processes including in Parkinson's disease [155, 156], anorexia nervosa [157], obesity [158], depression [159], chronic psychosocial stress [153, 160], schizophrenia and autism spectrum disorder [161]. It is thought that SCFAs might affect psychological function either through G-protein coupled receptors or as epigenetic modulators via histone deacetylases (HDACs), directly affecting brain function via humoral mechanisms and indirectly via hormonal, immune and neural pathways [154, 162, 163]

Immunomodulation by the gut microbiota is another crucial pathway that orchestrates microbiota-gut-brain communication [164, 165]. Mouse models of human disease, such as multiple sclerosis and stroke, have provided substantial insights into the role of the gut microbiota in peripheral immune cell responses and subsequent CNS effects [166, 167]. Experimental autoimmune encephalomyelitis (EAE; a mouse model of multiple sclerosis), GF and antibiotic treated mice all exhibit decreased inflammation and diseased scores compared to their conventional controls, indicating a role for the microbiota in neuroinflammation [168, 169]. Interestingly, colonization of GF mice with segmented filamentous bacteria promotes Th17 production and an EAE phenotype [170]. The hosts microbiota also plays a crucial role in controlling the maturation and function of CNS resident immune cells, particularly microglia [52, 149, 171]. Relative to controls, GF mice demonstrate altered microglia development and maturation in an age- and sex-dependent manner [172]. These effects can be normalized by colonization of the gut microbiota or treatment with SCFAs [52, 149]. Besides regulating intestinal barrier function, inflammatory signals can alter neuronal activity indirectly via the vagus nerve and HPA axis, or directly via cytokines in the brain [173].

A variety of cells including neurons and glia can produce cytokines, immunomodulatory signaling molecules that can cross the BBB. The lymphatic system may also facilitate the immunomodulatory potential of the gut microbiota [174]. Multiple lines of evidence have demonstrated that microbiota manipulations alter circulating and central cytokine levels [175–177]. Once in the CNS, cytokines can affect the metabolism of numerous neurotransmitters, thus altered cytokine levels can disrupt critical brain neurocircuitry [178]. For instance, in humans elevated levels of circulating pro-inflammatory cytokines interleukin IL-1β, IL-6 and TNFα are associated with neuropsychiatric disorders [179, 180]. However, cellular sources of these cytokines, and whether they are a cause or result of these disorders requires further analysis. Moreover, abnormal production of cytokines increases the risk for neurodevelopmental disorders by activating growth factor signal transduction pathways [181]. Finally, the role of microbiota-immune interactions is clearly portrayed in a myriad of CNS disorders implicated including multiple sclerosis [182], autism spectrum disorder [183], anxiety and depression [184], Alzheimer's disease [185], schizophrenia, and Parkinson's disease [186, 187]. However, the mechanisms underlying microbially mediated crosstalk between the gut microbiota, the immune system, and the CNS, merits further exploration.

One of the main neuroendocrine systems in the human body, the HPA axis, is also actively involved in microbiota-gut-brain communication [188]. The primary function of the HPA axis is regulation of the "fight or flight" stress response [189]. Research highlighting the association between the microbiota and the HPA axis has primarily come from studies in GF mice. For instance, a study showed that in response to restraint stress, adult male GF mice demonstrated an exaggerated HPA response, including elevated levels of corticosterone, aldosterone and adrenocorticotropic hormone (ACTH) [51, 190]. Additionally, this phenotype was rescued by colonization with a specific probiotic or pathogenic bacterium [190]. A recent study also found that GF mice exhibit altered serotonergic responses to acute stress exposure in a sex-dependent manner [191]. The gut microbiota also mediates the effects of maternal separation, a form of early-life stress that results in chronic activation of the HPA axis [192]. Using this model, conventionally colonized mice demonstrated increased anxiety and depressive-like behaviors, whereas GF mice showed no behavioral effects [193]. The effects of early-life stress on microbial composition is thought to impact the development of behavioral and emotional network systems, resulting in behavioral and psychiatric disorders [194]. Thus, it is not surprising that stress-related effects on the HPA axis, and subsequently the microbiota, can be seen in several pathological conditions such as irritable bowel syndrome (IBS) [195].

4 The Gut-Brain Axis and Disease

A growing body of evidence has implicated the microbiota-gut-brain axis in a myriad of physiological and neurological diseases and disease processes (Fig. 3). However, there is still extensive research required to investigate the mechanistic links that underlie microbiota-gut-brain communication and whether changes in the microbiota are central to the pathophysiology, or merely epiphenomenal [196].

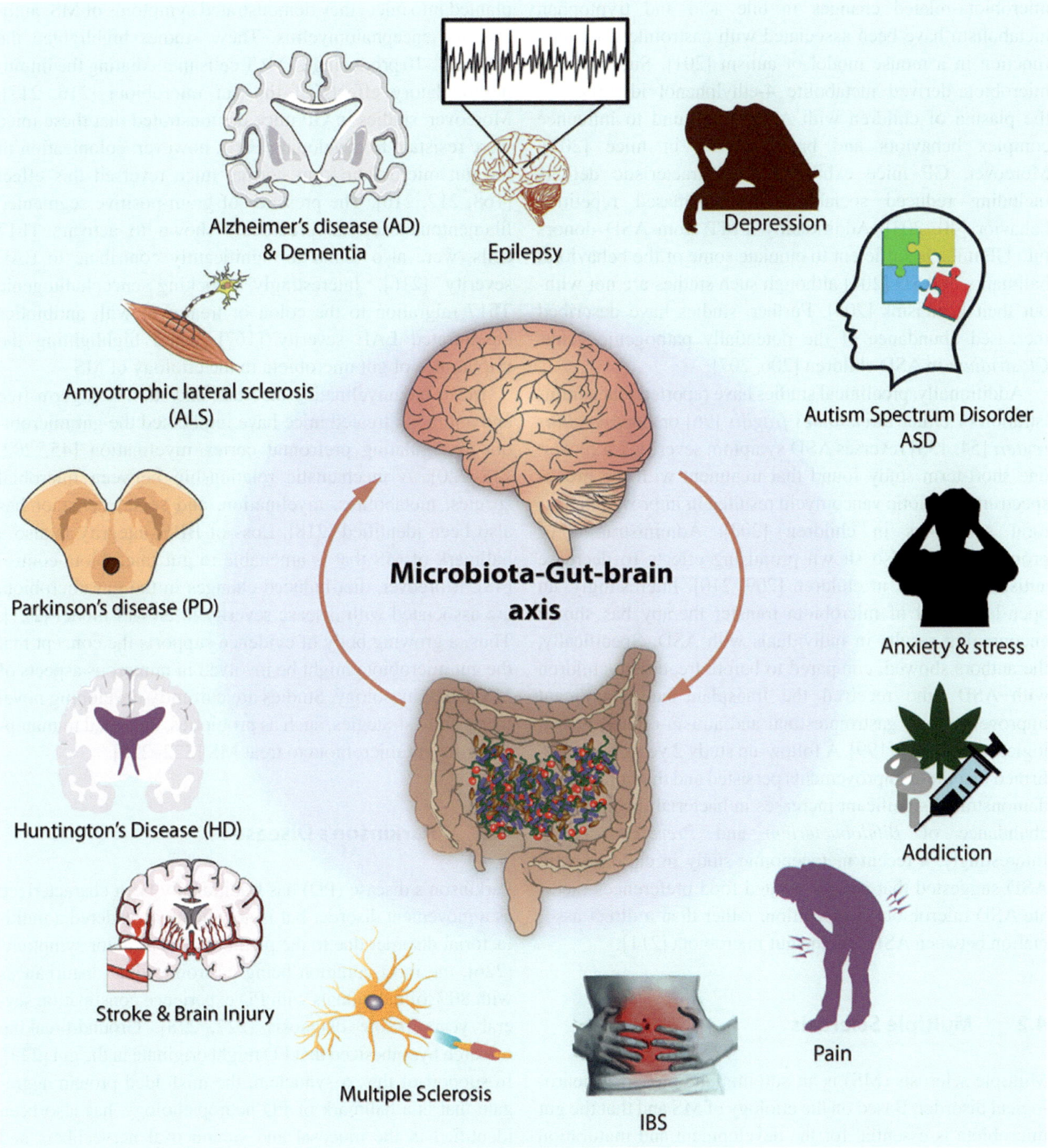

Fig. 3 The microbiota-gut-brain axis in disease. The microbiota-gut-brain axis has been implicated in a myriad of diseases and disease processes including psychiatric, neurodegenerative, neurodevelopmental, pain, stroke, and addiction

4.1 Autism Spectrum Disorder

Autism spectrum disorder (ASD) is a complex group of neurodevelopmental disorders characterized by profound deficits in social behavior and communication in association with repetitive and restrictive behaviors. Whilst ASD may be described primarily as a neurological disorder, influenced by both genetic and environmental factors, an often under-recognized feature is its marked comorbidity with gastrointestinal disturbances [197, 198]. Indeed, experimental studies, albeit relatively few, have shown alterations in the gut microbiota composition in ASD [161, 183, 199, 200]. Animal models have provided researchers with an insight into the role of the gut microbiota in ASD. For instance, microbiota-related changes in bile acid and tryptophan metabolism have been associated with gastrointestinal dysfunction in a mouse model of autism [201]. Similarly the microbiota-derived metabolite 4-ethylphenol identified in the plasma of children with ASD was found to influence complex behaviors and brain activity in mice [202]. Moreover, GF mice exhibit ASD characteristic deficits including reduced sociability and increased repetitive behavior [40, 203]. Additionally, FMT from ASD donors into GF mice is sufficient to emulate some of the behavioral hallmarks of ASD [204] although such studies are not without their criticisms [205]. Further, studies have described increased abundance of the potentially pathogenic genus *Clostridium* in ASD children [206, 207].

Additionally, preclinical studies have reported that administration of either *Bacteroides fragilis* [96] or *Lactobacillus reuteri* [54, 133] reverses ASD symptom severity. Moreover, one short-term study found that treatment with the broad-spectrum antibiotic vancomycin resulted in improved behavioral symptoms in children [208]. Administration of probiotics have also shown promising effects to decrease autism symptoms in children [209, 210]. Interestingly, an open-label trial of microbiota transfer therapy has shown encouraging results in individuals with ASD. Specifically, the authors showed, compared to before treatment, children with ASD who received the transplant had significant improvements in gastrointestinal and autism-related neurological symptoms [199]. A follow-up study 2 years later confirmed that these improvements persisted and that individuals demonstrated significant increases in bacterial diversity and abundance of *Bifidobacterium* and *Prevotella* [183]. Interestingly, a recent metagenomic study in children with ASD suggested that autistic-related food preferences mediate ASD microbiota composition, rather than a direct association between ASD and the gut microbiota [211].

4.2 Multiple Sclerosis

Multiple sclerosis (MS) is an autoimmune-mediated neurological disorder. Based on the etiology of MS and that the gut microbiota is essential for the development and maturation of the immune system, it is not surprising that MS-related therapies are now focused on targeting the gut microbiota as a potential therapeutic intervention [168, 212, 213]. Cross-sectional studies have also shown subtle changes in the gut microbiota of people with pediatric MS at lower taxonomic levels [214]. A recent longitudinal multi-omics study identified a number of alterations in the gut microbiota, immune and blood metabolome of individuals with MS [215].

Numerous studies have utilized EAE mouse models to investigate the role of the gut microbiota in MS. Interestingly, when fecal content from individuals with MS was transplanted into mice, they demonstrated symptoms of MS, autoimmune encephalomyelitis. These studies highlighted the role of IL-10 producing $CD4^+$ T cells in mediating the immunomodulatory effects of the gut microbiota [216, 217]. Moreover, studies in GF mice demonstrated that these mice were resistant to developing EAE; however, colonization of the gut microbiota from normal mice reversed this effect [168, 212, 216]. The presence of gram-positive segmented filamentous bacteria, previously shown to activate Th17 cells, were also found to significantly contribute to EAE severity [216]. Interestingly, blocking encephalitogenic Th17 migration to the colon or treatment with antibiotics ameliorated EAE severity [167], again highlighting the importance of gut microbiota in the etiology of MS.

MS is a demyelinating disorder, and studies in germ-free and antibiotic treated mice have implicated the gut microbiota in regulating prefrontal cortex myelination [45, 202, 218–220]. A mechanistic relationship between microbial species, metabolites, myelination, and social behavior has also been identified [218]. Loss of BBB integrity is also a hallmark of MS that is amenable to gut microbiota control [48]. Moreover, diet-induced changes in the gut microbiota are associated with disease severity in an MS model [221]. Thus, a growing body of evidence supports the concept that the gut microbiota might be involved in numerous aspects of MS pathophysiology. Studies are currently developing novel therapeutic strategies, such as probiotics, designed to manipulate the gut microbiota to treat MS [222–225].

4.3 Parkinson's Disease

Parkinson's disease (PD) has traditionally been characterized as a movement disorder but is now being considered a multifactorial disorder due to the plethora of non-motor symptoms [226], the most common being gastrointestinal disturbances with 80% of individuals with PD experience constipation several years before diagnosis [227, 228]. Ground-breaking research hypothesized that PD might originate in the gut [229]. In support of this, α-synuclein, the misfolded protein aggregate that is a hallmark of PD neuropathology, has also been identified in the mucosal and submucosal nerve fibers and ganglia of PD individuals, even prior to brain detection [230, 231]. Notably, α-synuclein aggregation appears to be regulated by the gut microbial metabolite lipopolysaccharide [232]. There is also preclinical evidence to suggest that α-synuclein may travel by retrograde transport to the brain via the vagus nerve [233]. Epidemiological studies have shown that individuals who underwent truncal vagotomy have a decreased risk of developing PD later in life [128, 129]. Replicated in mice, truncal vagotomy prevented gut-brain propagation of pathologic α-synuclein and associated neurodegeneration and behavioral deficits [234].

Moreover, the presence of an altered gut microbiota composition in individuals with PD has been extensively documented [235–239]. Whether or not a specific microbial signature for PD exists remains to be confirmed. However, emerging preclinical [240] and metagenomic studies [241] are providing strong evidence that support the role of the microbiota in PD pathophysiology. Transfer of fecal matter from individuals with PD into GF mice resulted in two hallmark symptoms of PD, neuroinflammation and motor impairment [242]. Moreover, antibiotic treatment improved these symptoms, providing further evidence for a role for the gut microbiota in PD. This study implicated SCFAs as neuroinflammatory activators in a mouse model of PD [242]. However, it is unknown if this translates to human studies as many individuals with PD exhibit decreased SCFA levels [243]. Further evidence demonstrated that amyloid producing *Escherichia coli* promotes α-synuclein pathology in the gut and the brain [244]. *Helicobacter pylori* infection has also been implicated in the development of PD symptomatology [245, 246]. A recent meta-analysis determined beneficial effects for *H. pylori* eradication by enhancing levodopa onset time and improving PD clinical severity [247]. Conversely, evidence has demonstrated that microbial tyrosine decarboxylases reduce the effectiveness of levodopa [248]. Indeed, mechanistic studies are merited to further understand how the gut microbiota influences not only PD symptomatology and pathophysiology but also its effects on drug safety and efficacy.

4.4 Alzheimer's Disease and Dementia

Alzheimer's disease (AD) is the most common neurodegenerative disorder and leading cause of dementia. The concept that AD is caused by alterations in the gut microbiota has been discussed for years. Moreover, new evidence is emerging further linking the pathogenic microbes in the gut with the development and progression of AD [249–251]. Influential experimental evidence also supports the concept that β-amyloid (Aβ) is acting as an innate antimicrobial peptide in the brain. However, it is ethically and logistically difficult to prove that there is an infective etiology in AD pathophysiology [252]. Interestingly, amyloid-like proteins can be produced by gut bacteria and can promote α-synucleinopathy in aged vagotomized rats. However, further research is required to validate this observation in humans [249].

While the neuropathology of AD primarily focuses on the accumulation of Aβ and abnormal tau tangles, it does not infer causality [253]. Therefore, recent research is exploring the role of viruses and bacteria in AD pathogenesis. In particular many studies have implicated herpes simplex virus type 1 as an important risk factor for developing AD [254, 255]. Moreover, GF transgenic AD mouse models have decreased Aβ pathology compared with conventional mice [256, 257]. Similarly, chronic treatment of transgenic mice with an antibiotic cocktail reduced Aβ plaque pathology and associated gliosis in the hippocampus and decreased insoluble Aβ plaques [258, 259]. Further highlighting that the microbiota plays a central role in regulating Aβ accumulation and hence AD pathology.

Cross-sectional studies have identified alterations in the gut microbiota profiles of individuals with AD, with a reduction in richness and diversity found in very mild to mild AD [247]. Individuals exhibited taxonomic differences with alterations in the Firmicutes: Bacteroidetes ratio (a correlational and observational output of microbiome analysis that provides information based on the direction of change compared to control) [260], which has also been described in obesity and provides an interesting association between AD and type 2 diabetes mellitus [261]. Moreover, AD and mild-cognitively impaired individuals exhibited altered bile acid profiles and this was associated with brain pathology [262]. Furthermore, one study determined that individuals with AD have a higher abundance of pro-inflammatory *Escherichia/Shigella* and decreased abundances of anti-inflammatory *E. rectale* compared to healthy controls. Alterations in microbiota composition are also associated with pro-inflammatory cytokine concentrations, suggesting a peripheral pro-inflammatory state in individuals with AD [185]. It is thought that mis-wired signaling between the gut microbiota and systemic inflammation might contribute to neurodegeneration that is observed in AD pathogenesis [263]. While this field of research is making extensive advancements, therapeutic applications are still far away.

4.5 Stroke and Brain Injury

Peripheral and systemic factors, including neuroinflammation, have been shown to exacerbate pathophysiological outcomes after stroke and brain injury [264]. Thus, a large field of research is focused on investigating the potential role of microbiota-related inflammation on stroke and brain injury outcomes. Cross-sectional studies have described alterations in microbiota composition in individuals presenting with stroke or transient ischemic attack compared with asymptomatic individuals. Decreased abundance of beneficial commensals such as *Bacteroides, Prevotella,* and *Faecalibacterium* was observed along with increased abundance of opportunistic pathogens *Enterobacter, Megasphaera* and *Desulfovibrio* [265]. Further, trimethylamine-N-oxide, a gut microbial metabolite with proatherogenic and prothrombotic effects, may act as a potential biomarker for stroke and possible therapeutic target [266, 267].

In preclinical mouse models, stroke severity associated with an altered microbiota composition and gastrointestinal function of gut permeability and motility [264, 268]. Additionally, fecal microbiota transfer from a cerebral ischemia model into GF mice [264], or from stroke individuals into antibiotic treated mice [269], resulted in increased abundance of pro-inflammatory T-leukocytes, exacerbation of ischemia-induced cerebral lesion volume, and stroke outcome. Pretreatment of broad-spectrum antibiotics also worsened stroke outcomes in mouse models [270]. Moreover, antibiotic induced alterations in the intestinal flora resulted in decreased pro-inflammatory IL-17 $\gamma\partial$ T-cell trafficking and chemokine expression [166]. Thus, it appears that the gut microbiota mediates post-stroke neuroinflammation by modulating T-cell trafficking to the brain. Conversely, probiotic supplementation has demonstrated beneficial effects for brain injury individuals. *Clostridium butyricum* showed neuroprotective effects in a murine model of traumatic brain injury [271]. The role of the gut microbiota in stroke and brain injury is only just being understood. Further clinical trials are required to fully understand the therapeutic potential of microbiota-based interventions for the prevention and treatment of stroke and traumatic brain injuries.

4.6 Epilepsy

Recent evidence from both human and animal studies suggest a potential role for the gut microbiota in certain forms of epilepsy [272]. Numerous studies in GF mice have implicated the gut microbiota in regulating synaptic changes in key brain regions involved in epileptogenesis [45, 47, 176, 273]. Interestingly, the ketogenic diet is a well-established, non-pharmacological treatment used for many years in children with drug-resistant epilepsy. Further, there is a strong association between the ketogenic diet, the gut microbiota composition of infants [274, 275] and animals [276], with epilepsy. While the mechanisms underlying the effects of the ketogenic diet have not been fully determined, many studies have proposed several potential mechanisms [277]. More recently, studies in GF mice have shown that the beneficial effects of the ketogenic diet are dependent on the gut microbiota [278]. Additionally, one small-scale study in 6 drug-resistant epilepsy patients showed that antibiotic treatment reduced seizures by 90% in 1 patient and ablated seizures in the remaining 5. These effects were abolished within 2 weeks of antibiotic cessation, perhaps due to recovery of the gut microbiota [279]. Thus, it is thought that future research will focus on the potential role of the gut microbiota as a mediator of epileptogenesis and investigate it as a possible therapeutic target.

4.7 Huntington Disease

Huntington Disease (HD) is a heritable neurodegenerative disease caused by an autosomal dominant gene [280]. Symptoms include motor alterations, cognitive decline and psychiatric decline which can lead to dementia [281]. There is a lack of research investigating the relationship between the gut microbiota and HD, perhaps because it is predominantly considered a genetic disorder. However, intrinsic and extrinsic factors can affect the progression of HD [282]. One metabolomic study observed altered microbial metabolites in the serum of HD patients who were early-stage or premorbid compared to control groups [283]. Recent work showed that absence of gut microbiota in a mouse model of HD (BACHD-GF) resulted in alterations in myelin thickness and plasticity in the corpus callosum [284]. Moreover, a preclinical study reported for the first-time significant differences in microbiota composition in a transgenic mouse model of HD. Specifically, the authors observed an increase in Bacteroidetes and decreased Firmicutes abundance in the HD mice compared to controls. However, no differences in diversity were observed in female mice, suggesting a sex-specific effect [285]. Another recent study investigated the brain microbiome of HD patients and identified *Pseudomonas*, *Acinetobacter*, and *Burkholderia* as prevalent bacterial genera [286].

4.8 Amyotrophic Lateral Sclerosis

Amyotrophic lateral sclerosis (ALS) is a progressive neurodegenerative disorder that is characterized by the progressive death of motor neurons and an average survival time of 5 years from diagnosis [287]. Mouse models of ALS have implicated changes in the gut microbiota with the pathogenesis of ALS. For instance, these models demonstrate altered gut barrier permeability, microbiota composition and decreased abundance of butyrate producing bacteria [288, 289]. Moreover, ALS-prone Sod1 transgenic mice have been shown to have altered microbiota and metabolite composition. This phenotype was enhanced under GF conditions or after antibiotic treatment [290]. However, cross-sectional studies in humans have not provided sufficient evidence to support a relationship between the gut microbiota and ALS pathogenesis. Two small-scale studies have reported alterations in gut microbiota composition in ALS patients compared to controls [291, 292], whereas another study reported no difference between 25 patients with ALS and 32 healthy controls [293].

4.9 Depression

There is extensive evidence implicating an altered microbiota profile in MDD, with individuals demonstrating differences in a variety of genera and decreased microbial richness when compared to controls [122, 294–298]. For instance, individuals with MDD tend to have higher abundances of phyla associated with pro-inflammatory states such as Bacteroidetes and decreased abundance of Firmicutes, which are associated with decreased inflammation. However, there is considerable variability between studies, perhaps due to small sample size or different sampling/analysis methodologies.

A recent large-scale study, the Flemish Gut Flora project, provides one of the most convincing datasets for the role of the microbiota-gut-brain axis in mental health [299]. In this study the authors determined that depressed individuals and individuals who reported a lower quality of life segregated into a certain enterotype (classification of a living organism based on its bacteriological ecosystem in the gut microbiome) compared to healthy controls. This enterotype was characterized as having a decreased microbial load and abundance of *Faecalibacterium*. Interestingly, the authors also associated increased abundance of butyrate-producing bacteria with a higher quality of life score. Furthermore, they curated 56 gut-brain modules potentially involved in brain function and behavior [299]. Interestingly, the HELIUS (Healthy Life in an Urban Setting) study examined microbiome-depression associations across 6 ethnic groups. Analyses demonstrated robust associations between the gut microbiota and depressive symptoms and this association generalizes across ethnic groups [300, 301]. Additional support for a causal role of the microbiota in MDD comes from preclinical studies in which fecal content from individuals with MDD was transplanted into murine models. Animals that received an FMT from depressed individuals displayed a heightened inflammatory state along with depressed and anxiety-like phenotypes compared to FMT from healthy controls [122, 298].

There is a growing body of data supporting that interventions targeting the gut microbiota confers brain health benefits [4, 302, 303]. For instance, probiotic administration resulted in improved depressive scores and elicited beneficial metabolic effects in an MDD cohort [304]. Further, *L. rhamnosus* HN001 supplementation during pregnancy reduced postpartum depression and anxiety-like behaviors [305]. Moreover, a placebo-controlled parallel study investigated the effects of synbiotic (co-administration of prebiotic and probiotic) supplementation in mild-moderate MDD individuals. Probiotic supplementation demonstrated beneficial effects by lowering depression scores and improved tryptophan signaling [306]. Despite the great potential of microbiota-targeted interventions, it is clear that more longitudinal large-scale studies are required to maximize the potential of psychobiotic interventions (i.e., live bacteria that confer mental health benefits).

4.10 Anxiety

Anxiety and depression frequently occur together; therefore, most studies investigating the microbiota-gut-brain axis and psychiatric disorders will often measure both behaviors. To date, there has been clear preclinical evidence to support the relationship between the gut microbiota and anxiety, whereby an animal's anxiety-like phenotype may be determined by its behavior in certain environments (open field test, light/dark box, elevated plus maze), or by its response to specific stressors [307, 308]. Studies utilizing GF mice highlight the casual role of the gut microbiota in anxiety related disorders. For instance, GF mice display decreased anxiety-like behavior and an increased corticosterone stress response implicating HPA axis dysfunction when compared to normal conventionally colonized mice [51, 309, 310]. Conversely, GF rats exhibit anxiety-like behavior and exacerbated neuroendocrine responses compared to controls [311]. Additionally, anxiety-related microbial differences are strain-specific, with the anxious BALB/c mice having a unique microbiota profile compared to the more resilient Swiss Webster mouse [312]. Further, fecal transfer from donor to recipient mice could transfer the behavioral phenotype of the donor. More recently, preclinical studies have also highlighted the therapeutic potential of prebiotic and probiotic treatments for stress-related disorders such as anxiety and depression [132, 175].

Conversely, very few studies have investigated the relationship between the gut microbiota and anxiety in clinical scenarios [302, 313]. However, a study recently demonstrated a relationship between the gut microbiota and depression and anxiety symptoms in individuals, where severity negatively associated with microbial richness, alpha diversity and specific bacterial taxa [314]. A longitudinal study also found a significant relationship between risky lifestyle clusters and risk for anxiety, depression, and stress [315]. Further two small interventional studies which report that probiotics reduce anxiety in healthy individuals [316, 317] and in clinical participants with chronic fatigue syndrome [318]. Further, a recent placebo-controlled study identified *Lactobacillus plantarum DR7* as a potential psychobiotic for the treatment of stress and anxiety [319]. However, it must be noted that further translational studies investigating the effects of psychobiotics for the treatment of stress-related disorders such as anxiety and depression are merited.

4.11 Schizophrenia

Schizophrenia is a debilitating psychiatric disorder characterized by chronic physical and psychosocial deterioration. Its complexity and heterogeneity make its etiology difficult to define. Further, mainstream treatments fail to address all aspects of the disorder and are associated with significant side-effects. Recently, there has been growing interest in the role of the gut microbiota in schizophrenia. However, detangling this relationship is difficult due to the use of antipsychotic medications, which are known to affect gut-microbiota composition [320, 321]. Nonetheless, a few studies have identified oral and fecal microbiota differences between healthy individuals and schizophrenics [322–324]. Transplant of fecal content from individuals with schizophrenia into GF mice induced hyperactivity and heightened responses in addition to changes in glutamate transmission in the hippocampus. However, there were no observed changes in social behavior and pre-pulse inhibition suggesting that the microbiome may not affect all aspects of schizophrenia, or that the results are not fully indicative of a schizophrenia-like phenotype [323].

Additionally, there is significant discord in studies investigating microbial composition in individuals with schizophrenia. While the majority of studies display no significant changes in microbial diversity two recent studies suggest there is an increase in both α and beta (β) diversity (alpha diversity is a statistic describing community variation within a community (sample), whereas beta diversity quantifies similarities between communities (samples)) [324, 325]. However, the considerable heterogeneity in the etiology and manifestation of schizophrenia may be the cause of this observation. Moreover, there are significant differences in the abundance of taxa that correlate with disease severity. A recent preliminary study of individuals with schizophrenia reported alterations in microbiota composition, with reduced abundance of *Lactobacillus* and *Bifidobacteria* species compared to healthy controls [326]. Notably, differences in microbiota composition were linked with negative symptom severity and increased risk for remission. In another study, *Proteobacteria* and *Clostridia* were found to be decreased in individuals with schizophrenia compared to healthy controls. Moreover, the abundance of *Ruminococcacea* was associated with lower severity of negative symptoms, while B*acteroides* was associated with worse depressive symptoms [327]. Interestingly, a recent metagenome-wide association study identified differences in SCFA synthesis, tryptophan metabolism and neurotransmitter synthesis/degradation in treatment-naïve schizophrenia individuals [324]. Moreover, the authors showed that transplantation of *S. vestibularis* from schizophrenia individuals induced deficits in social behaviors and altered neurotransmitter levels in peripheral tissues in recipient mice. However, the existing studies implicating the microbiota in schizophrenia are all cross-sectional and of a small sample size; therefore, causal relationships cannot be directly inferred.

Whilst probiotics have been proposed as a non-invasive therapeutic for the treatment of schizophrenia, there is thus far no probiotic that demonstrates beneficial effects on the core aspects of the disorder. Some studies have reported a benefit of probiotics in alleviating the gastrointestinal problems commonly observed in schizophrenia individuals [328, 329]. Recently, a study found that administration of a probiotic supplement with vitamin D to individuals with schizophrenia resulted in improved PANSS (Positive and Negative Symptom Scale) scores, along with symptomatic improvements and reduced inflammation [330]. However, it is unknown what component of the intervention resulted in this effect. In another study, consumption of *Bifidobacterium breve* for 4 weeks improved PANSS and anxiety/depression scores [331].

4.12 Irritable Bowel Syndrome

Irritable bowel syndrome (IBS) is a common gastrointestinal disorder now viewed as a disorder of gut-brain axis interactions in which pain or discomfort related to bowel habits are clinical hallmarks [332]. Symptoms of IBS commonly present co-morbidly with behavioral, psychological, psychosocial and environmental factors [333]. Over the years, dysmotility, visceral hypersensitivity, intestinal permeability, immune activation have all been implicated in the pathogenesis of IBS [334, 335]. However, despite considerable research efforts the pathophysiology of IBS remains poorly understood.

Numerous studies have implicated the gut microbiota as an important potential pathophysiological factor in IBS [336, 337]. IBS is associated with a reduction in microbial diversity and increased instability of the microbiota [338–340]. Moreover, different IBS subtypes show different microbiota compositions, with the majority demonstrating an increased abundance of *Firmicutes* and reduced amounts of *Bacteroides* and *Bifidobacteria* [339, 341, 342]. A recent meta-analysis identified *Lachnospiraceae* strains to be consistently decreased in individuals with IBD and correlated these changes with immune-related pathways [343]. Another pathway linking the microbiota-gut-brain axis with IBS etiology is serotonin signaling. Serotonin is produced, stored and released by enterochromaffin cells in the gut and it controls many gastrointestinal functions including peristalsis, vasodilation, secretion in addition to pain and sensory perception [344]. Disturbances to this signaling pathway contribute to altered gut function. Serotonergic function and tryptophan metabolism are also known to be changed in individuals with IBS [345–348]. Further, altered bile acid absorption, immune activation, pain, and an altered stress response have been shown to contribute to the pathophysiology of IBS [349–351].

Microbiota-directed therapies have shown promising effects for treating IBS [352, 353]. In one clinical study, probiotic supplementation (*L. acidophilus* Rosell-52 and *B. longum* Rosell-175) was successful in ameliorating stress-related GI disturbances [354]. Another study found beneficial effects of a *B. longum* probiotic in reducing depressive symptoms [355]. A recent meta-analysis determined that particular combinations of probiotics, or specific species and strains, seemed to have beneficial effects on global IBS symptoms and abdominal pain; however, it could not define their efficacy [353]. Further, it has been proposed that probiotics elicit their effects on IBS symptomology indirectly via anti-inflammatory mechanisms [356]. These beneficial effects may also be observed during dietary restriction such as the low FODMAP diet in IBS [357]. Furthermore, FMT has become a popular treatment for IBS; however, its long-term success is questionable [358, 359]. Finally, antibiotic treatment is also showing positive responses in a subset of individuals with IBS [360].

4.13 Pain Disorders

Outside of abdominal pain, studies investigating the role of the gut microbiota in nociceptive (somatic and visceral), inflammatory and neuropathic pain remain sparse [176, 361]. To date, the majority of clinical evidence implicating the microbiota in pain responses have come from nociceptive disorders of the gut [362], while preclinical studies have investigated pain responses throughout the body. Chronic visceral pain is a predominant symptom observed in IBD [363] and FGIDs [362]; however, it is unclear as to what extent the gut microbiota plays in mediating the associated visceral pain behavior. Numerous human studies have identified alterations in the gut microbiota composition in individuals with chronic or recurrent visceral pain in IBD [364, 365] and IBS [339, 366, 367]. Individuals exhibit decreased relative abundance of *Bifidobacterium* and *Lactobacillus* genera, and an increased Firmicutes: Bacteroidetes ratio [352]. Moreover, FMT from IBS individuals into GF rats resulted in hypersensitivity to colorectal distension compared to control animals [368]. Despite the preclinical evidence, therapeutic interventions in humans have proven challenging with varied success [369–374]. Thus, it remains unclear if these microbiota changes are a causative or a secondary effect of visceral pain.

One of the first studies associating the gut microbiota with inflammatory pain involved GF mice, where they demonstrated a reduced pain response following administration of a range of inflammatory agents [375]. Interestingly, a recent study demonstrated that an exopolysaccharide produced by a putative probiotic strain *Lactobacillus paraplantarum* was capable of reducing inflammatory hyperalgesia in rats [376]. In humans, antibiotics are the preferred treatment for *C. difficile* infections; however, FMT is well accepted as an alternative therapeutic strategy [377]. A recent review highlights the efficacy of FMT for the treatment of IBD with studies showing 20–33% reduction in pain index in ulcerative colitis individuals, 50–75% in Crohn's disease and varying effects in 0–80% in pouchitis [378]. As of yet, there are no effective probiotics for the treatment of Crohn's disease, [373]. However, there are mixed reports for probiotic interventions in treating ulcerative colitis and pouchitis [373, 379]. Arthritic pain is also associated with inflammation. While evidence has implicated a link between the gut microbiota, specifically *Prevotella copri* [380], in untreated early-onset rheumatoid arthritis, human studies investigating the effects of microbiota-based therapeutics remain relatively low [381].

Studies investigating the role of the gut microbiota in neuropathic pain (migraine and headache, stroke-related pain, cancer pain, diabetic neuropathy, MS-related pain) are also limited. A recent animal study linked the gut microbiota with chemotherapy-induced neuropathic pain [382]. Further, a clinical study with probiotic administration reported an improvement in post-chemotherapy abdominal pain in colorectal cancer patients [383]. This effect is due to chemotherapy-induced GI toxicity of the microbiome-immune system. One case study also reported positive effects of FMT treatment in diabetic neuropathy [384]. Whilst there is no conclusive evidence providing a role for the gut microbiota in migraine, clinical studies have reported that probiotic administration improves migraine symptoms [385].

4.14 Addiction

Most studies investigating the microbiota-gut-brain axis in relation to addiction have focused on the impact of drug use [386]. Addictive drugs not only have a direct effect on the composition of the gut microbiota but are also associated with a myriad of comorbidities that are already associated with the microbiota-gut-brain axis, including depression, stress and anxiety, and chronic inflammation [387]. Extensive research has studied the effect of alcohol abuse on the gut microbiota [388]. A recent preclinical study linked alcohol consumption to alterations in microbiota composition, increased intestinal permeability and hepatic inflammation [389]. Clinical research has associated the negative effects of alcohol-withdrawal with increased gut permeability, inflammation, and altered microbiota composition [390]. In support of this, a recent study found that binge drinking resulted in a distinct microbiome composition that associated with alterations in craving, social cognition, and impulsivity [391]. Interestingly, a reduced gut microbiome has been shown to protect against alcohol-induced neuroinflammation [392]. Moreover, preclinical supplementation of *L. rhamnosus GG* was shown to decrease the severity of alcoholic hepatitis [393].

Furthermore, other substances of abuse tend to demonstrate similar effects to alcohol. For instance, clinical studies investigating the effects of tobacco smoke have associated smoking and cessation with changes in microbiota composition and inflammation [394, 395]. Opioid use studies have also implicated the gut microbiota in the progression of morphine dependence, GI dysfunction, altered inflammation and disrupted metabolic processes [396]. Moreover, antibiotic treatment in alcohol and opioid studies reduced drug-associated increases in gut permeability, inflammation and enteric neuron signaling [199, 397–399]. Moreover, chronic Δ9-tetrahydrocannabino (THC) treatment, a component of cannabis, altered microbiota composition in a mouse model [400]. These effects resulted in reduced energy consumption and prevented high-fat diet-induced weight gain and adiposity. Regarding cocaine dependence, studies have shown that alterations in the gut microbiota enhanced an animal's behavioral responses to cocaine by affecting the brain's reward circuitry [401, 402]. A recent preclinical study demonstrated that depletion of the gut microbiota with antibiotics altered reward responses to both cocaine and social stimuli [403]. This preliminary research highlights the connection between the gut microbiota, social behavior, and addiction. Further longitudinal studies are required to investigate the role of the gut microbiota in addictive behaviors and to define "at risk" addictive phenotypes.

5 Conclusions

Extensive research over the last few years has suggested that the gut microbiota is essential for normal development and maintenance of brain function. Our understanding of how the microbiota influences gut-brain communication is beginning to gain traction. Accumulating evidence from preclinical and clinical studies have implicated the microbiota in a variety of psychiatric, neurological, and neurodegenerative disorders. However, it remains to be determined if changes in the microbiota are the cause of the pathophysiology or merely correlational. Thus, caution should be taken when interpreting data from cross-sectional observational clinical studies. Most are of a small-sample size, often with participant selection bias, various confounders, as well as different sampling, sequencing, and analysis methods.

Thus, to further delineate the complex processes associated with the microbiota-gut-brain axis, more well-controlled and well-designed longitudinal studies are required. These studies should focus on the development of microbiota-based interventions such as probiotics/prebiotics strains, antibiotics, and fecal microbiota transplant therapies. Additionally, it will be imperative that these studies move away from correlational analysis and explore the molecular mechanisms underlying microbiota-gut-brain axis communication. Identifying and understanding the role of specific microbial metabolites and their effects on host physiology and health will be of tremendous benefit for the development of novel therapeutic approaches. While microbiota-based medicine is challenged by interindividual differences in microbiome composition, these differences may also provide an opportunity to develop patient-specific therapies for the treatment of an array of disorders.

Indeed, the field of microbiota-gut-brain axis research has made major advances in a short space of time. In the next few years, we hope to further our understanding as to what role the gut microbiota plays and translate this knowledge into novel therapeutics for a range of disorders.

References

1. Sender R, Fuchs S, Milo R. Are we really vastly outnumbered? Revisiting the ratio of bacterial to host cells in humans. Cell. 2016;164:337–40.
2. Gilbert JA, Blaser MJ, Caporaso JG, Jansson JK, Lynch SV, Knight R. Current understanding of the human microbiome. Nat Med. 2018;24:392–400.
3. Collins SM, Bercik P. Gut microbiota: intestinal bacteria influence brain activity in healthy humans. Nat Rev Gastroenterol Hepatol. 2013;10:326–7.
4. Cryan JF, Cryan JF, Cowan CSM, Sandhu KV, Bastiaanssen TFS, Boehme M, Codagnone MG, Cussotto S, Fulling C, Golubeva AV, Guzzetta KE, Jaggar M, Long-Smith CM, Lyte JM, Martin JA, Molinero-Perez A, Moloney G, Morelli E, Morillas E, O'Connor R, Cruz-Pereira JS, Peterson VL, Rea K, Ritz NL, Sherwin E, Spichak S, Teichman EM, van de Wouw M, Ventura-Silva AP, Wallace-Fitzsimons SE, Hyland N, Clarke G, Dinan TG. The microbiota-gut-brain axis. Physiol Rev. 2019;99:1877–2013.
5. Sherwin E, Dinan TG, Cryan JF. Recent developments in understanding the role of the gut microbiota in brain health and disease. Ann N Y Acad Sci. 2018;1420:5–25.
6. Human Microbiome Project Consortium. A framework for human microbiome research. Nature. 2012;486:215–21.
7. Human Microbiome Project Consortium. Structure, function and diversity of the healthy human microbiome. Nature. 2012;486:207–14.
8. Integrative HMP (iHMP) Research Network Consortium. The Integrative Human Microbiome Project. Nature. 2019;569:641–8.
9. Clemente JC, Ursell LK, Parfrey LW, Knight R. The impact of the gut microbiota on human health: an integrative view. Cell. 2012;148:1258–70.
10. Feng Q, Chen WD, Wang YD. Gut microbiota: an integral moderator in health and disease. Front Microbiol. 2018;9:151.
11. Sekirov I, Russell SL, Antunes LC, Finlay BB. Gut microbiota in health and disease. Physiol Rev. 2010;90:859–904.
12. Foster JA, Rinaman L, Cryan JF. Stress & the gut-brain axis: regulation by the microbiome. Neurobiol Stress. 2017;7:124–36.

13. Mayer EA. Gut feelings: the emerging biology of gut-brain communication. Nat Rev Neurosci. 2011;12:453–66.
14. Rhee SH, Pothoulakis C, Mayer EA. Principles and clinical implications of the brain-gut-enteric microbiota axis. Nat Rev Gastroenterol Hepatol. 2009;6:306–14.
15. Bordenstein SR, Theis KR. Host biology in light of the microbiome: ten principles of holobionts and hologenomes. PLoS Biol. 2015;13:e1002226.
16. Gaulke CA, Arnold HK, Humphreys IR, Kembel SW, O'Dwyer JP, Sharpton TJ. Ecophylogenetics clarifies the evolutionary association between mammals and their gut microbiota. mBio. 2018;9:e01348-18.
17. Suzuki TA, Fitzstevens JL, Schmidt VT, Enav H, Huus KE, Mbong Ngwese M, Griesshamme A, Pfleiderer A, Adegbite BR, Zinsou JF, Esen M, Velavan TP, Adegnika AA, Song LH, Spector TD, Muehlbauer AL, Marchi N, Kang H, Maier L, Blekhman R, Ségurel L, Ko G, Youngblut ND, Kremsner P, Ley RE. Codiversification of gut microbiota with humans. Science. 2022;377:1328–32.
18. Pasolli E, Asnicar F, Manara S, Zolfo M, Karcher N, Armanini F, Beghini F, Manghi P, Tett A, Ghensi P, Collado MC, Rice BL, Dulong C, Morgan XC, Golden CD, Quince C, Huttenhower C, Segata N. Extensive unexplored human microbiome diversity revealed by over 150,000 genomes from metagenomes spanning age, geography, and lifestyle. Cell. 2019;176:649–62.e20.
19. Costello EK, Lauber CL, Hamady M, Fierer N, Gordon JI, Knight R. Bacterial community variation in human body habitats across space and time. Science. 2009;326:1694–7.
20. Claesson MJ, Jeffery IB, Conde S, Power SE, O'Connor EM, Cusack S, Harris HM, Coakley M, Lakshminarayanan B, O'Sullivan O, Fitzgerald GF, Deane J, O'Connor M, Harnedy N, O'Connor K, O'Mahony D, van Sinderen D, Wallace M, Brennan L, Stanton C, Marchesi JR, Fitzgerald AP, Shanahan F, Hill C, Ross RP, O'Toole PW. Gut microbiota composition correlates with diet and health in the elderly. Nature. 2012;488:178–84.
21. Roswall J, Olsson LM, Kovatcheva-Datchary P, Nilsson S, Tremaroli V, Simon MC, Kiilerich P, Akrami R, Kramer M, Uhlen M, Gummesson A, Kristiansen K, Dahlgren J, Backhed F. Developmental trajectory of the healthy human gut microbiota during the first 5 years of life. Cell Host Microbe. 2021;29:765–76.e3.
22. Salazar N, Arboleya S, Valdes L, Stanton C, Ross P, Ruiz L, Gueimonde M, de Los Reyes-Gavilan CG. The human intestinal microbiome at extreme ages of life. Dietary intervention as a way to counteract alterations. Front Genet. 2014;5:406.
23. Hubel DH, Wiesel TN. The period of susceptibility to the physiological effects of unilateral eye closure in kittens. J Physiol. 1970;206:419–36.
24. Alberini CM, Travaglia A. Infantile amnesia: a critical period of learning to learn and remember. J Neurosci. 2017;37:5783–95.
25. Werker JF, Hensch TK. Critical periods in speech perception: new directions. Annu Rev Psychol. 2015;66:173–96.
26. Borre YE, O'Keeffe GW, Clarke G, Stanton C, Dinan TG, Cryan JF. Microbiota and neurodevelopmental windows: implications for brain disorders. Trends Mol Med. 2014;20:509–18.
27. Cowan CSM, Dinan TG, Cryan JF. Annual research review: critical windows—the microbiota-gut-brain axis in neurocognitive development. J Child Psychol Psychiatry. 2020;61:353–71.
28. Goyal MS, Venkatesh S, Milbrandt J, Gordon JI, Raichle ME. Feeding the brain and nurturing the mind: linking nutrition and the gut microbiota to brain development. Proc Natl Acad Sci U S A. 2015;112:14105–12.
29. Hensch TK. Critical period plasticity in local cortical circuits. Nat Rev Neurosci. 2005;6:877–88.
30. Collado MC, Segata N. Initial exploration of in utero microbial colonization. Nat Med. 2020;26(4):469–70.
31. Fujimura KE, Sitarik AR, Havstad S, Lin DL, Levan S, Fadrosh D, Panzer AR, Lamere B, Rackaityte E, Lukacs NW, Wegienka G, Boushey HA, Ownby DR, Zoratti EM, Levin AM, Johnson CC, Lynch SV. Neonatal gut microbiota associates with childhood multisensitized atopy and T cell differentiation. Nat Med. 2016;22:1187–91.
32. Kennedy KM, de Goffau MC, Perez-Muñoz ME, Arrieta M-C, Bäckhed F, Bork P, Braun T, Bushman FD, Dore J, de Vos WM, Earl AM, Eisen JA, Elovitz MA, Ganal-Vonarburg SC, Gänzle MG, Garrett WS, Hall LJ, Hornef MW, Huttenhower C, Konnikova L, Lebeer S, Macpherson AJ, Massey RC, McHardy AC, Koren O, Lawley TD, Ley RE, O'Mahony L, O'Toole PW, Pamer EG, Parkhill J, Raes J, Rattei T, Salonen A, Segal E, Segata N, Shanahan F, Sloboda DM, Smith GCS, Sokol H, Spector TD, Surette MG, Tannock GW, Walker AW, Yassour M, Walter J. Questioning the fetal microbiome illustrates pitfalls of low-biomass microbial studies. Nature. 2023;613:639–49.
33. Perez-Munoz ME, Arrieta MC, Ramer-Tait AE, Walter J. A critical assessment of the "sterile womb" and "in utero colonization" hypotheses: implications for research on the pioneer infant microbiome. Microbiome. 2017;5:48.
34. Rackaityte E, Halkias J, Fukui EM, Mendoza VF, Hayzelden C, Crawford ED, Fujimura KE, Burt TD, Lynch SV. Viable bacterial colonization is highly limited in the human intestine in utero. Nat Med. 2020;26(4):599–607.
35. Walker RW, Clemente JC, Peter I, Loos RJF. The prenatal gut microbiome: are we colonized with bacteria in utero? Pediatr Obes. 2017;12(Suppl 1):3–17.
36. Walter J, Hornef MW. A philosophical perspective on the prenatal in utero microbiome debate. Microbiome. 2021;9:5.
37. Ferretti P, Pasolli E, Tett A, Asnicar F, Gorfer V, Fedi S, Armanini F, Truong DT, Manara S, Zolfo M, Beghini F, Bertorelli R, De Sanctis V, Bariletti I, Canto R, Clementi R, Cologna M, Crifo T, Cusumano G, Gottardi S, Innamorati C, Mase C, Postai D, Savoi D, Duranti S, Lugli GA, Mancabelli L, Turroni F, Ferrario C, Milani C, Mangifesta M, Anzalone R, Viappiani A, Yassour M, Vlamakis H, Xavier R, Collado CM, Koren O, Tateo S, Soffiati M, Pedrotti A, Ventura M, Huttenhower C, Bork P, Segata N. Mother-to-infant microbial transmission from different body sites shapes the developing infant gut microbiome. Cell Host Microbe. 2018;24:133–45.e5.
38. Yassour M, Jason E, Hogstrom LJ, Arthur TD, Tripathi S, Siljander H, Selvenius J, Oikarinen S, Hyoty H, Virtanen SM, Ilonen J, Ferretti P, Pasolli E, Tett A, Asnicar F, Segata N, Vlamakis H, Lander ES, Huttenhower C, Knip M, Xavier RJ. Strain-level analysis of mother-to-child bacterial transmission during the first few months of life. Cell Host Microbe. 2018;24:146–54.e4.
39. Gareau MG. Microbiota-gut-brain axis and cognitive function. Adv Exp Med Biol. 2014;817:357–71.
40. Luczynski P, McVey Neufeld KA, Oriach CS, Clarke G, Dinan TG, Cryan JF. Growing up in a bubble: using germ-free animals to assess the influence of the gut microbiota on brain and behavior. Int J Neuropsychopharmacol. 2016;19:pyw020.
41. Sampson TR, Mazmanian SK. Control of brain development, function, and behavior by the microbiome. Cell Host Microbe. 2015;17:565–76.
42. Spichak S, Guzzetta KE, O'Leary OF, Clarke G, Dinan TG, Cryan JF. Without a bug's life: germ-free rodents to interrogate microbiota-gut-neuroimmune interactions. Drug Discov Today Dis Model. 2018;28:79–93.
43. Vuong HE, Pronovost GN, Williams DW, Coley EJL, Siegler EL, Qiu A, Kazantsev M, Wilson CJ, Rendon T, Hsiao EY. The maternal microbiome modulates fetal neurodevelopment in mice. Nature. 2020;586:281–6.
44. Ogbonnaya ES, Clarke G, Shanahan F, Dinan TG, Cryan JF, O'Leary OF. Adult hippocampal neurogenesis is regulated by the microbiome. Biol Psychiatry. 2015;78:e7–9.

45. Hoban AE, Stilling RM, Ryan FJ, Shanahan F, Dinan TG, Claesson MJ, Clarke G, Cryan JF. Regulation of prefrontal cortex myelination by the microbiota. Transl Psychiatry. 2016;6:e774.
46. Lynch CMK, Nagpal J, Clarke G, Cryan JF. Wrapping things up: recent developments in understanding the role of the microbiome in regulating myelination. Curr Opin Physiol. 2021;23:100468.
47. Luczynski P, Whelan SO, O'Sullivan C, Clarke G, Shanahan F, Dinan TG, Cryan JF. Adult microbiota-deficient mice have distinct dendritic morphological changes: differential effects in the amygdala and hippocampus. Eur J Neurosci. 2016;44: 2654–66.
48. Braniste V, Al-Asmakh M, Kowal C, Anuar F, Abbaspour A, Toth M, Korecka A, Bakocevic N, Ng LG, Kundu P, Gulyas B, Halldin C, Hultenby K, Nilsson H, Hebert H, Volpe BT, Diamond B, Pettersson S. The gut microbiota influences blood-brain barrier permeability in mice. Sci Transl Med. 2014;6:263ra158.
49. Knox EG, Aburto MR, Tessier C, Nagpal J, Clarke G, O'Driscoll CM, Cryan JF. Microbial-derived metabolites induce actin cytoskeletal rearrangement and protect blood-brain barrier function. iScience. 2022;25:105648.
50. Knox EG, Lynch CMK, Lee YS, O'Driscoll CM, Clarke G, Cryan JF, Aburto MR. The gut microbiota is important for the maintenance of blood-cerebrospinal fluid barrier integrity. Eur J Neurosci. 2023;57(2):233–41.
51. Clarke G, Grenham S, Scully P, Fitzgerald P, Moloney RD, Shanahan F, Dinan TG, Cryan JF. The microbiome-gut-brain axis during early life regulates the hippocampal serotonergic system in a sex-dependent manner. Mol Psychiatry. 2013;18:666–73.
52. Erny D, Hrabe de Angelis AL, Jaitin D, Wieghofer P, Staszewski O, David E, Keren-Shaul H, Mahlakoiv T, Jakobshagen K, Buch T, Schwierzeck V, Utermohlen O, Chun E, Garrett WS, McCoy KD, Diefenbach A, Staeheli P, Stecher B, Amit I, Prinz M. Host microbiota constantly control maturation and function of microglia in the CNS. Nat Neurosci. 2015;18:965–77.
53. Lynch CMK, Clarke G, Cryan JF. Powering up microbiome-microglia interactions. Cell Metab. 2021;33:2097–9.
54. Buffington SA, Di Prisco GV, Auchtung TA, Ajami NJ, Petrosino JF, Costa-Mattioli M. Microbial reconstitution reverses maternal diet-induced social and synaptic deficits in offspring. Cell. 2016;165:1762–75.
55. Cirulli F, Musillo C, Berry A. Maternal obesity as a risk factor for brain development and mental health in the offspring. Neuroscience. 2020;447:122–35.
56. Jasarevic E, Bale TL. Prenatal and postnatal contributions of the maternal microbiome on offspring programming. Front Neuroendocrinol. 2019;55:100797.
57. Jasarevic E, Howard CD, Misic AM, Beiting DP, Bale TL. Stress during pregnancy alters temporal and spatial dynamics of the maternal and offspring microbiome in a sex-specific manner. Sci Rep. 2017;7:44182.
58. Jasarevic E, Howard CD, Morrison K, Misic A, Weinkopff T, Scott P, Hunter C, Beiting D, Bale TL. The maternal vaginal microbiome partially mediates the effects of prenatal stress on offspring gut and hypothalamus. Nat Neurosci. 2018;21:1061–71.
59. Rincel M, Aubert P, Chevalier J, Grohard PA, Basso L, Monchaux de Oliveira C, Helbling JC, Levy E, Chevalier G, Leboyer M, Eberl G, Laye S, Capuron L, Vergnolle N, Neunlist M, Boudin H, Lepage P, Darnaudery M. Multi-hit early life adversity affects gut microbiota, brain and behavior in a sex-dependent manner. Brain Behav Immun. 2019;80:179–92.
60. Zijlmans MA, Korpela K, Riksen-Walraven JM, de Vos WM, de Weerth C. Maternal prenatal stress is associated with the infant intestinal microbiota. Psychoneuroendocrinology. 2015;53:233–45.
61. Korpela K, de Vos WM. Early life colonization of the human gut: microbes matter everywhere. Curr Opin Microbiol. 2018;44:70–8.
62. Stewart CJ, Ajami NJ, O'Brien JL, Hutchinson DS, Smith DP, Wong MC, Ross MC, Lloyd RE, Doddapaneni H, Metcalf GA, Muzny D, Gibbs RA, Vatanen T, Huttenhower C, Xavier RJ, Rewers M, Hagopian W, Toppari J, Ziegler AG, She JX, Akolkar B, Lernmark A, Hyoty H, Vehik K, Krischer JP, Petrosino JF. Temporal development of the gut microbiome in early childhood from the TEDDY study. Nature. 2018;562:583–8.
63. Derrien M, Alvarez AS, de Vos WM. The gut microbiota in the first decade of life. Trends Microbiol. 2019;27:997–1010.
64. Hollister EB, Riehle K, Luna RA, Weidler EM, Rubio-Gonzales M, Mistretta TA, Raza S, Doddapaneni HV, Metcalf GA, Muzny DM, Gibbs RA, Petrosino JF, Shulman RJ, Versalovic J. Structure and function of the healthy pre-adolescent pediatric gut microbiome. Microbiome. 2015;3:36.
65. Dominguez-Bello MG, Costello EK, Contreras M, Magris M, Hidalgo G, Fierer N, Knight R. Delivery mode shapes the acquisition and structure of the initial microbiota across multiple body habitats in newborns. Proc Natl Acad Sci U S A. 2010;107:11971–5.
66. Fouhy F, Watkins C, Hill CJ, O'Shea CA, Nagle B, Dempsey EM, O'Toole PW, Ross RP, Ryan CA, Stanton C. Perinatal factors affect the gut microbiota up to four years after birth. Nat Commun. 2019;10:1517.
67. Combellick JL, Shin H, Shin D, Cai Y, Hagan H, Lacher C, Lin DL, McCauley K, Lynch SV, Dominguez-Bello MG. Author correction: differences in the fecal microbiota of neonates born at home or in the hospital. Sci Rep. 2019;9:9044.
68. Ho NT, Li F, Lee-Sarwar KA, Tun HM, Brown BP, Pannaraj PS, Bender JM, Azad MB, Thompson AL, Weiss ST, Azcarate-Peril MA, Litonjua AA, Kozyrskyj AL, Jaspan HB, Aldrovandi GM, Kuhn L. Meta-analysis of effects of exclusive breastfeeding on infant gut microbiota across populations. Nat Commun. 2018;9:4169.
69. Callaghan BL, Fields A, Gee DG, Gabard-Durnam L, Caldera C, Humphreys KL, Goff B, Flannery J, Telzer EH, Shapiro M, Tottenham N. Mind and gut: associations between mood and gastrointestinal distress in children exposed to adversity. Dev Psychopathol. 2020;32:309–28.
70. Gur TL, Worly BL, Bailey MT. Stress and the commensal microbiota: importance in parturition and infant neurodevelopment. Front Psychiatry. 2015;6:5.
71. Korpela K, Salonen A, Saxen H, Nikkonen A, Peltola V, Jaakkola T, de Vos W, Kolho KL. Antibiotics in early life associate with specific gut microbiota signatures in a prospective longitudinal infant cohort. Pediatr Res. 2020;88(3):438–43.
72. O'Connor R, Moloney GM, Fulling C, O'Riordan KJ, Fitzgerald P, Bastiaanssen TFS, Schellekens H, Dinan TG, Cryan JF. Maternal antibiotic administration during a critical developmental window has enduring neurobehavioural effects in offspring mice. Behav Brain Res. 2021;404:113156.
73. Tun HM, Konya T, Takaro TK, Brook JR, Chari R, Field CJ, Guttman DS, Becker AB, Mandhane PJ, Turvey SE, Subbarao P, Sears MR, Scott JA, Kozyrskyj AL, CHILD Study Investigators. Exposure to household furry pets influences the gut microbiota of infant at 3-4 months following various birth scenarios. Microbiome. 2017;5:40.
74. Kan JM, Cowan CSM, Ooi CY, Kasparian NA. What can the gut microbiome teach us about the connections between child physical and mental health? A systematic review. Dev Psychobiol. 2019;61:700–13.
75. Codagnone MG, Spichak S, O'Mahony SM, O'Leary OF, Clarke G, Stanton C, Dinan TG, Cryan JF. Programming bugs: microbiota and the developmental origins of brain health and disease. Biol Psychiatry. 2019;85:150–63.
76. Milani C, Duranti S, Bottacini F, Casey E, Turroni F, Mahony J, Belzer C, Delgado Palacio S, Arboleya Montes S, Mancabelli

L, Lugli GA, Rodriguez JM, Bode L, de Vos W, Gueimonde M, Margolles A, van Sinderen D, Ventura M. The first microbial colonizers of the human gut: composition, activities, and health implications of the infant gut microbiota. Microbiol Mol Biol Rev. 2017;81:e00036-17.
77. Lynch CMK, Cowan CSM, Bastiaanssen TFS, Moloney GM, Theune N, van de Wouw M, Florensa Zanuy E, Ventura-Silva AP, Codagnone MG, Villalobos-Manriquez F, Segalla M, Koc F, Stanton C, Ross P, Dinan TG, Clarke G, Cryan JF. Critical windows of early-life microbiota disruption on behaviour, neuroimmune function, and neurodevelopment. Brain Behav Immun. 2022;108:309–27.
78. O'Mahony SM, Felice VD, Nally K, Savignac HM, Claesson MJ, Scully P, Woznicki J, Hyland NP, Shanahan F, Quigley EM, Marchesi JR, O'Toole PW, Dinan TG, Cryan JF. Disturbance of the gut microbiota in early-life selectively affects visceral pain in adulthood without impacting cognitive or anxiety-related behaviors in male rats. Neuroscience. 2014;277:885–901.
79. Verdu EF, Bercik P, Verma-Gandhu M, Huang XX, Blennerhassett P, Jackson W, Mao Y, Wang L, Rochat F, Collins SM. Specific probiotic therapy attenuates antibiotic induced visceral hypersensitivity in mice. Gut. 2006;55:182–90.
80. Moloney G, O'Connor R, Fulling C, O'Riordan KJ, Fitzgerald P, Bastiaanssen TFS, Schellekens H, Dinan TG, Cryan JF. Maternal antibiotic administration during a critical developmental window has enduring neurobehavioural effects in offspring mice. Behav Brain Res. 2021;404:113156.
81. O'Mahony SM, Clarke G, Dinan TG, Cryan JF. Early-life adversity and brain development: is the microbiome a missing piece of the puzzle? Neuroscience. 2017;342:37–54.
82. Carlson AL, Xia K, Azcarate-Peril MA, Goldman BD, Ahn M, Styner MA, Thompson AL, Geng X, Gilmore JH, Knickmeyer RC. Infant gut microbiome associated with cognitive development. Biol Psychiatry. 2018;83:148–59.
83. Gao W, Salzwedel AP, Carlson AL, Xia K, Azcarate-Peril MA, Styner MA, Thompson AL, Geng X, Goldman BD, Gilmore JH, Knickmeyer RC. Gut microbiome and brain functional connectivity in infants—a preliminary study focusing on the amygdala. Psychopharmacology. 2019;236:1641–51.
84. Aatsinki AK, Lahti L, Uusitupa HM, Munukka E, Keskitalo A, Nolvi S, O'Mahony S, Pietila S, Elo LL, Eerola E, Karlsson H, Karlsson L. Gut microbiota composition is associated with temperament traits in infants. Brain Behav Immun. 2019;80:849–58.
85. Christian LM, Galley JD, Hade EM, Schoppe-Sullivan S, Kamp Dush C, Bailey MT. Gut microbiome composition is associated with temperament during early childhood. Brain Behav Immun. 2015;45:118–27.
86. Loughman A, Ponsonby AL, O'Hely M, Symeonides C, Collier F, Tang MLK, Carlin J, Ranganathan S, Allen K, Pezic A, Saffery R, Jacka F, Harrison LC, Sly PD, Vuillermin P, BIS Investigator Group. Gut microbiota composition during infancy and subsequent behavioural outcomes. EBioMedicine. 2020;52:102640.
87. Firestein MR, Myers MM, Austin J, Stark RI, Barone JL, Ludwig RJ, Welch MG. Perinatal antibiotics alter preterm infant EEG and neurobehavior in the Family Nurture Intervention trial. Dev Psychobiol. 2019;61:661–9.
88. Sjogren YM, Tomicic S, Lundberg A, Bottacher MF, Bjorksten B, Sverremark-Ekstrom E, Jenmalm MC. Influence of early gut microbiota on the maturation of childhood mucosal and systemic immune responses. Clin Exp Allergy. 2009;39:1842–51.
89. Sugino KY, Paneth N, Comstock SS. Michigan cohorts to determine associations of maternal pre-pregnancy body mass index with pregnancy and infant gastrointestinal microbial communities: late pregnancy and early infancy. PLoS One. 2019;14:e0213733.
90. Indrio F, Di Mauro A, Riezzo G, Civardi E, Intini C, Corvaglia L, Ballardini E, Bisceglia M, Cinquetti M, Brazzoduro E, Del Vecchio A, Tafuri S, Francavilla R. Prophylactic use of a probiotic in the prevention of colic, regurgitation, and functional constipation: a randomized clinical trial. JAMA Pediatr. 2014;168: 228–33.
91. Partty A, Kalliomaki M, Wacklin P, Salminen S, Isolauri E. A possible link between early probiotic intervention and the risk of neuropsychiatric disorders later in childhood: a randomized trial. Pediatr Res. 2015;77:823–8.
92. Callaghan BL, Cowan CS, Richardson R. Treating generational stress: effect of paternal stress on development of memory and extinction in offspring is reversed by probiotic treatment. Psychol Sci. 2016;27:1171–80.
93. Cowan CS, Callaghan BL, Richardson R. The effects of a probiotic formulation (Lactobacillus rhamnosus and L. helveticus) on developmental trajectories of emotional learning in stressed infant rats. Transl Psychiatry. 2016;6:e823.
94. Cowan CSM, Stylianakis AA, Richardson R. Early-life stress, microbiota, and brain development: probiotics reverse the effects of maternal separation on neural circuits underpinning fear expression and extinction in infant rats. Dev Cogn Neurosci. 2019;37:100627.
95. Fukui H, Oshima T, Tanaka Y, Oikawa Y, Makizaki Y, Ohno H, Tomita T, Watari J, Miwa H. Effect of probiotic Bifidobacterium bifidum G9-1 on the relationship between gut microbiota profile and stress sensitivity in maternally separated rats. Sci Rep. 2018;8:12384.
96. Hsiao EY, McBride SW, Hsien S, Sharon G, Hyde ER, McCue T, Codelli JA, Chow J, Reisman SE, Petrosino JF, Patterson PH, Mazmanian SK. Microbiota modulate behavioral and physiological abnormalities associated with neurodevelopmental disorders. Cell. 2013;155:1451–63.
97. Korpela K, Salonen A, Vepsalainen O, Suomalainen M, Kolmeder C, Varjosalo M, Miettinen S, Kukkonen K, Savilahti E, Kuitunen M, de Vos WM. Probiotic supplementation restores normal microbiota composition and function in antibiotic-treated and in caesarean-born infants. Microbiome. 2018;6:182.
98. O'Mahony SM, McVey Neufeld KA, Waworuntu RV, Pusceddu MM, Manurung S, Murphy K, Strain C, Laguna MC, Peterson VL, Stanton C, Berg BM, Dinan TG, Cryan JF. The enduring effects of early-life stress on the microbiota-gut-brain axis are buffered by dietary supplementation with milk fat globule membrane and a prebiotic blend. Eur J Neurosci. 2020;51:1042–58.
99. Guerville F, De Souto Barreto P, Ader I, Andrieu S, Casteilla L, Dray C, Fazilleau N, Guyonnet S, Langin D, Liblau R, Parini A, Valet P, Vergnolle N, Rolland Y, Vellas B. Revisiting the hallmarks of aging to identify markers of biological age. J Prev Alzheimers Dis. 2020;7:56–64.
100. Konturek PC, Haziri D, Brzozowski T, Hess T, Heyman S, Kwiecien S, Konturek SJ, Koziel J. Emerging role of fecal microbiota therapy in the treatment of gastrointestinal and extra-gastrointestinal diseases. J Physiol Pharmacol. 2015;66:483–91.
101. Lopez-Otin C, Blasco MA, Partridge L, Serrano M, Kroemer G. The hallmarks of aging. Cell. 2013;153:1194–217.
102. Glimstedt G. The germfree animal as a research tool. Ann N Y Acad Sci. 1959;78:281–4.
103. Gustafsson B. Germ-free rearing of rats. Acta Anat (Basel). 1946;2:376–91.
104. Teng Y, Mu J, Xu F, Zhang X, Sriwastva MK, Liu QM, Li X, Lei C, Sundaram K, Hu X, Zhang L, Park JW, Hwang JY, Rouchka EC, Zhang X, Yan J, Merchant ML, Zhang H-G. Gut bacterial isoamylamine promotes age-related cognitive dysfunction by promoting microglial cell death. Cell Host Microbe. 2022;30:944–60. e8.
105. Ghosh TS, Shanahan F, O'Toole PW. The gut microbiome as a modulator of healthy ageing. Nat Rev Gastroenterol Hepatol. 2022;19:565–84.

106. O'Toole PW, Jeffery IB. Gut microbiota and aging. Science. 2015;350:1214–5.
107. Vauzour D, Camprubi-Robles M, Miquel-Kergoat S, Andres-Lacueva C, Banati D, Barberger-Gateau P, Bowman GL, Caberlotto L, Clarke R, Hogervorst E, Kiliaan AJ, Lucca U, Manach C, Minihane AM, Mitchell ES, Perneczky R, Perry H, Roussel AM, Schuermans J, Sijben J, Spencer JP, Thuret S, van de Rest O, Vandewoude M, Wesnes K, Williams RJ, Williams RS, Ramirez M. Nutrition for the ageing brain: towards evidence for an optimal diet. Ageing Res Rev. 2017;35:222–40.
108. Claesson MJ, Cusack S, O'Sullivan O, Greene-Diniz R, de Weerd H, Flannery E, Marchesi JR, Falush D, Dinan T, Fitzgerald G, Stanton C, van Sinderen D, O'Connor M, Harnedy N, O'Connor K, Henry C, O'Mahony D, Fitzgerald AP, Shanahan F, Twomey C, Hill C, Ross RP, O'Toole PW. Composition, variability, and temporal stability of the intestinal microbiota of the elderly. Proc Natl Acad Sci U S A. 2011;108(Suppl 1):4586–91.
109. Ghosh TS, Das M, Jeffery IB, O'Toole PW. Adjusting for age improves identification of gut microbiome alterations in multiple diseases. elife. 2020;9:e50240.
110. Wilmanski T, Diener C, Rappaport N, Patwardhan S, Wiedrick J, Lapidus J, Earls JC, Zimmer A, Glusman G, Robinson M, Yurkovich JT, Kado DM, Cauley JA, Zmuda J, Lane NE, Magis AT, Lovejoy JC, Hood L, Gibbons SM, Orwoll ES, Price ND. Gut microbiome pattern reflects healthy ageing and predicts survival in humans. Nat Metab. 2021;3:274–86.
111. Boehme M, van de Wouw M, Bastiaanssen TFS, Olavarria-Ramirez L, Lyons K, Fouhy F, Golubeva AV, Moloney GM, Minuto C, Sandhu KV, Scott KA, Clarke G, Stanton C, Dinan TG, Schellekens H, Cryan JF. Mid-life microbiota crises: middle age is associated with pervasive neuroimmune alterations that are reversed by targeting the gut microbiome. Mol Psychiatry. 2020;25(10):2567–83.
112. Ostlund-Lagerstrom L, Kihlgren A, Repsilber D, Bjorksten B, Brummer RJ, Schoultz I. Probiotic administration among free-living older adults: a double blinded, randomized, placebo-controlled clinical trial. Nutr J. 2016;15:80.
113. Ghosh TS, Rampelli S, Jeffery IB, Santoro A, Neto M, Capri M, Giampieri E, Jennings A, Candela M, Turroni S, Zoetendal EG, Hermes GDA, Elodie C, Meunier N, Brugere CM, Pujos-Guillot E, Berendsen AM, De Groot L, Feskins EJM, Kaluza J, Pietruszka B, Bielak MJ, Comte B, Maijo-Ferre M, Nicoletti C, De Vos WM, Fairweather-Tait S, Cassidy A, Brigidi P, Franceschi C, O'Toole PW. Mediterranean diet intervention alters the gut microbiome in older people reducing frailty and improving health status: the NU-AGE 1-year dietary intervention across five European countries. Gut. 2020;69(7):1218–28.
114. Cox LM, Schafer MJ, Sohn J, Vincentini J, Weiner HL, Ginsberg SD, Blaser MJ. Calorie restriction slows age-related microbiota changes in an Alzheimer's disease model in female mice. Sci Rep. 2019;9:17904.
115. Mattison JA, Roth GS, Beasley TM, Tilmont EM, Handy AM, Herbert RL, Longo DL, Allison DB, Young JE, Bryant M, Barnard D, Ward WF, Qi W, Ingram DK, de Cabo R. Impact of caloric restriction on health and survival in rhesus monkeys from the NIA study. Nature. 2012;489:318–21.
116. Rusli F, Boekschoten MV, Borelli V, Sun C, Lute C, Menke AL, van den Heuvel J, Salvioli S, Franceschi C, Muller M, Steegenga WT. Plasticity of lifelong calorie-restricted C57BL/6J mice in adapting to a medium-fat diet intervention at old age. Aging Cell. 2018;17:e12696.
117. Boehme M, Guzzetta KE, Bastiaanssen TFS, van de Wouw M, Moloney GM, Gual-Grau A, Spichak S, Olavarría-Ramírez L, Fitzgerald P, Morillas E, Ritz NL, Jaggar M, Cowan CSM, Crispie F, Donoso F, Halitzki E, Neto MC, Sichetti M, Golubeva AV, Fitzgerald RS, Claesson MJ, Cotter PD, O'Leary OF, Dinan TG, Cryan JF. Microbiota from young mice counteracts selective age-associated behavioral deficits. Nature Aging. 2021;1:666–76.
118. Franceschi C, Garagnani P, Parini P, Giuliani C, Santoro A. Inflammaging: a new immune-metabolic viewpoint for age-related diseases. Nat Rev Endocrinol. 2018;14:576–90.
119. Prenderville JA, Kennedy PJ, Dinan TG, Cryan JF. Adding fuel to the fire: the impact of stress on the ageing brain. Trends Neurosci. 2015;38:13–25.
120. Thevaranjan N, Puchta A, Schulz C, Naidoo A, Szamosi JC, Verschoor CP, Loukov D, Schenck LP, Jury J, Foley KP, Schertzer JD, Larche MJ, Davidson DJ, Verdu EF, Surette MG, Bowdish DME. Age-associated microbial dysbiosis promotes intestinal permeability, systemic inflammation, and macrophage dysfunction. Cell Host Microbe. 2018;23:570.
121. Esposito P, Chandler N, Kandere K, Basu S, Jacobson S, Connolly R, Tutor D, Theoharides TC. Corticotropin-releasing hormone and brain mast cells regulate blood-brain-barrier permeability induced by acute stress. J Pharmacol Exp Ther. 2002;303:1061–6.
122. Kelly JR, Borre Y, O'Brien C, Patterson E, El Aidy S, Deane J, Kennedy PJ, Beers S, Scott K, Moloney G, Hoban AE, Scott L, Fitzgerald P, Ross P, Stanton C, Clarke G, Cryan JF, Dinan TG. Transferring the blues: depression-associated gut microbiota induces neurobehavioural changes in the rat. J Psychiatr Res. 2016;82:109–18.
123. Montagne A, Barnes SR, Sweeney MD, Halliday MR, Sagare AP, Zhao Z, Toga AW, Jacobs RE, Liu CY, Amezcua L, Harrington MG, Chui HC, Law M, Zlokovic BV. Blood-brain barrier breakdown in the aging human hippocampus. Neuron. 2015;85:296–302.
124. Erny D, Prinz M. How microbiota shape microglial phenotypes and epigenetics. Glia. 2020;68(8):1655–72.
125. Parker A, Romano S, Ansorge R, Aboelnour A, Le Gall G, Savva GM, Pontifex MG, Telatin A, Baker D, Jones E, Vauzour D, Rudder S, Blackshaw LA, Jeffery G, Carding SR. Fecal microbiota transfer between young and aged mice reverses hallmarks of the aging gut, eye, and brain. Microbiome. 2022;10:68.
126. Fulling C, Dinan TG, Cryan JF. Gut microbe to brain signaling: what happens in vagus. Neuron. 2019;101:998–1002.
127. Whitlock FA. Some psychiatric consequences of gastreotomy. Br Med J. 1961;1:1560–4.
128. Liu B, Fang F, Pedersen NL, Tillander A, Ludvigsson JF, Ekbom A, Svenningsson P, Chen H, Wirdefeldt K. Vagotomy and Parkinson disease: a Swedish register-based matched-cohort study. Neurology. 2017;88:1996–2002.
129. Svensson E, Horvath-Puho E, Thomsen RW, Djurhuus JC, Pedersen L, Borghammer P, Sorensen HT. Vagotomy and subsequent risk of Parkinson's disease. Ann Neurol. 2015;78:522–9.
130. Borovikova LV, Ivanova S, Zhang M, Yang H, Botchkina GI, Watkins LR, Wang H, Abumrad N, Eaton JW, Tracey KJ. Vagus nerve stimulation attenuates the systemic inflammatory response to endotoxin. Nature. 2000;405:458–62.
131. Breit S, Kupferberg A, Rogler G, Hasler G. Vagus nerve as modulator of the brain-gut axis in psychiatric and inflammatory disorders. Front Psychiatry. 2018;9:44.
132. Bravo JA, Forsythe P, Chew MV, Escaravage E, Savignac HM, Dinan TG, Bienenstock J, Cryan JF. Ingestion of Lactobacillus strain regulates emotional behavior and central GABA receptor expression in a mouse via the vagus nerve. Proc Natl Acad Sci U S A. 2011;108:16050–5.
133. Sgritta M, Dooling SW, Buffington SA, Momin EN, Francis MB, Britton RA, Costa-Mattioli M. Mechanisms underlying microbial-mediated changes in social behavior in mouse models of autism spectrum disorder. Neuron. 2019;101:246–59.e6.
134. Ye L, Bae M, Cassilly CD, Jabba SV, Thorpe DW, Martin AM, Lu H-Y, Wang J, Thompson JD, Lickwar CR. Enteroendocrine cells sense bacterial tryptophan catabolites to activate enteric and vagal neuronal pathways. Cell Host Microbe. 2021;29:179–96.e9.

135. Muller PA, Schneeberger M, Matheis F, Wang P, Kerner Z, Ilanges A, Pellegrino K, Del Mármol J, Castro TBR, Furuichi M, Perkins M, Han W, Rao A, Pickard AJ, Cross JR, Honda K, de Araujo I, Mucida D. Microbiota modulate sympathetic neurons via a gut–brain circuit. Nature. 2020;583:441–6.
136. Jaglin M, Rhimi M, Philippe C, Pons N, Bruneau A, Goustard B, Daugé V, Maguin E, Naudon L, Rabot S. Indole, a signaling molecule produced by the gut microbiota, negatively impacts emotional behaviors in rats. Front Neurosci. 2018;12:216.
137. Han W, Tellez LA, Perkins MH, Perez IO, Qu T, Ferreira J, Ferreira TL, Quinn D, Liu ZW, Gao XB, Kaelberer MM, Bohorquez DV, Shammah-Lagnado SJ, de Lartigue G, de Araujo IE. A neural circuit for gut-induced reward. Cell. 2018;175:665–78.e23.
138. Suarez AN, Hsu TM, Liu CM, Noble EE, Cortella AM, Nakamoto EM, Hahn JD, de Lartigue G, Kanoski SE. Gut vagal sensory signaling regulates hippocampus function through multi-order pathways. Nat Commun. 2018;9:2181.
139. Chen Y, Xu J, Chen Y. Regulation of neurotransmitters by the gut microbiota and effects on cognition in neurological disorders. Nutrients. 2021;13:2099.
140. Gheorghe CE, Martin JA, Manriquez FV, Dinan TG, Cryan JF, Clarke G. Focus on the essentials: tryptophan metabolism and the microbiome-gut-brain axis. Curr Opin Pharmacol. 2019;48:137–45.
141. Desbonnet L, Garrett L, Clarke G, Kiely B, Cryan JF, Dinan TG. Effects of the probiotic Bifidobacterium infantis in the maternal separation model of depression. Neuroscience. 2010;170:1179–88.
142. Yano JM, Yu K, Donaldson GP, Shastri GG, Ann P, Ma L, Nagler CR, Ismagilov RF, Mazmanian SK, Hsiao EY. Indigenous bacteria from the gut microbiota regulate host serotonin biosynthesis. Cell. 2015;161:264–76.
143. Kaelberer MM, Buchanan KL, Klein ME, Barth BB, Montoya MM, Shen X, Bohorquez DV. A gut-brain neural circuit for nutrient sensory transduction. Science. 2018;361:eaat5236.
144. Lyte M. Microbial endocrinology and the microbiota-gut-brain axis. Adv Exp Med Biol. 2014;817:3–24.
145. Lyte M. Microbial endocrinology: host-microbiota neuroendocrine interactions influencing brain and behavior. Gut Microbes. 2014;5:381–9.
146. Strandwitz P, Kim KH, Terekhova D, Liu JK, Sharma A, Levering J, McDonald D, Dietrich D, Ramadhar TR, Lekbua A, Mroue N, Liston C, Stewart EJ, Dubin MJ, Zengler K, Knight R, Gilbert JA, Clardy J, Lewis K. GABA-modulating bacteria of the human gut microbiota. Nat Microbiol. 2019;4:396–403.
147. Wan L, Ge WR, Zhang S, Sun YL, Wang B, Yang G. Case-control study of the effects of gut microbiota composition on neurotransmitter metabolic pathways in children with attention deficit hyperactivity disorder. Front Neurosci. 2020;14:127.
148. Gill PA, van Zelm MC, Muir JG, Gibson PR. Review article: short chain fatty acids as potential therapeutic agents in human gastrointestinal and inflammatory disorders. Aliment Pharmacol Ther. 2018;48:15–34.
149. Erny D, Dokalis N, Mezö C, Castoldi A, Mossad O, Staszewski O, Frosch M, Villa M, Fuchs V, Mayer A, Neuber J, Sosat J, Tholen S, Schilling O, Vlachos A, Blank T, Gomez de Agüero M, Macpherson AJ, Pearce EJ, Prinz M. Microbiota-derived acetate enables the metabolic fitness of the brain innate immune system during health and disease. Cell Metab. 2021;33:2260–76.e7.
150. Erny D, Hrabe de Angelis AL, Prinz M. Communicating systems in the body: how microbiota and microglia cooperate. Immunology. 2017;150:7–15.
151. Tahara Y, Yamazaki M, Sukigara H, Motohashi H, Sasaki H, Miyakawa H, Haraguchi A, Ikeda Y, Fukuda S, Shibata S. Gut microbiota-derived short chain fatty acids induce circadian clock entrainment in mouse peripheral tissue. Sci Rep. 2018;8:1395.
152. Canfora EE, Jocken JW, Blaak EE. Short-chain fatty acids in control of body weight and insulin sensitivity. Nat Rev Endocrinol. 2015;11:577–91.
153. van de Wouw M, Boehme M, Lyte JM, Wiley N, Strain C, O'Sullivan O, Clarke G, Stanton C, Dinan TG, Cryan JF. Short-chain fatty acids: microbial metabolites that alleviate stress-induced brain-gut axis alterations. J Physiol. 2018;596:4923–44.
154. Dalile B, Van Oudenhove L, Vervliet B, Verbeke K. The role of short-chain fatty acids in microbiota-gut-brain communication. Nat Rev Gastroenterol Hepatol. 2019;16:461–78.
155. Hopfner F, Kunstner A, Muller SH, Kunzel S, Zeuner KE, Margraf NG, Deuschl G, Baines JF, Kuhlenbaumer G. Gut microbiota in Parkinson disease in a northern German cohort. Brain Res. 2017;1667:41–5.
156. Li W, Wu X, Hu X, Wang T, Liang S, Duan Y, Jin F, Qin B. Structural changes of gut microbiota in Parkinson's disease and its correlation with clinical features. Sci China Life Sci. 2017;60:1223–33.
157. Morita C, Tsuji H, Hata T, Gondo M, Takakura S, Kawai K, Yoshihara K, Ogata K, Nomoto K, Miyazaki K, Sudo N. Gut dysbiosis in patients with anorexia nervosa. PLoS One. 2015;10:e0145274.
158. Schwiertz A, Taras D, Schafer K, Beijer S, Bos NA, Donus C, Hardt PD. Microbiota and SCFA in lean and overweight healthy subjects. Obesity (Silver Spring). 2010;18:190–5.
159. Skonieczna-Zydecka K, Grochans E, Maciejewska D, Szkup M, Schneider-Matyka D, Jurczak A, Loniewski I, Kaczmarczyk M, Marlicz W, Czerwinska-Rogowska M, Pelka-Wysiecka J, Dec K, Stachowska E. Faecal short chain fatty acids profile is changed in Polish depressive women. Nutrients. 2018;10:1939.
160. Michels N, Van de Wiele T, De Henauw S. Chronic psychosocial stress and gut health in children: associations with calprotectin and fecal short-chain fatty acids. Psychosom Med. 2017;79:927–35.
161. Liu S, Li E, Sun Z, Fu D, Duan G, Jiang M, Yu Y, Mei L, Yang P, Tang Y, Zheng P. Altered gut microbiota and short chain fatty acids in Chinese children with autism spectrum disorder. Sci Rep. 2019;9:287.
162. Paul B, Barnes S, Demark-Wahnefried W, Morrow C, Salvador C, Skibola C, Tollefsbol TO. Influences of diet and the gut microbiome on epigenetic modulation in cancer and other diseases. Clin Epigenetics. 2015;7:112.
163. Stilling RM, Dinan TG, Cryan JF. Microbial genes, brain & behaviour—epigenetic regulation of the gut-brain axis. Genes Brain Behav. 2014;13:69–86.
164. Ratsika A, Cruz Pereira JS, Lynch CMK, Clarke G, Cryan JF. Microbiota-immune-brain interactions: a lifespan perspective. Curr Opin Neurobiol. 2023;78:102652.
165. Zheng D, Liwinski T, Elinav E. Interaction between microbiota and immunity in health and disease. Cell Res. 2020;30:492–506.
166. Benakis C, Brea D, Caballero S, Faraco G, Moore J, Murphy M, Sita G, Racchumi G, Ling L, Pamer EG, Iadecola C, Anrather J. Commensal microbiota affects ischemic stroke outcome by regulating intestinal gammadelta T cells. Nat Med. 2016;22:516–23.
167. Duc D, Vigne S, Bernier-Latmani J, Yersin Y, Ruiz F, Gaia N, Leo S, Lazarevic V, Schrenzel J, Petrova TV, Pot C. Disrupting myelin-specific Th17 cell gut homing confers protection in an adoptive transfer experimental autoimmune encephalomyelitis. Cell Rep. 2019;29:378–90.e4.
168. Lee YK, Menezes JS, Umesaki Y, Mazmanian SK. Proinflammatory T-cell responses to gut microbiota promote experimental autoimmune encephalomyelitis. Proc Natl Acad Sci U S A. 2011;108(Suppl 1):4615–22.
169. Ochoa-Reparaz J, Mielcarz DW, Ditrio LE, Burroughs AR, Foureau DM, Haque-Begum S, Kasper LH. Role of gut commensal microflora in the development of experimental autoimmune encephalomyelitis. J Immunol. 2009;183:6041–50.

170. Gaboriau-Routhiau V, Rakotobe S, Lecuyer E, Mulder I, Lan A, Bridonneau C, Rochet V, Pisi A, De Paepe M, Brandi G, Eberl G, Snel J, Kelly D, Cerf-Bensussan N. The key role of segmented filamentous bacteria in the coordinated maturation of gut helper T cell responses. Immunity. 2009;31:677–89.
171. Low D, Ginhoux F. Recent advances in the understanding of microglial development and homeostasis. Cell Immunol. 2018;330:68–78.
172. Thion MS, Low D, Silvin A, Chen J, Grisel P, Schulte-Schrepping J, Blecher R, Ulas T, Squarzoni P, Hoeffel G, Coulpier F, Siopi E, David FS, Scholz C, Shihui F, Lum J, Amoyo AA, Larbi A, Poidinger M, Buttgereit A, Lledo PM, Greter M, Chan JKY, Amit I, Beyer M, Schultze JL, Schlitzer A, Pettersson S, Ginhoux F, Garel S. Microbiome influences prenatal and adult microglia in a sex-specific manner. Cell. 2018;172:500–16.e6.
173. El Aidy S, Dinan TG, Cryan JF. Immune modulation of the brain-gut-microbe axis. Front Microbiol. 2014;5:146.
174. Da Mesquita S, Fu Z, Kipnis J. The meningeal lymphatic system: a new player in neurophysiology. Neuron. 2018;100:375–88.
175. Burokas A, Arboleya S, Moloney RD, Peterson VL, Murphy K, Clarke G, Stanton C, Dinan TG, Cryan JF. Targeting the microbiota-gut-brain axis: prebiotics have anxiolytic and antidepressant-like effects and reverse the impact of chronic stress in mice. Biol Psychiatry. 2017;82:472–87.
176. Luczynski P, Tramullas M, Viola M, Shanahan F, Clarke G, O'Mahony S, Dinan TG, Cryan JF. Microbiota regulates visceral pain in the mouse. elife. 2017;6:e25887.
177. O'Mahony L, McCarthy J, Kelly P, Hurley G, Luo F, Chen K, O'Sullivan GC, Kiely B, Collins JK, Shanahan F, Quigley EM. Lactobacillus and bifidobacterium in irritable bowel syndrome: symptom responses and relationship to cytokine profiles. Gastroenterology. 2005;128:541–51.
178. Miller AH, Haroon E, Raison CL, Felger JC. Cytokine targets in the brain: impact on neurotransmitters and neurocircuits. Depress Anxiety. 2013;30:297–306.
179. Chu AL, Stochl J, Lewis G, Zammit S, Jones PB, Khandaker GM. Longitudinal association between inflammatory markers and specific symptoms of depression in a prospective birth cohort. Brain Behav Immun. 2019;76:74–81.
180. Khandaker GM, Zammit S, Burgess S, Lewis G, Jones PB. Association between a functional interleukin 6 receptor genetic variant and risk of depression and psychosis in a population-based birth cohort. Brain Behav Immun. 2018;69:264–72.
181. van Sadelhoff JHJ, Perez Pardo P, Wu J, Garssen J, van Bergenhenegouwen J, Hogenkamp A, Hartog A, Kraneveld AD. The gut-immune-brain axis in autism spectrum disorders; a focus on amino acids. Front Endocrinol (Lausanne). 2019;10:247.
182. Cosorich I, Dalla-Costa G, Sorini C, Ferrarese R, Messina MJ, Dolpady J, Radice E, Mariani A, Testoni PA, Canducci F, Comi G, Martinelli V, Falcone M. High frequency of intestinal TH17 cells correlates with microbiota alterations and disease activity in multiple sclerosis. Sci Adv. 2017;3:e1700492.
183. Kang DW, Adams JB, Coleman DM, Pollard EL, Maldonado J, McDonough-Means S, Caporaso JG, Krajmalnik-Brown R. Long-term benefit of Microbiota Transfer Therapy on autism symptoms and gut microbiota. Sci Rep. 2019;9:5821.
184. McGaughey KD, Yilmaz-Swenson T, Elsayed NM, Cruz DA, Rodriguiz RM, Kritzer MD, Peterchev AV, Roach J, Wetsel WC, Williamson DE. Relative abundance of Akkermansia spp. and other bacterial phylotypes correlates with anxiety- and depressive-like behavior following social defeat in mice. Sci Rep. 2019;9:3281.
185. Cattaneo A, Cattane N, Galluzzi S, Provasi S, Lopizzo N, Festari C, Ferrari C, Guerra UP, Paghera B, Muscio C, Bianchetti A, Volta GD, Turla M, Cotelli MS, Gennuso M, Prelle A, Zanetti O, Lussignoli G, Mirabile D, Bellandi D, Gentile S, Belotti G, Villani D, Harach T, Bolmont T, Padovani A, Boccardi M, Frisoni GB, INDIA-FBP Group. Association of brain amyloidosis with pro-inflammatory gut bacterial taxa and peripheral inflammation markers in cognitively impaired elderly. Neurobiol Aging. 2017;49:60–8.
186. Perez-Pardo P, Dodiya HB, Engen PA, Forsyth CB, Huschens AM, Shaikh M, Voigt RM, Naqib A, Green SJ, Kordower JH, Shannon KM, Garssen J, Kraneveld AD, Keshavarzian A. Role of TLR4 in the gut-brain axis in Parkinson's disease: a translational study from men to mice. Gut. 2019;68:829–43.
187. Peter I, Dubinsky M, Bressman S, Park A, Lu C, Chen N, Wang A. Anti-tumor necrosis factor therapy and incidence of Parkinson disease among patients with inflammatory bowel disease. JAMA Neurol. 2018;75:939–46.
188. Frankiensztajn LM, Elliott E, Koren O. The microbiota and the hypothalamus-pituitary-adrenocortical (HPA) axis, implications for anxiety and stress disorders. Curr Opin Neurobiol. 2020;62:76–82.
189. Mayer EA. The neurobiology of stress and gastrointestinal disease. Gut. 2000;47:861–9.
190. Sudo N, Chida Y, Aiba Y, Sonoda J, Oyama N, Yu XN, Kubo C, Koga Y. Postnatal microbial colonization programs the hypothalamic-pituitary-adrenal system for stress response in mice. J Physiol. 2004;558:263–75.
191. Lyte JM, Gheorghe CE, Goodson MS, Kelley-Loughnane N, Dinan TG, Cryan JF, Clarke G. Gut-brain axis serotonergic responses to acute stress exposure are microbiome-dependent. Neurogastroenterol Motil. 2020;32:e13881.
192. O'Mahony SM, Hyland NP, Dinan TG, Cryan JF. Maternal separation as a model of brain-gut axis dysfunction. Psychopharmacology. 2011;214:71–88.
193. De Palma G, Blennerhassett P, Lu J, Deng Y, Park AJ, Green W, Denou E, Silva MA, Santacruz A, Sanz Y, Surette MG, Verdu EF, Collins SM, Bercik P. Microbiota and host determinants of behavioural phenotype in maternally separated mice. Nat Commun. 2015;6:7735.
194. De Weerth C. Do bacteria shape our development? Crosstalk between intestinal microbiota and HPA axis. Neurosci Biobehav Rev. 2017;83:458–71.
195. Dinan TG, Quigley EM, Ahmed SM, Scully P, O'Brien S, O'Mahony L, O'Mahony S, Shanahan F, Keeling PW. Hypothalamic-pituitary-gut axis dysregulation in irritable bowel syndrome: plasma cytokines as a potential biomarker? Gastroenterology. 2006;130:304–11.
196. Cryan JF, O'Riordan KJ, Sandhu K, Peterson V, Dinan TG. The gut microbiome in neurological disorders. Lancet Neurol. 2020;19:179–94.
197. Chaidez V, Hansen RL, Hertz-Picciotto I. Gastrointestinal problems in children with autism, developmental delays or typical development. J Autism Dev Disord. 2014;44:1117–27.
198. Hodges H, Fealko C, Soares N. Autism spectrum disorder: definition, epidemiology, causes, and clinical evaluation. Transl Pediatr. 2020;9:S55–65.
199. Kang DW, Adams JB, Gregory AC, Borody T, Chittick L, Fasano A, Khoruts A, Geis E, Maldonado J, McDonough-Means S, Pollard EL, Roux S, Sadowsky MJ, Lipson KS, Sullivan MB, Caporaso JG, Krajmalnik-Brown R. Microbiota Transfer Therapy alters gut ecosystem and improves gastrointestinal and autism symptoms: an open-label study. Microbiome. 2017;5:10.
200. Mayer EA, Padua D, Tillisch K. Altered brain-gut axis in autism: comorbidity or causative mechanisms? BioEssays. 2014;36:933–9.
201. Golubeva AV, Joyce SA, Moloney G, Burokas A, Sherwin E, Arboleya S, Flynn I, Khochanskiy D, Moya-Perez A, Peterson V, Rea K, Murphy K, Makarova O, Buravkov S, Hyland NP, Stanton C, Clarke G, Gahan CGM, Dinan TG, Cryan JF. Microbiota-related changes in bile acid & tryptophan metabolism are associ-

ated with gastrointestinal dysfunction in a mouse model of autism. EBioMedicine. 2017;24:166–78.
202. Needham BD, Funabashi M, Adame MD, Wang Z, Boktor JC, Haney J, Wu W-L, Rabut C, Ladinsky MS, Hwang S-J, Guo Y, Zhu Q, Griffiths JA, Knight R, Bjorkman PJ, Shapiro MG, Geschwind DH, Holschneider DP, Fischbach MA, Mazmanian SK. A gut-derived metabolite alters brain activity and anxiety behaviour in mice. Nature. 2022;602:647–53.
203. Desbonnet L, Clarke G, Shanahan F, Dinan TG, Cryan JF. Microbiota is essential for social development in the mouse. Mol Psychiatry. 2014;19:146–8.
204. Sharon G, Cruz NJ, Kang DW, Gandal MJ, Wang B, Kim YM, Zink EM, Casey CP, Taylor BC, Lane CJ, Bramer LM, Isern NG, Hoyt DW, Noecker C, Sweredoski MJ, Moradian A, Borenstein E, Jansson JK, Knight R, Metz TO, Lois C, Geschwind DH, Krajmalnik-Brown R, Mazmanian SK. Human gut microbiota from autism spectrum disorder promote behavioral symptoms in mice. Cell. 2019;177:1600–18.e17.
205. Brock J. Can gut bacteria cause autism (in mice)? June 14, 2019. medium.com. https://medium.com/dr-jon-brock/can-gut-bacteria-cause-autism-in-mice-582306fd7235.
206. Kandeel WA, Meguid NA, Bjorklund G, Eid EM, Farid M, Mohamed SK, Wakeel KE, Chirumbolo S, Elsaeid A, Hammad DY. Impact of Clostridium bacteria in children with autism spectrum disorder and their anthropometric measurements. J Mol Neurosci. 2020;70(6):897–907.
207. Kim S, Kim H, Yim YS, Ha S, Atarashi K, Tan TG, Longman RS, Honda K, Littman DR, Choi GB, Huh JR. Maternal gut bacteria promote neurodevelopmental abnormalities in mouse offspring. Nature. 2017;549:528–32.
208. Sandler RH, Finegold SM, Bolte ER, Buchanan CP, Maxwell AP, Vaisanen ML, Nelson MN, Wexler HM. Short-term benefit from oral vancomycin treatment of regressive-onset autism. J Child Neurol. 2000;15:429–35.
209. Russo AJ. Decreased plasma myeloperoxidase associated with probiotic therapy in autistic children. Clin Med Insights Pediatr. 2015;9:13–7.
210. Tomova A, Husarova V, Lakatosova S, Bakos J, Vlkova B, Babinska K, Ostatnikova D. Gastrointestinal microbiota in children with autism in Slovakia. Physiol Behav. 2015;138:179–87.
211. Yap CX, Henders AK, Alvares GA, Wood DL, Krause L, Tyson GW, Restuadi R, Wallace L, McLaren T, Hansell NK. Autism-related dietary preferences mediate autism-gut microbiome associations. Cell. 2021;184:5916–31.e17.
212. Berer K, Mues M, Koutrolos M, Rasbi ZA, Boziki M, Johner C, Wekerle H, Krishnamoorthy G. Commensal microbiota and myelin autoantigen cooperate to trigger autoimmune demyelination. Nature. 2011;479:538–41.
213. Tremlett H, Waubant E. Gut microbiome and pediatric multiple sclerosis. Mult Scler. 2018;24:64–8.
214. Tremlett H, Fadrosh DW, Faruqi AA, Zhu F, Hart J, Roalstad S, Graves J, Lynch S, Waubant E, US Network of Pediatric MS Centers Centers. Gut microbiota in early pediatric multiple sclerosis: a case-control study. Eur J Neurol. 2016;23:1308–21.
215. Cantoni C, Lin Q, Dorsett Y, Ghezzi L, Liu Z, Pan Y, Chen K, Han Y, Li Z, Xiao H, Gormley M, Liu Y, Bokoliya S, Panier H, Suther C, Evans E, Deng L, Locca A, Mikesell R, Obert K, Newland P, Wu Y, Salter A, Cross AH, Tarr PI, Lovett-Racke A, Piccio L, Zhou Y. Alterations of host-gut microbiome interactions in multiple sclerosis. EBioMedicine. 2022;76:103798.
216. Berer K, Gerdes LA, Cekanaviciute E, Jia X, Xiao L, Xia Z, Liu C, Klotz L, Stauffer U, Baranzini SE, Kumpfel T, Hohlfeld R, Krishnamoorthy G, Wekerle H. Gut microbiota from multiple sclerosis patients enables spontaneous autoimmune encephalomyelitis in mice. Proc Natl Acad Sci U S A. 2017;114:10719–24.
217. Cekanaviciute E, Yoo BB, Runia TF, Debelius JW, Singh S, Nelson CA, Kanner R, Bencosme Y, Lee YK, Hauser SL, Crabtree-Hartman E, Sand IK, Gacias M, Zhu Y, Casaccia P, Cree BAC, Knight R, Mazmanian SK, Baranzini SE. Gut bacteria from multiple sclerosis patients modulate human T cells and exacerbate symptoms in mouse models. Proc Natl Acad Sci U S A. 2017;114:10713–8.
218. Gacias M, Gaspari S, Santos PM, Tamburini S, Andrade M, Zhang F, Shen N, Tolstikov V, Kiebish MA, Dupree JL, Zachariou V, Clemente JC, Casaccia P. Microbiota-driven transcriptional changes in prefrontal cortex override genetic differences in social behavior. elife. 2016;5:e13442.
219. Keogh CE, Kim DHJ, Pusceddu MM, Knotts TA, Rabasa G, Sladek JA, Hsieh MT, Honeycutt M, Brust-Mascher I, Barboza M, Gareau MG. Myelin as a regulator of development of the microbiota-gut-brain axis. Brain Behav Immun. 2021;91:437–50.
220. Lu J, Synowiec S, Lu L, Yu Y, Bretherick T, Takada S, Yarnykh V, Caplan J, Caplan M, Claud EC, Drobyshevsky A. Microbiota influence the development of the brain and behaviors in C57BL/6J mice. PLoS One. 2018;13:e0201829.
221. Libbey JE, Sanchez JM, Doty DJ, Sim JT, Cusick MF, Cox JE, Fischer KF, Round JL, Fujinami RS. Variations in diet cause alterations in microbiota and metabolites that follow changes in disease severity in a multiple sclerosis model. Benefic Microbes. 2018;9:495–513.
222. Dargahi N, Matsoukas J, Apostolopoulos V. Streptococcus thermophilus ST285 alters pro-inflammatory to anti-inflammatory cytokine secretion against multiple sclerosis peptide in mice. Brain Sci. 2020;10:126.
223. Kouchaki E, Tamtaji OR, Salami M, Bahmani F, Daneshvar Kakhaki R, Akbari E, Tajabadi-Ebrahimi M, Jafari P, Asemi Z. Clinical and metabolic response to probiotic supplementation in patients with multiple sclerosis: a randomized, double-blind, placebo-controlled trial. Clin Nutr. 2017;36:1245–9.
224. Tamtaji OR, Kouchaki E, Salami M, Aghadavod E, Akbari E, Tajabadi-Ebrahimi M, Asemi Z. The effects of probiotic supplementation on gene expression related to inflammation, insulin, and lipids in patients with multiple sclerosis: a randomized, double-blind, placebo-controlled trial. J Am Coll Nutr. 2017;36:660–5.
225. Tankou SK, Regev K, Healy BC, Tjon E, Laghi L, Cox LM, Kivisakk P, Pierre IV, Hrishikesh L, Gandhi R, Cook S, Glanz B, Stankiewicz J, Weiner HL. A probiotic modulates the microbiome and immunity in multiple sclerosis. Ann Neurol. 2018;83:1147–61.
226. Greenland JC, Williams-Gray CH, Barker RA. The clinical heterogeneity of Parkinson's disease and its therapeutic implications. Eur J Neurosci. 2019;49:328–38.
227. Felice VD, Quigley EM, Sullivan AM, O'Keeffe GW, O'Mahony SM. Microbiota-gut-brain signalling in Parkinson's disease: implications for non-motor symptoms. Parkinsonism Relat Disord. 2016;27:1–8.
228. Knudsen K, Krogh K, Ostergaard K, Borghammer P. Constipation in Parkinson's disease: subjective symptoms, objective markers, and new perspectives. Mov Disord. 2017;32:94–105.
229. Braak H, de Vos RA, Bohl J, Del Tredici K. Gastric alpha-synuclein immunoreactive inclusions in Meissner's and Auerbach's plexuses in cases staged for Parkinson's disease-related brain pathology. Neurosci Lett. 2006;396:67–72.
230. Forsyth CB, Shannon KM, Kordower JH, Voigt RM, Shaikh M, Jaglin JA, Estes JD, Dodiya HB, Keshavarzian A. Increased intestinal permeability correlates with sigmoid mucosa alpha-synuclein staining and endotoxin exposure markers in early Parkinson's disease. PLoS One. 2011;6:e28032.
231. Hilton D, Stephens M, Kirk L, Edwards P, Potter R, Zajicek J, Broughton E, Hagan H, Carroll C. Accumulation of alpha-synuclein in the bowel of patients in the pre-clinical phase of Parkinson's disease. Acta Neuropathol. 2014;127:235–41.

232. Bhattacharyya D, Mohite GM, Krishnamoorthy J, Gayen N, Mehra S, Navalkar A, Kotler SA, Ratha BN, Ghosh A, Kumar R, Garai K, Mandal AK, Maji SK, Bhunia A. Lipopolysaccharide from gut microbiota modulates alpha-synuclein aggregation and alters its biological function. ACS Chem Neurosci. 2019;10:2229–36.
233. Holmqvist S, Chutna O, Bousset L, Aldrin-Kirk P, Li W, Bjorklund T, Wang ZY, Roybon L, Melki R, Li JY. Direct evidence of Parkinson pathology spread from the gastrointestinal tract to the brain in rats. Acta Neuropathol. 2014;128:805–20.
234. Kim S, Kwon SH, Kam TI, Panicker N, Karuppagounder SS, Lee S, Lee JH, Kim WR, Kook M, Foss CA, Shen C, Lee H, Kulkarni S, Pasricha PJ, Lee G, Pomper MG, Dawson VL, Dawson TM, Ko HS. Transneuronal propagation of pathologic alpha-synuclein from the gut to the brain models Parkinson's disease. Neuron. 2019;103:627–41.e7.
235. Aho VTE, Pereira PAB, Voutilainen S, Paulin L, Pekkonen E, Auvinen P, Scheperjans F. Gut microbiota in Parkinson's disease: temporal stability and relations to disease progression. EBioMedicine. 2019;44:691–707.
236. Heintz-Buschart A, Pandey U, Wicke T, Sixel-Doring F, Janzen A, Sittig-Wiegand E, Trenkwalder C, Oertel WH, Mollenhauer B, Wilmes P. The nasal and gut microbiome in Parkinson's disease and idiopathic rapid eye movement sleep behavior disorder. Mov Disord. 2018;33:88–98.
237. Keshavarzian A, Green SJ, Engen PA, Voigt RM, Naqib A, Forsyth CB, Mutlu E, Shannon KM. Colonic bacterial composition in Parkinson's disease. Mov Disord. 2015;30:1351–60.
238. Scheperjans F, Aho V, Pereira PA, Koskinen K, Paulin L, Pekkonen E, Haapaniemi E, Kaakkola S, Eerola-Rautio J, Pohja M, Kinnunen E, Murros K, Auvinen P. Gut microbiota are related to Parkinson's disease and clinical phenotype. Mov Disord. 2015;30:350–8.
239. Sun MF, Zhu YL, Zhou ZL, Jia XB, Xu YD, Yang Q, Cui C, Shen YQ. Neuroprotective effects of fecal microbiota transplantation on MPTP-induced Parkinson's disease mice: gut microbiota, glial reaction and TLR4/TNF-alpha signaling pathway. Brain Behav Immun. 2018;70:48–60.
240. Challis C, Hori A, Sampson TR, Yoo BB, Challis RC, Hamilton AM, Mazmanian SK, Volpicelli-Daley LA, Gradinaru V. Gut-seeded alpha-synuclein fibrils promote gut dysfunction and brain pathology specifically in aged mice. Nat Neurosci. 2020;23:327–36.
241. Wallen ZD, Demirkan A, Twa G, Cohen G, Dean MN, Standaert DG, Sampson TR, Payami H. Metagenomics of Parkinson's disease implicates the gut microbiome in multiple disease mechanisms. Nat Commun. 2022;13:6958.
242. Sampson TR, Debelius JW, Thron T, Janssen S, Shastri GG, Ilhan ZE, Challis C, Schretter CE, Rocha S, Gradinaru V, Chesselet MF, Keshavarzian A, Shannon KM, Krajmalnik-Brown R, Wittung-Stafshede P, Knight R, Mazmanian SK. Gut microbiota regulate motor deficits and neuroinflammation in a model of Parkinson's disease. Cell. 2016;167:1469–80.e12.
243. Unger MM, Spiegel J, Dillmann KU, Grundmann D, Philippeit H, Burmann J, Fassbender K, Schwiertz A, Schafer KH. Short chain fatty acids and gut microbiota differ between patients with Parkinson's disease and age-matched controls. Parkinsonism Relat Disord. 2016;32:66–72.
244. Sampson TR, Challis C, Jain N, Moiseyenko A, Ladinsky MS, Shastri GG, Thron T, Needham BD, Horvath I, Debelius JW, Janssen S, Knight R, Wittung-Stafshede P, Gradinaru V, Chapman M, Mazmanian SK. A gut bacterial amyloid promotes alpha-synuclein aggregation and motor impairment in mice. elife. 2020;9:e53111.
245. Dobbs SM, Dobbs RJ, Weller C, Charlett A, Bjarnason IT, Lawson AJ, Letley D, Harbin L, Price AB, Ibrahim MA, Oxlade NL, Bowthorpe J, Leckstroem D, Smee C, Plant JM, Peterson DW. Differential effect of Helicobacter pylori eradication on time-trends in brady/hypokinesia and rigidity in idiopathic parkinsonism. Helicobacter. 2010;15:279–94.
246. Shen X, Yang H, Wu Y, Zhang D, Jiang H. Meta-analysis: association of Helicobacter pylori infection with Parkinson's diseases. Helicobacter. 2017;22
247. Vogt NM, Kerby RL, Dill-Mcfarland KA, Harding SJ, Merluzzi AP, Johnson SC, Carlsson CM, Asthana S, Zetterberg H, Blennow K, Bendlin BB, Rey FE. Gut microbiome alterations in Alzheimer's disease. Sci Rep. 2017;7:13537.
248. van Kessel SP, Frye AK, El-Gendy AO, Castejon M, Keshavarzian A, van Dijk G, El Aidy S. Gut bacterial tyrosine decarboxylases restrict levels of levodopa in the treatment of Parkinson's disease. Nat Commun. 2019;10:310.
249. Friedland RP. Mechanisms of molecular mimicry involving the microbiota in neurodegeneration. J Alzheimers Dis. 2015;45:349–62.
250. Hill JM, Clement C, Pogue AI, Bhattacharjee S, Zhao Y, Lukiw WJ. Pathogenic microbes, the microbiome, and Alzheimer's disease (AD). Front Aging Neurosci. 2014;6:127.
251. Itzhaki RF, Lathe R, Balin BJ, Ball MJ, Bearer EL, Braak H, Bullido MJ, Carter C, Clerici M, Cosby SL, Del Tredici K, Field H, Fulop T, Grassi C, Griffin WS, Haas J, Hudson AP, Kamer AR, Kell DB, Licastro F, Letenneur L, Lovheim H, Mancuso R, Miklossy J, Otth C, Palamara AT, Perry G, Preston C, Pretorius E, Strandberg T, Tabet N, Taylor-Robinson SD, Whittum-Hudson JA. Microbes and Alzheimer's disease. J Alzheimers Dis. 2016;51:979–84.
252. Soscia SJ, Kirby JE, Washicosky KJ, Tucker SM, Ingelsson M, Hyman B, Burton MA, Goldstein LE, Duong S, Tanzi RE, Moir RD. The Alzheimer's disease-associated amyloid beta-protein is an antimicrobial peptide. PLoS One. 2010;5:e9505.
253. Braak H, Braak E. Neuropathological stageing of Alzheimer-related changes. Acta Neuropathol. 1991;82:239–59.
254. Eimer WA, Vijaya Kumar DK, Navalpur Shanmugam NK, Rodriguez AS, Mitchell T, Washicosky KJ, Gyorgy B, Breakefield XO, Tanzi RE, Moir RD. Alzheimer's disease-associated beta-amyloid is rapidly seeded by herpesviridae to protect against brain infection. Neuron. 2018;99:56–63.e3.
255. Itzhaki RF. Herpes simplex virus type 1 and Alzheimer's disease: possible mechanisms and signposts. FASEB J. 2017;31:3216–26.
256. Harach T, Marungruang N, Duthilleul N, Cheatham V, McCoy KD, Frisoni G, Neher JJ, Fak F, Jucker M, Lasser T, Bolmont T. Reduction of abeta amyloid pathology in APPPS1 transgenic mice in the absence of gut microbiota. Sci Rep. 2017;7:41802.
257. Mezo C, Dokalis N, Mossad O, Staszewski O, Neuber J, Yilmaz B, Schnepf D, de Aguero MG, Ganal-Vonarburg SC, Macpherson AJ, Meyer-Luehmann M, Staeheli P, Blank T, Prinz M, Erny D. Different effects of constitutive and induced microbiota modulation on microglia in a mouse model of Alzheimer's disease. Acta Neuropathol Commun. 2020;8:119.
258. Dodiya HB, Kuntz T, Shaik SM, Baufeld C, Leibowitz J, Zhang X, Gottel N, Zhang X, Butovsky O, Gilbert JA, Sisodia SS. Sex-specific effects of microbiome perturbations on cerebral Abeta amyloidosis and microglia phenotypes. J Exp Med. 2019;216:1542–60.
259. Minter MR, Hinterleitner R, Meisel M, Zhang C, Leone V, Zhang X, Oyler-Castrillo P, Zhang X, Musch MW, Shen X, Jabri B, Chang EB, Tanzi RE, Sisodia SS. Antibiotic-induced perturbations in microbial diversity during post-natal development alters amyloid pathology in an aged APPSWE/PS1DeltaE9 murine model of Alzheimer's disease. Sci Rep. 2017;7:10411.
260. Zhuang ZQ, Shen LL, Li WW, Fu X, Zeng F, Gui L, Lu Y, Cai M, Zhu C, Tan YL, Zheng P, Li HY, Zhu J, Zhou HD, Bu XL, Wang YJ. Gut microbiota is altered in patients with Alzheimer's disease. J Alzheimers Dis. 2018;63:1337–46.

261. Arnold SE, Arvanitakis Z, Macauley-Rambach SL, Koenig AM, Wang HY, Ahima RS, Craft S, Gandy S, Buettner C, Stoeckel LE, Holtzman DM, Nathan DM. Brain insulin resistance in type 2 diabetes and Alzheimer disease: concepts and conundrums. Nat Rev Neurol. 2018;14:168–81.
262. Nho K, Kueider-Paisley A, Mahmoudiandehkordi S, Arnold M, Risacher SL, Louie G, Blach C, Baillie R, Han X, Kastenmuller G, Jia W, Xie G, Ahmad S, Hankemeier T, van Duijn CM, Trojanowski JQ, Shaw LM, Weiner MW, Doraiswamy PM, Saykin AJ, Kaddurah-Daouk R, Alzheimer's Disease Neuroimaging Initiative and the Alzheimer Disease Metabolomics Consortium. Altered bile acid profile in mild cognitive impairment and Alzheimer's disease: relationship to neuroimaging and CSF biomarkers. Alzheimers Dement. 2019;15:232–44.
263. Connell E, Le Gall G, Pontifex MG, Sami S, Cryan JF, Clarke G, Müller M, Vauzour D. Microbial-derived metabolites as a risk factor of age-related cognitive decline and dementia. Mol Neurodegener. 2022;17:43.
264. Singh V, Roth S, Llovera G, Sadler R, Garzetti D, Stecher B, Dichgans M, Liesz A. Microbiota dysbiosis controls the neuroinflammatory response after stroke. J Neurosci. 2016;36:7428–40.
265. Yin J, Liao SX, He Y, Wang S, Xia GH, Liu FT, Zhu JJ, You C, Chen Q, Zhou L, Pan SY, Zhou HW. Dysbiosis of gut microbiota with reduced trimethylamine-N-oxide level in patients with large-artery atherosclerotic stroke or transient ischemic attack. J Am Heart Assoc. 2015;4:e002699.
266. Tan C, Wang H, Gao X, Xu R, Zeng X, Cui Z, Zhu J, Wu Q, Xia G, Zhou H, He Y, Yin J. Dynamic changes and prognostic value of gut microbiota-dependent trimethylamine-N-oxide in acute ischemic stroke. Front Neurol. 2020;11:29.
267. Zhu C, Li G, Lv Z, Li J, Wang X, Kang J, Zhan C. Association of plasma trimethylamine-N-oxide levels with post-stroke cognitive impairment: a 1-year longitudinal study. Neurol Sci. 2020;41:57–63.
268. Houlden A, Goldrick M, Brough D, Vizi ES, Lenart N, Martinecz B, Roberts IS, Denes A. Brain injury induces specific changes in the caecal microbiota of mice via altered autonomic activity and mucoprotein production. Brain Behav Immun. 2016;57:10–20.
269. Xia GH, You C, Gao XX, Zeng XL, Zhu JJ, Xu KY, Tan CH, Xu RT, Wu QH, Zhou HW, He Y, Yin J. Stroke dysbiosis index (SDI) in gut microbiome are associated with brain injury and prognosis of stroke. Front Neurol. 2019;10:397.
270. Winek K, Engel O, Koduah P, Heimesaat MM, Fischer A, Bereswill S, Dames C, Kershaw O, Gruber AD, Curato C, Oyama N, Meisel C, Meisel A, Dirnagl U. Depletion of cultivatable gut microbiota by broad-spectrum antibiotic pretreatment worsens outcome after murine stroke. Stroke. 2016;47:1354–63.
271. Sun J, Ling Z, Wang F, Chen W, Li H, Jin J, Zhang H, Pang M, Yu J, Liu J. Clostridium butyricum pretreatment attenuates cerebral ischemia/reperfusion injury in mice via anti-oxidation and anti-apoptosis. Neurosci Lett. 2016;613:30–5.
272. De Caro C, Iannone LF, Citraro R, Striano P, De Sarro G, Constanti A, Cryan JF, Russo E. Can we 'seize' the gut microbiota to treat epilepsy? Neurosci Biobehav Rev. 2019;107:750–64.
273. Cowan CSM, Hoban AE, Ventura-Silva AP, Dinan TG, Clarke G, Cryan JF. Gutsy moves: the amygdala as a critical node in microbiota to brain signaling. BioEssays. 2018;40.
274. Lindefeldt M, Eng A, Darban H, Bjerkner A, Zetterstrom CK, Allander T, Andersson B, Borenstein E, Dahlin M, Prast-Nielsen S. The ketogenic diet influences taxonomic and functional composition of the gut microbiota in children with severe epilepsy. NPJ Biofilms Microbiomes. 2019;5:5.
275. Xie G, Zhou Q, Qiu CZ, Dai WK, Wang HP, Li YH, Liao JX, Lu XG, Lin SF, Ye JH, Ma ZY, Wang WJ. Ketogenic diet poses a significant effect on imbalanced gut microbiota in infants with refractory epilepsy. World J Gastroenterol. 2017;23:6164–71.
276. Newell C, Bomhof MR, Reimer RA, Hittel DS, Rho JM, Shearer J. Ketogenic diet modifies the gut microbiota in a murine model of autism spectrum disorder. Mol Autism. 2016; 7:37.
277. Rogawski MA, Loscher W, Rho JM. Mechanisms of action of antiseizure drugs and the ketogenic diet. Cold Spring Harb Perspect Med. 2016;6:a022780.
278. Olson CA, Vuong HE, Yano JM, Liang QY, Nusbaum DJ, Hsiao EY. The gut microbiota mediates the anti-seizure effects of the ketogenic diet. Cell. 2018;174:497.
279. Braakman HMH, van Ingen J. Can epilepsy be treated by antibiotics? J Neurol. 2018;265:1934–6.
280. Testa CM, Jankovic J. Huntington disease: a quarter century of progress since the gene discovery. J Neurol Sci. 2019;396:52–68.
281. Jimenez-Sanchez M, Licitra F, Underwood BR, Rubinsztein DC. Huntington's disease: mechanisms of pathogenesis and therapeutic strategies. Cold Spring Harb Perspect Med. 2017;7(7):a024240.
282. van Dellen A, Hannan AJ. Genetic and environmental factors in the pathogenesis of Huntington's disease. Neurogenetics. 2004;5:9–17.
283. Rosas HD, Doros G, Bhasin S, Thomas B, Gevorkian S, Malarick K, Matson W, Hersch SM. A systems-level "misunderstanding": the plasma metabolome in Huntington's disease. Ann Clin Transl Neurol. 2015;2:756–68.
284. Radulescu CI, Garcia-Miralles M, Sidik H, Bardile CF, Yusof N, Lee HU, Ho EXP, Chu CW, Layton E, Low D, De Sessions PF, Pettersson S, Ginhoux F, Pouladi MA. Manipulation of microbiota reveals altered callosal myelination and white matter plasticity in a model of Huntington disease. Neurobiol Dis. 2019;127: 65–75.
285. Kong G, Cao KL, Judd LM, Li S, Renoir T, Hannan AJ. Microbiome profiling reveals gut dysbiosis in a transgenic mouse model of Huntington's disease. Neurobiol Dis. 2020;135:104268.
286. Alonso R, Pisa D, Carrasco L. Brain microbiota in Huntington's disease patients. Front Microbiol. 2019;10:2622.
287. Turner MR, Hardiman O, Benatar M, Brooks BR, Chio A, de Carvalho M, Ince PG, Lin C, Miller RG, Mitsumoto H, Nicholson G, Ravits J, Shaw PJ, Swash M, Talbot K, Traynor BJ, Van den Berg LH, Veldink JH, Vucic S, Kiernan MC. Controversies and priorities in amyotrophic lateral sclerosis. Lancet Neurol. 2013;12:310–22.
288. Wu S, Yi J, Zhang YG, Zhou J, Sun J. Leaky intestine and impaired microbiome in an amyotrophic lateral sclerosis mouse model. Physiol Rep. 2015;3:e12356.
289. Zhang YG, Wu S, Yi J, Xia Y, Jin D, Zhou J, Sun J. Target intestinal microbiota to alleviate disease progression in amyotrophic lateral sclerosis. Clin Ther. 2017;39:322–36.
290. Blacher E, Bashiardes S, Shapiro H, Rothschild D, Mor U, Dori-Bachash M, Kleimeyer C, Moresi C, Harnik Y, Zur M, Zabari M, Brik RB, Kviatcovsky D, Zmora N, Cohen Y, Bar N, Levi I, Amar N, Mehlman T, Brandis A, Biton I, Kuperman Y, Tsoory M, Alfahel L, Harmelin A, Schwartz M, Israelson A, Arike L, Johansson MEV, Hansson GC, Gotkine M, Segal E, Elinav E. Potential roles of gut microbiome and metabolites in modulating ALS in mice. Nature. 2019;572:474–80.
291. Fang X. Potential role of gut microbiota and tissue barriers in Parkinson's disease and amyotrophic lateral sclerosis. Int J Neurosci. 2016;126:771–6.
292. Zhai CD, Zheng JJ, An BC, Huang HF, Tan ZC. Intestinal microbiota composition in patients with amyotrophic lateral sclerosis: establishment of bacterial and archaeal communities analyses. Chin Med J. 2019;132:1815–22.
293. Brenner D, Hiergeist A, Adis C, Mayer B, Gessner A, Ludolph AC, Weishaupt JH. The fecal microbiome of ALS patients. Neurobiol Aging. 2018;61:132–7.

294. Bastiaanssen TFS, Cussotto S, Claesson MJ, Clarke G, Dinan TG, Cryan JF. Gutted! Unraveling the role of the microbiome in major depressive disorder. Harv Rev Psychiatry. 2020;28:26–39.
295. Huang Y, Shi X, Li Z, Shen Y, Shi X, Wang L, Li G, Yuan Y, Wang J, Zhang Y, Zhao L, Zhang M, Kang Y, Liang Y. Possible association of Firmicutes in the gut microbiota of patients with major depressive disorder. Neuropsychiatr Dis Treat. 2018;14:3329–37.
296. Milaneschi Y, Kappelmann N, Ye Z, Lamers F, Moser S, Jones PB, Burgess S, Penninx B, Khandaker GM. Association of inflammation with depression and anxiety: evidence for symptom-specificity and potential causality from UK Biobank and NESDA cohorts. Mol Psychiatry. 2021;26(12):7393–402.
297. Naseribafrouei A, Hestad K, Avershina E, Sekelja M, Linlokken A, Wilson R, Rudi K. Correlation between the human fecal microbiota and depression. Neurogastroenterol Motil. 2014;26:1155–62.
298. Zheng P, Zeng B, Zhou C, Liu M, Fang Z, Xu X, Zeng L, Chen J, Fan S, Du X, Zhang X, Yang D, Yang Y, Meng H, Li W, Melgiri ND, Licinio J, Wei H, Xie P. Gut microbiome remodeling induces depressive-like behaviors through a pathway mediated by the host's metabolism. Mol Psychiatry. 2016;21:786–96.
299. Valles-Colomer M, Falony G, Darzi Y, Tigchelaar EF, Wang J, Tito RY, Schiweck C, Kurilshikov A, Joossens M, Wijmenga C, Claes S, Van Oudenhove L, Zhernakova A, Vieira-Silva S, Raes J. The neuroactive potential of the human gut microbiota in quality of life and depression. Nat Microbiol. 2019;4:623–32.
300. Bosch JA, Nieuwdorp M, Zwinderman AH, Deschasaux M, Radjabzadeh D, Kraaij R, Davids M, de Rooij SR, Lok A. The gut microbiota and depressive symptoms across ethnic groups. Nat Commun. 2022;13:7129.
301. Radjabzadeh D, Bosch JA, Uitterlinden AG, Zwinderman AH, Ikram MA, Van Meurs JBJ, Luik AI, Nieuwdorp M, Lok A, Van Duijn CM, Kraaij R, Amin N. Gut microbiome-wide association study of depressive symptoms. Nat Commun. 2022;13:7128.
302. Long-Smith C, O'Riordan KJ, Clarke G, Stanton C, Dinan TG, Cryan JF. Microbiota-gut-brain axis: new therapeutic opportunities. Annu Rev Pharmacol Toxicol. 2020;60:477–502.
303. Sarkar A, Lehto SM, Harty S, Dinan TG, Cryan JF, Burnet PWJ. Psychobiotics and the manipulation of bacteria-gut-brain signals. Trends Neurosci. 2016;39:763–81.
304. Akkasheh G, Kashani-Poor Z, Tajabadi-Ebrahimi M, Jafari P, Akbari H, Taghizadeh M, Memarzadeh MR, Asemi Z, Esmaillzadeh A. Clinical and metabolic response to probiotic administration in patients with major depressive disorder: a randomized, double-blind, placebo-controlled trial. Nutrition. 2016;32:315–20.
305. Slykerman RF, Hood F, Wickens K, Thompson JMD, Barthow C, Murphy R, Kang J, Rowden J, Stone P, Crane J, Stanley T, Abels P, Purdie G, Maude R, Mitchell EA, Probiotic in Pregnancy Study Group. Effect of Lactobacillus rhamnosus HN001 in pregnancy on postpartum symptoms of depression and anxiety: a randomised double-blind placebo-controlled trial. EBioMedicine. 2017;24:159–65.
306. Kazemi A, Soltani S, Ghorabi S, Keshtkar A, Daneshzad E, Nasri F, Mazloomi SM. Effect of probiotic and synbiotic supplementation on inflammatory markers in health and disease status: a systematic review and meta-analysis of clinical trials. Clin Nutr. 2020;39:789–819.
307. Foster JA, Mcvey Neufeld KA. Gut-brain axis: how the microbiome influences anxiety and depression. Trends Neurosci. 2013;36:305–12.
308. Malan-Muller S, Valles-Colomer M, Raes J, Lowry CA, Seedat S, Hemmings SMJ. The gut microbiome and mental health: implications for anxiety- and trauma-related disorders. OMICS. 2018;22:90–107.
309. Diaz Heijtz R, Wang S, Anuar F, Qian Y, Bjorkholm B, Samuelsson A, Hibberd ML, Forssberg H, Pettersson S. Normal gut microbiota modulates brain development and behavior. Proc Natl Acad Sci U S A. 2011;108:3047–52.
310. Neufeld KM, Kang N, Bienenstock J, Foster JA. Reduced anxiety-like behavior and central neurochemical change in germ-free mice. Neurogastroenterol Motil. 2011;23:255–64.e119.
311. Crumeyrolle-Arias M, Jaglin M, Bruneau A, Vancassel S, Cardona A, Dauge V, Naudon L, Rabot S. Absence of the gut microbiota enhances anxiety-like behavior and neuroendocrine response to acute stress in rats. Psychoneuroendocrinology. 2014;42:207–17.
312. Bercik P, Denou E, Collins J, Jackson W, Lu J, Jury J, Deng Y, Blennerhassett P, Macri J, McCoy KD, Verdu EF, Collins SM. The intestinal microbiota affect central levels of brain-derived neurotropic factor and behavior in mice. Gastroenterology. 2011;141:599–609, 609 e1–3.
313. McGuinness AJ, Davis JA, Dawson SL, Loughman A, Collier F, O'Hely M, Simpson CA, Green J, Marx W, Hair C, Guest G, Mohebbi M, Berk M, Stupart D, Watters D, Jacka FN. A systematic review of gut microbiota composition in observational studies of major depressive disorder, bipolar disorder and schizophrenia. Mol Psychiatry. 2022;27:1920–35.
314. Madan A, Thompson D, Fowler JC, Ajami NJ, Salas R, Frueh BC, Bradshaw MR, Weinstein BL, Oldham JM, Petrosino JF. The gut microbiota is associated with psychiatric symptom severity and treatment outcome among individuals with serious mental illness. J Affect Disord. 2020;264:98–106.
315. Collins S, Hoare E, Allender S, Olive L, Leech RM, Winpenny EM, Jacka F, Lotfalian M. A longitudinal study of lifestyle behaviours in emerging adulthood and risk for symptoms of depression, anxiety, and stress. J Affect Disord. 2023;327:244–53.
316. Colica C, Avolio E, Bollero P, Costa de Miranda R, Ferraro S, Sinibaldi Salimei P, De Lorenzo A, Di Renzo L. Evidences of a new psychobiotic formulation on body composition and anxiety. Mediators Inflamm. 2017;2017:5650627.
317. Messaoudi M, Violle N, Bisson JF, Desor D, Javelot H, Rougeot C. Beneficial psychological effects of a probiotic formulation (Lactobacillus helveticus R0052 and Bifidobacterium longum R0175) in healthy human volunteers. Gut Microbes. 2011;2:256–61.
318. Rao AV, Bested AC, Beaulne TM, Katzman MA, Iorio C, Berardi JM, Logan AC. A randomized, double-blind, placebo-controlled pilot study of a probiotic in emotional symptoms of chronic fatigue syndrome. Gut Pathog. 2009;1:6.
319. Chong HX, Yusoff NAA, Hor YY, Lew LC, Jaafar MH, Choi SB, Yusoff MSB, Wahid N, Abdullah M, Zakaria N, Ong KL, Park YH, Liong MT. Lactobacillus plantarum DR7 alleviates stress and anxiety in adults: a randomised, double-blind, placebo-controlled study. Benefic Microbes. 2019;10:355–73.
320. Cussotto S, Sandhu KV, Dinan TG, Cryan JF. The neuroendocrinology of the microbiota-gut-brain axis: a behavioural perspective. Front Neuroendocrinol. 2018;51:80–101.
321. Maier L, Pruteanu M, Kuhan M, Zeller G, Telzerow A, Anderson EE, Brochado AR, Fernandez KC, Dose H, Mori H, Patil KR, Bork P, Typas A. Extensive impact of non-antibiotic drugs on human gut bacteria. Nature. 2018;555:623–8.
322. Castro-Nallar E, Bendall ML, Perez-Losada M, Sabuncyan S, Severance EG, Dickerson FB, Schroeder JR, Yolken RH, Crandall KA. Composition, taxonomy and functional diversity of the oropharynx microbiome in individuals with schizophrenia and controls. PeerJ. 2015;3:e1140.
323. Zheng P, Zeng B, Liu M, Chen J, Pan J, Han Y, Liu Y, Cheng K, Zhou C, Wang H, Zhou X, Gui S, Perry SW, Wong ML, Licinio J, Wei H, Xie P. The gut microbiome from patients with schizophrenia modulates the glutamate-glutamine-GABA cycle and schizophrenia-relevant behaviors in mice. Sci Adv. 2019;5:eaau8317.

324. Zhu F, Ju Y, Wang W, Wang Q, Guo R, Ma Q, Sun Q, Fan Y, Xie Y, Yang Z, Jie Z, Zhao B, Xiao L, Yang L, Zhang T, Feng J, Guo L, He X, Chen Y, Chen C, Gao C, Xu X, Yang H, Wang J, Dang Y, Madsen L, Brix S, Kristiansen K, Jia H, Ma X. Metagenome-wide association of gut microbiome features for schizophrenia. Nat Commun. 2020;11:1612.
325. Olde Loohuis LM, Mangul S, Ori APS, Jospin G, Koslicki D, Yang HT, Wu T, Boks MP, Lomen-Hoerth C, Wiedau-Pazos M, Cantor RM, de Vos WM, Kahn RS, Eskin E, Ophoff RA. Transcriptome analysis in whole blood reveals increased microbial diversity in schizophrenia. Transl Psychiatry. 2018;8:96.
326. Schwarz E, Maukonen J, Hyytiainen T, Kieseppa T, Oresic M, Sabunciyan S, Mantere O, Saarela M, Yolken R, Suvisaari J. Analysis of microbiota in first episode psychosis identifies preliminary associations with symptom severity and treatment response. Schizophr Res. 2018;192:398–403.
327. Nguyen TT, Kosciolek T, Maldonado Y, Daly RE, Martin AS, McDonald D, Knight R, Jeste DV. Differences in gut microbiome composition between persons with chronic schizophrenia and healthy comparison subjects. Schizophr Res. 2019;204:23–9.
328. Dickerson FB, Stallings C, Origoni A, Katsafanas E, Savage CL, Schweinfurth LA, Goga J, Khushalani S, Yolken RH. Effect of probiotic supplementation on schizophrenia symptoms and association with gastrointestinal functioning: a randomized, placebo-controlled trial. Prim Care Companion CNS Disord. 2014;16:PCC.13m01579.
329. Severance EG, Gressitt KL, Stallings CR, Katsafanas E, Schweinfurth LA, Savage CLG, Adamos MB, Sweeney KM, Origoni AE, Khushalani S, Dickerson FB, Yolken RH. Probiotic normalization of Candida albicans in schizophrenia: a randomized, placebo-controlled, longitudinal pilot study. Brain Behav Immun. 2017;62:41–5.
330. Ghaderi A, Banafshe HR, Mirhosseini N, Moradi M, Karimi MA, Mehrzad F, Bahmani F, Asemi Z. Clinical and metabolic response to vitamin D plus probiotic in schizophrenia patients. BMC Psychiatry. 2019;19:77.
331. Okubo R, Koga M, Katsumata N, Odamaki T, Matsuyama S, Oka M, Narita H, Hashimoto N, Kusumi I, Xiao J, Matsuoka YJ. Effect of bifidobacterium breve A-1 on anxiety and depressive symptoms in schizophrenia: a proof-of-concept study. J Affect Disord. 2019;245:377–85.
332. Enck P, Aziz Q, Barbara G, Farmer AD, Fukudo S, Mayer EA, Niesler B, Quigley EM, Rajilic-Stojanovic M, Schemann M, Schwille-Kiuntke J, Simren M, Zipfel S, Spiller RC. Irritable bowel syndrome. Nat Rev Dis Primers. 2016;2:16014.
333. Rodino-Janeiro BK, Vicario M, Alonso-Cotoner C, Pascua-Garcia R, Santos J. A review of microbiota and irritable bowel syndrome: future in therapies. Adv Ther. 2018;35:289–310.
334. Chong PP, Chin VK, Looi CY, Wong WF, Madhavan P, Yong VC. The microbiome and irritable bowel syndrome—a review on the pathophysiology, current research and future therapy. Front Microbiol. 2019;10:1136.
335. Quigley EM. Bugs on the brain; brain in the gut—seeking explanations for common gastrointestinal symptoms. Ir J Med Sci. 2013;182:1–6.
336. El-Salhy M, Valeur J, Hausken T, Gunnar Hatlebakk J. Changes in fecal short-chain fatty acids following fecal microbiota transplantation in patients with irritable bowel syndrome. Neurogastroenterol Motil. 2021;33:e13983.
337. Rajilic-Stojanovic M, Jonkers DM, Salonen A, Hanevik K, Raes J, Jalanka J, de Vos WM, Manichanh C, Golic N, Enck P, Philippou E, Iraqi FA, Clarke G, Spiller RC, Penders J. Intestinal microbiota and diet in IBS: causes, consequences, or epiphenomena? Am J Gastroenterol. 2015;110:278–87.
338. Durban A, Abellan JJ, Jimenez-Hernandez N, Artacho A, Garrigues V, Ortiz V, Ponce J, Latorre A, Moya A. Instability of the faecal microbiota in diarrhoea-predominant irritable bowel syndrome. FEMS Microbiol Ecol. 2013;86:581–9.
339. Jeffery IB, O'Toole PW, Ohman L, Claesson MJ, Deane J, Quigley EM, Simren M. An irritable bowel syndrome subtype defined by species-specific alterations in faecal microbiota. Gut. 2012;61:997–1006.
340. Matto J, Maunuksela L, Kajander K, Palva A, Korpela R, Kassinen A, Saarela M. Composition and temporal stability of gastrointestinal microbiota in irritable bowel syndrome—a longitudinal study in IBS and control subjects. FEMS Immunol Med Microbiol. 2005;43:213–22.
341. Jalanka-Tuovinen J, Salojarvi J, Salonen A, Immonen O, Garsed K, Kelly FM, Zaitoun A, Palva A, Spiller RC, De Vos WM. Faecal microbiota composition and host-microbe cross-talk following gastroenteritis and in postinfectious irritable bowel syndrome. Gut. 2014;63:1737–45.
342. Kerckhoffs AP, Samsom M, van der Rest ME, de Vogel J, Knol J, Ben-Amor K, Akkermans LM. Lower Bifidobacteria counts in both duodenal mucosa-associated and fecal microbiota in irritable bowel syndrome patients. World J Gastroenterol. 2009;15:2887–92.
343. Ravichandar JD, Rutherford E, Chow C-ET, Han A, Yamamoto ML, Narayan N, Kaplan GG, Beck PL, Claesson MJ, Dabbagh K, Iwai S, Desantis TZ. Strain level and comprehensive microbiome analysis in inflammatory bowel disease via multi-technology meta-analysis identifies key bacterial influencers of disease. Front Microbiol. 2022;13:961020.
344. Mawe GM, Coates MD, Moses PL. Review article: intestinal serotonin signalling in irritable bowel syndrome. Aliment Pharmacol Ther. 2006;23:1067–76.
345. Camilleri M. Serotonin in the gastrointestinal tract. Curr Opin Endocrinol Diabetes Obes. 2009;16:53–9.
346. Clarke G, Fitzgerald P, Cryan JF, Cassidy EM, Quigley EM, Dinan TG. Tryptophan degradation in irritable bowel syndrome: evidence of indoleamine 2,3-dioxygenase activation in a male cohort. BMC Gastroenterol. 2009;9:6.
347. Clarke G, McKernan DP, Gaszner G, Quigley EM, Cryan JF, Dinan TG. A distinct profile of tryptophan metabolism along the kynurenine pathway downstream of toll-like receptor activation in irritable bowel syndrome. Front Pharmacol. 2012;3:90.
348. Fitzgerald P, Cassidy Eugene M, Clarke G, Scully P, Barry S, Quigley Eamonn MM, Shanahan F, Cryan J, Dinan Timothy G. Tryptophan catabolism in females with irritable bowel syndrome: relationship to interferon-gamma, severity of symptoms and psychiatric co-morbidity. Neurogastroenterol Motil. 2008;20:1291–7.
349. Jeffery IB, Das A, O'Herlihy E, Coughlan S, Cisek K, Moore M, Bradley F, Carty T, Pradhan M, Dwibedi C, Shanahan F, O'Toole PW. Differences in fecal microbiomes and metabolomes of people with vs without irritable bowel syndrome and bile acid malabsorption. Gastroenterology. 2020;158:1016–28.e8.
350. Kennedy PJ, Clarke G, Quigley EM, Groeger JA, Dinan TG, Cryan JF. Gut memories: towards a cognitive neurobiology of irritable bowel syndrome. Neurosci Biobehav Rev. 2012;36:310–40.
351. Kennedy PJ, Cryan JF, Quigley EM, Dinan TG, Clarke G. A sustained hypothalamic-pituitary-adrenal axis response to acute psychosocial stress in irritable bowel syndrome. Psychol Med. 2014;44:3123–34.
352. Clarke G, Cryan JF, Dinan TG, Quigley EM. Review article: probiotics for the treatment of irritable bowel syndrome—focus on lactic acid bacteria. Aliment Pharmacol Ther. 2012;35: 403–13.
353. Ford AC, Harris LA, Lacy BE, Quigley EMM, Moayyedi P. Systematic review with meta-analysis: the efficacy of prebiotics, probiotics, synbiotics and antibiotics in irritable bowel syndrome. Aliment Pharmacol Ther. 2018;48:1044–60.

354. Diop L, Guillou S, Durand H. Probiotic food supplement reduces stress-induced gastrointestinal symptoms in volunteers: a double-blind, placebo-controlled, randomized trial. Nutr Res. 2008;28:1–5.
355. Pinto-Sanchez MI, Hall GB, Ghajar K, Nardelli A, Bolino C, Lau JT, Martin FP, Cominetti O, Welsh C, Rieder A, Traynor J, Gregory C, De Palma G, Pigrau M, Ford AC, Macri J, Berger B, Bergonzelli G, Surette MG, Collins SM, Moayyedi P, Bercik P. Probiotic Bifidobacterium longum NCC3001 reduces depression scores and alters brain activity: a pilot study in patients with irritable bowel syndrome. Gastroenterology. 2017;153:448–59.e8.
356. Ticinesi A, Milani C, Lauretani F, Nouvenne A, Mancabelli L, Lugli GA, Turroni F, Duranti S, Mangifesta M, Viappiani A, Ferrario C, Magagio M, Ventura M, Meschi T. Gut microbiota composition is associated with polypharmacy in elderly hospitalized patients. Sci Rep. 2017;7:11102.
357. Dinan TG, Cryan JF. Gut-brain axis in 2016: brain-gut-microbiota axis—mood, metabolism and behaviour. Nat Rev Gastroenterol Hepatol. 2017;14:69–70.
358. Halkjaer SI, Christensen AH, Lo BZS, Browne PD, Gunther S, Hansen LH, Petersen AM. Faecal microbiota transplantation alters gut microbiota in patients with irritable bowel syndrome: results from a randomised, double-blind placebo-controlled study. Gut. 2018;67:2107–15.
359. Mizuno S, Masaoka T, Naganuma M, Kishimoto T, Kitazawa M, Kurokawa S, Nakashima M, Takeshita K, Suda W, Mimura M, Hattori M, Kanai T. Bifidobacterium-rich fecal donor may be a positive predictor for successful fecal microbiota transplantation in patients with irritable bowel syndrome. Digestion. 2017;96:29–38.
360. Pimentel M, Morales W, Chua K, Barlow G, Weitsman S, Kim G, Amichai MM, Pokkunuri V, Rook E, Mathur R, Marsh Z. Effects of rifaximin treatment and retreatment in nonconstipated IBS subjects. Dig Dis Sci. 2011;56:2067–72.
361. Rea K, O'Mahony S, Dinan TG, Cryan JF. Pain bugs: gut microbiota and pain disorders. Curr Opin Physiol. 2019;11:97–102.
362. Pusceddu MM, Gareau MG. Visceral pain: gut microbiota, a new hope? J Biomed Sci. 2018;25:73.
363. Holzer P, Farzi A, Hassan AM, Zenz G, Jacan A, Reichmann F. Visceral inflammation and immune activation stress the brain. Front Immunol. 2017;8:1613.
364. Conte MP, Schippa S, Zamboni I, Penta M, Chiarini F, Seganti L, Osborn J, Falconieri P, Borrelli O, Cucchiara S. Gut-associated bacterial microbiota in paediatric patients with inflammatory bowel disease. Gut. 2006;55:1760–7.
365. Manichanh C, Rigottier-Gois L, Bonnaud E, Gloux K, Pelletier E, Frangeul L, Nalin R, Jarrin C, Chardon P, Marteau P, Roca J, Dore J. Reduced diversity of faecal microbiota in Crohn's disease revealed by a metagenomic approach. Gut. 2006;55:205–11.
366. Noor SO, Ridgway K, Scovell L, Kemsley EK, Lund EK, Jamieson C, Johnson IT, Narbad A. Ulcerative colitis and irritable bowel patients exhibit distinct abnormalities of the gut microbiota. BMC Gastroenterol. 2010;10:134.
367. Shankar V, Homer D, Rigsbee L, Khamis HJ, Michail S, Raymer M, Reo NV, Paliy O. The networks of human gut microbe-metabolite associations are different between health and irritable bowel syndrome. ISME J. 2015;9:1899–903.
368. Crouzet L, Gaultier E, Del'Homme C, Cartier C, Delmas E, Dapoigny M, Fioramonti J, Bernalier-Donadille A. The hypersensitivity to colonic distension of IBS patients can be transferred to rats through their fecal microbiota. Neurogastroenterol Motil. 2013;25:e272–82.
369. Bennet JD, Brinkman M. Treatment of ulcerative colitis by implantation of normal colonic flora. Lancet. 1989;1:164.
370. Ding C, Fan W, Gu L, Tian H, Ge X, Gong J, Nie Y, Li N. Outcomes and prognostic factors of fecal microbiota transplantation in patients with slow transit constipation: results from a prospective study with long-term follow-up. Gastroenterol Rep (Oxf). 2018;6:101–7.
371. Holvoet T, Joossens M, Wang J, Boelens J, Verhasselt B, Laukens D, van Vlierberghe H, Hindryckx P, De Vos M, De Looze D, Raes J. Assessment of faecal microbial transfer in irritable bowel syndrome with severe bloating. Gut. 2017;66: 980–2.
372. Kellermayer R, Nagy-Szakal D, Harris RA, Luna RA, Pitashny M, Schady D, Mir SA, Lopez ME, Gilger MA, Belmont J, Hollister EB, Versalovic J. Serial fecal microbiota transplantation alters mucosal gene expression in pediatric ulcerative colitis. Am J Gastroenterol. 2015;110:604–6.
373. Knox NC, Forbes JD, Van Domselaar G, Bernstein CN. The gut microbiome as a target for IBD treatment: are we there yet? Curr Treat Options Gastroenterol. 2019;17:115–26.
374. Naseer M, Poola S, Ali S, Samiullah S, Tahan V. Prebiotics and probiotics in inflammatory bowel disease (IBD): where are we now and where are we going? Curr Clin Pharmacol. 2020;15(3):216–33.
375. Amaral FA, Sachs D, Costa VV, Fagundes CT, Cisalpino D, Cunha TM, Ferreira SH, Cunha FQ, Silva TA, Nicoli JR, Vieira LQ, Souza DG, Teixeira MM. Commensal microbiota is fundamental for the development of inflammatory pain. Proc Natl Acad Sci U S A. 2008;105:2193–7.
376. Dinic M, Pecikoza U, Djokic J, Stepanovic-Petrovic R, Milenkovic M, Stevanovic M, Filipovic N, Begovic J, Golic N, Lukic J. Exopolysaccharide produced by probiotic strain Lactobacillus paraplantarum BGCG11 reduces inflammatory hyperalgesia in rats. Front Pharmacol. 2018;9:1.
377. Culligan EP, Sleator RD. Advances in the microbiome: applications to Clostridium difficile infection. J Clin Med. 2016;5:83.
378. Levy AN, Allegretti JR. Insights into the role of fecal microbiota transplantation for the treatment of inflammatory bowel disease. Ther Adv Gastroenterol. 2019;12:1756284819836893.
379. Derwa Y, Gracie DJ, Hamlin PJ, Ford AC. Systematic review with meta-analysis: the efficacy of probiotics in inflammatory bowel disease. Aliment Pharmacol Ther. 2017;46:389–400.
380. Scher JU, Sczesnak A, Longman RS, Segata N, Ubeda C, Bielski C, Rostron T, Cerundolo V, Pamer EG, Abramson SB, Huttenhower C, Littman DR. Expansion of intestinal Prevotella copri correlates with enhanced susceptibility to arthritis. elife. 2013;2:e01202.
381. Lowe J, Briggs A, Whittle S, Hoon E, Stephenson M. Effectiveness of probiotics in the management of inflammatory arthritis: a systematic review protocol. JBI Database System Rev Implement Rep. 2018;16:2295–303.
382. Shen X, Yang H, Wu Y, Zhang D, Jiang H. Meta-analysis: association of Helicobacter pylori infection with Parkinson's diseases. Helicobacter. 2017;22.
383. Osterlund P, Ruotsalainen T, Korpela R, Saxelin M, Ollus A, Valta P, Kouri M, Elomaa I, Joensuu H. Lactobacillus supplementation for diarrhoea related to chemotherapy of colorectal cancer: a randomised study. Br J Cancer. 2007;97:1028–34.
384. Cai TT, Ye XL, Yong HJ, Song B, Zheng XL, Cui BT, Zhang FM, Lu YB, Miao H, Ding DF. Fecal microbiota transplantation relieve painful diabetic neuropathy: a case report. Medicine (Baltimore). 2018;97:e13543.
385. De Roos NM, Van Hemert S, Rovers JMP, Smits MG, Witteman BJM. The effects of a multispecies probiotic on migraine and markers of intestinal permeability-results of a randomized placebo-controlled study. Eur J Clin Nutr. 2017;71: 1455–62.
386. Meckel KR, Kiraly DD. A potential role for the gut microbiome in substance use disorders. Psychopharmacology. 2019;236:1513–30.

387. Garcia-Cabrerizo R, Carbia C, Riordan KJO, Schellekens H, Cryan JF. Microbiota-gut-brain axis as a regulator of reward processes. J Neurochem. 2021;157:1495–524.
388. Peterson VL, Jury NJ, Cabrera-Rubio R, Draper LA, Crispie F, Cotter PD, Dinan TG, Holmes A, Cryan JF. Drunk bugs: chronic vapour alcohol exposure induces marked changes in the gut microbiome in mice. Behav Brain Res. 2017;323:172–6.
389. Bajaj JS. Alcohol, liver disease and the gut microbiota. Nat Rev Gastroenterol Hepatol. 2019;16:235–46.
390. Leclercq S, Matamoros S, Cani PD, Neyrinck AM, Jamar F, Starkel P, Windey K, Tremaroli V, Backhed F, Verbeke K, de Timary P, Delzenne NM. Intestinal permeability, gut-bacterial dysbiosis, and behavioral markers of alcohol-dependence severity. Proc Natl Acad Sci U S A. 2014;111:E4485–93.
391. Carbia C, Bastiaanssen TFS, Iannone LF, García-Cabrerizo R, Boscaini S, Berding K, Strain CR, Clarke G, Stanton C, Dinan TG, Cryan JF. The Microbiome-Gut-Brain axis regulates social cognition & craving in young binge drinkers. EBioMedicine. 2023;89:104442.
392. Lowe PP, Gyongyosi B, Satishchandran A, Iracheta-Vellve A, Cho Y, Ambade A, Szabo G. Reduced gut microbiome protects from alcohol-induced neuroinflammation and alters intestinal and brain inflammasome expression. J Neuroinflammation. 2018;15:298.
393. Bull-Otterson L, Feng W, Kirpich I, Wang Y, Qin X, Liu Y, Gobejishvili L, Joshi-Barve S, Ayvaz T, Petrosino J, Kong M, Barker D, McClain C, Barve S. Metagenomic analyses of alcohol induced pathogenic alterations in the intestinal microbiome and the effect of Lactobacillus rhamnosus GG treatment. PLoS One. 2013;8:e53028.
394. Allais L, Kerckhof FM, Verschuere S, Bracke KR, De Smet R, Laukens D, Van den Abbeele P, De Vos M, Boon N, Brusselle GG, Cuvelier CA, Van de Wiele T. Chronic cigarette smoke exposure induces microbial and inflammatory shifts and mucin changes in the murine gut. Environ Microbiol. 2016;18:1352–63.
395. Biedermann L, Brulisauer K, Zeitz J, Frei P, Scharl M, Vavricka SR, Fried M, Loessner MJ, Rogler G, Schuppler M. Smoking cessation alters intestinal microbiota: insights from quantitative investigations on human fecal samples using FISH. Inflamm Bowel Dis. 2014;20:1496–501.
396. Lee K, Vuong HE, Nusbaum DJ, Hsiao EY, Evans CJ, Taylor AMW. The gut microbiota mediates reward and sensory responses associated with regimen-selective morphine dependence. Neuropsychopharmacology. 2018;43:2606–14.
397. Lowe PP, Gyongyosi B, Satishchandran A, Iracheta-Vellve A, Ambade A, Kodys K, Catalano D, Ward DV, Szabo G. Alcohol-related changes in the intestinal microbiome influence neutrophil infiltration, inflammation and steatosis in early alcoholic hepatitis in mice. PLoS One. 2017;12:e0174544.
398. Plein LM, Rittner HL. Opioids and the immune system—friend or foe. Br J Pharmacol. 2018;175:2717–25.
399. Wang F, Meng J, Zhang L, Johnson T, Chen C, Roy S. Morphine induces changes in the gut microbiome and metabolome in a morphine dependence model. Sci Rep. 2018;8:3596.
400. Cluny NL, Keenan CM, Reimer RA, Le Foll B, Sharkey KA. Prevention of diet-induced obesity effects on body weight and gut microbiota in mice treated chronically with delta9-tetrahydrocannabinol. PLoS One. 2015;10:e0144270.
401. Fields CT, Sampson TR, Bruce-Keller AJ, Kiraly DD, Hsiao EY, de Vries GJ. Defining dysbiosis in disorders of movement and motivation. J Neurosci. 2018;38:9414–22.
402. Kiraly DD, Walker DM, Calipari ES, Labonte B, Issler O, Pena CJ, Ribeiro EA, Russo SJ, Nestler EJ. Alterations of the host microbiome affect behavioral responses to cocaine. Sci Rep. 2016;6:35455.
403. García-Cabrerizo R, Barros-Santos T, Campos D, Cryan JF. The gut microbiota alone and in combination with a social stimulus regulates cocaine reward in the mouse. Brain Behav Immun. 2023;107:286–91.

Changes in the Gut Microbiome as Seen in Diabetes and Obesity

Gillian M. Barlow, Shreya Celly, and Ruchi Mathur

Highlights

- Perturbations in the balance of microbial populations can precipitate shifts toward inflammatory states and metabolic dysfunction, and contribute to the development of both Type 2 diabetes (T2D) and obesity.
- Alterations in gut microbial populations can impact intestinal barrier function and host metabolic and signaling pathways associated with insulin resistance. Microbial modulation of lipid accumulation, bacterial lipopolysaccharide (LPS), and short chain fatty production acids (SCFAs) also influences inflammation and/or or insulin signaling.
- Relative abundances of the genera *Clostridium* and *Phascolarctobacterium* positively correlate with insulin sensitivity in overweight/obese adults, and *Dialister* positively correlates with fasting insulin and insulin area-under-the-curve (AUC). Relative abundances of *Akkermansia, Faecalibacterium, Oscillibacter,* and *Alistipes* have been associated with obesity.
- Increased levels of methanogenic archaea in the gut, e.g. *Methanobrevibacter smithii*, are associated with impaired glucose tolerance and obesity. These effects may be mediated through SCFA production and stimulation of the incretin glucagon-like peptide 1 (GLP-1).
- Food additives including artificial sweeteners and emulsifiers may also influence the development of T2D and obesity. For example, aspartame inhibits intestinal alkaline phosphatase (IAP), resulting in elevated fasting glucose levels and impaired insulin-stimulated glucose disposal in rodent models, and saccharin and the emulsifier polysorbate-80 induce microbial encroachment and increase levels of bacterial LPS and flagellin, which act on toll-like receptors (TLR) 4 and 5 to induce inflammation.

G. M. Barlow · S. Celly
Medically Associated Science and Technology (MAST) Program, Cedars-Sinai Medical Center, West Hollywood, CA, USA
e-mail: gillian.barlow@cshs.org

R. Mathur (✉)
Medically Associated Science and Technology (MAST) Program, Cedars-Sinai Medical Center, West Hollywood, CA, USA

Division of Endocrinology, Diabetes, and Metabolism, Department of Medicine, Cedars-Sinai Medical Center, Los Angeles, CA, USA
e-mail: ruchi.mathur@cshs.org

1 Two Global Pandemics: Diabetes and Obesity

1.1 Diabetes

Definitions, Prevalence and Costs

The World Health Organization (WHO) defines diabetes as a "chronic metabolic disease characterized by elevated levels of blood glucose, which leads over time to serious damage to the heart, blood vessels, eyes, kidneys, and nerves" [1]. Based on 2021 estimates from the International Diabetes Federation (IDF), the number of adults aged 20–79 years with diabetes currently stands at 537 million worldwide [2] (Fig. 1). This is projected to increase to 643 million by 2030, and to 783 million by 2045, with the largest increases occurring in economies that are progressing from low- to middle-income status, such that 3 in 4 adults with diabetes live in low- and middle-income countries [2]. In the US, 37.3 million people (11.3% of the US population) were estimated to have diabetes, of whom an estimated 8.5 million were undiagnosed [3] (see Fig. 2). These escalating rates are associated with dramatic increases in health care costs. In 2021, global diabetes-related health expenditure was approximately $966 billion, 42.9% of which occurred in the North America and Caribbean Region, and projections for 2030 and 2045 are $1028 billion and $1054 billion, respectively [2].

While there are newly categorized subtypes of diabetes, the principal types of diabetes remain Type 1 diabetes (T1D) and Type 2 diabetes (T2D). T1D is due to autoimmune β-cell

M. Pimentel et al. (eds.), *Clinical Understanding of the Human Gut Microbiome*, https://doi.org/10.1007/978-3-031-46712-7_5

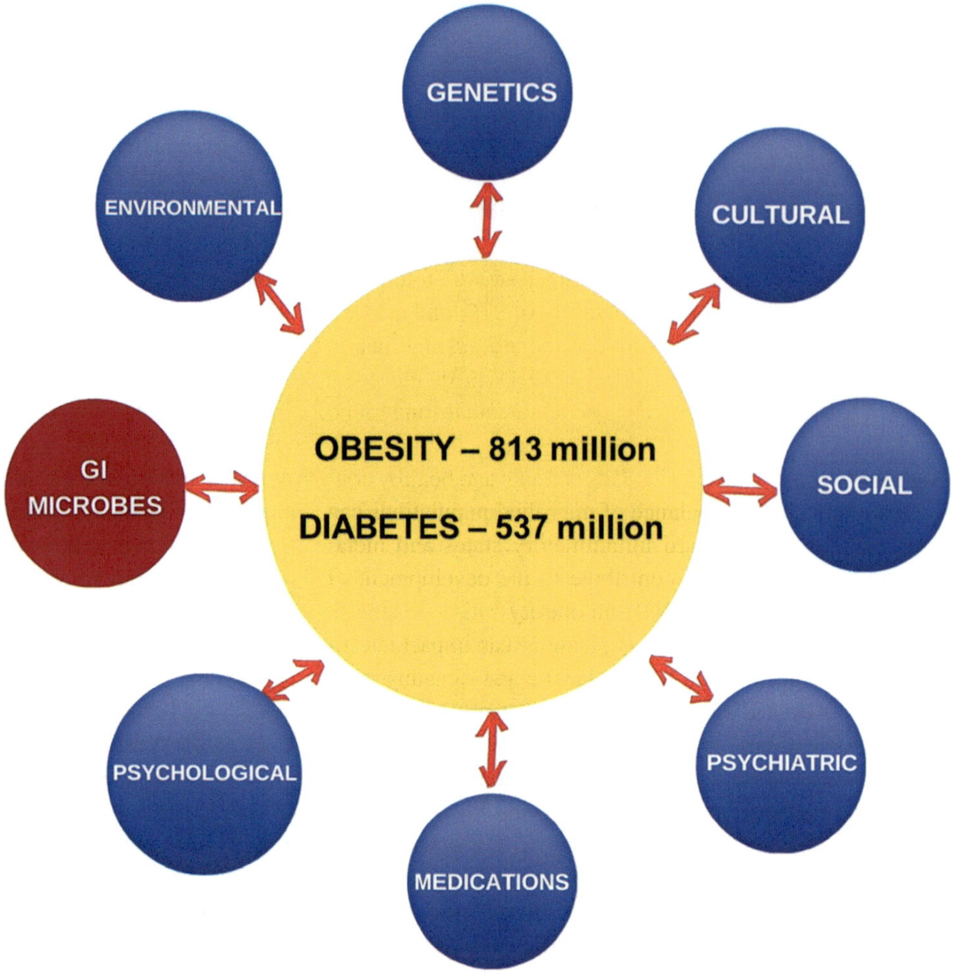

Fig. 1 Factors influencing the prevalence of obesity and diabetes. Prevalences shown are for adults worldwide in 2020–2021

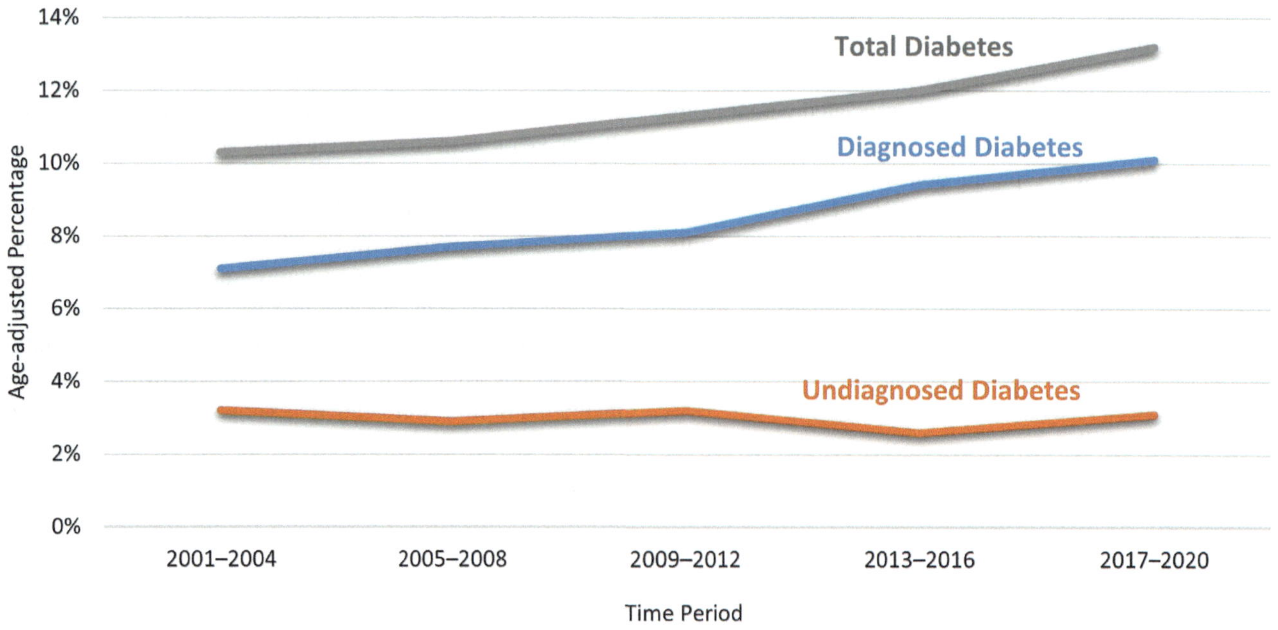

Fig. 2 Trends in Prevalence of Diagnosed Diabetes, Undiagnosed Diabetes, and Total Diabetes Among Adults Aged 18 Years or Older, United States, 2001–2004 to 2017–2020. (Source: Centers for Disease Control and Prevention (CDC))
Publisher's notes: Diagnosed diabetes was based on self-report. Undiagnosed diabetes was based on fasting plasma glucose and HbA1c levels among people self-reporting no diabetes

Use of this material does not imply endorsement by the Centers for Disease Control and Prevention (CDC), the Agency for Toxic Substances and Disease Registry (ATSDR), United States Department of Health and Human Services (HHS), or the United States Government. This material is available on the CDC website at no charge destruction, leading to insulin deficiency [4]. As β-cells are depleted, the pancreas becomes unable to produce enough insulin to regulate blood sugar levels. This causes hyperglycemia (elevated blood glucose) and associated symptoms such as polydipsia, polyphagia, and polyuria [5]. T2D is precipitated by insulin resistance leading to increased demand on the insulin producing beta cells. Once demand exceeds supply, hyperglycemia ensues [4]. T2D is a multifactorial disease: diet, genetics, and environmental factors are significant contributing factors [6, 7] (Fig. 1). Both T1D and T2D are diagnosed by fasting plasma glucose levels ≥126 mg/dL; a 2-h plasma glucose value of 200 mg/dL after a 75 gram oral glucose tolerance test [8, 9] or hemoglobin A_{1c} (HbA_{1c}) levels ≥6.5% [9].

In addition, the Centers for Disease Control and Prevention (CDC) suggests that 38% of American adults are living with prediabetes [3], defined as a state of intermediate hyperglycemia using three specific parameters: fasting plasma glucose of 5.6–6.9 mmol/L (100 to 125 mg/dL) [4]; impaired glucose tolerance, which is defined as 2 h plasma glucose of 7.8–11.0 mmol/L (140–199 mg/dL) after ingestion of 75 g of oral glucose, or a combination of the two based on a 2 h oral glucose tolerance test; and an HbA_{1c} of 5.7–6.4% also meets the criteria for prediabetes [4].

Individuals with diabetes have increased risk for cardiovascular disease (seen in up to 21.6% of adults with diabetes) [10], as hyperglycemia and insulin resistance increase oxidative stress, disturb protein kinase C signaling and increase levels of advanced glycation end-products [10]. These in turn result in vascular inflammation, vasoconstriction, thrombosis, atherogenesis [11], and atherosclerosis [12]. Other common conditions in individuals with T2D include hypertension, hyperlipidemia, chronic kidney disease, neuropathy, retinopathy, and Alzheimer's Disease [10].

Prediabetes and T2D are linked to obesity through insulin resistance [13]. Insulin resistance occurs when muscle, fat, or liver cells do not respond to insulin and cannot take up glucose from the blood, causing the pancreas to produce more insulin to allow glucose to enter these cells [14]. Insulin resistance is central to the development of T2D, and is also linked to hypertension, hyperlipidemia, atherosclerosis, and polycystic ovarian disease [13]. Insulin resistance is also associated with changes in the gut microbiome, specifically increases in the phyla Firmicutes and Actinobacteria and a decrease in Bacteroidetes, and with altered intestinal permeability [15]. This can increase the absorption of bacterial lipopolysaccharide (LPS), leading in turn to increased activation of inflammatory pathways [15].

1.2 Obesity

Definitions, Prevalence, and Costs

Obesity is defined as an excess of body fat resulting in a body mass index (BMI) above the normal range [16]. BMI measures weight in relation to a person's height: a normal BMI in the Caucasian population is between 18.5 and 25 kg/m^2, underweight less than18.5 kg/m^2, overweight between 25 kg/m^2 and 29.9 kg/m^2, and obese above 30 kg/m^2 [17]. However, these values may differ in various ethnic groups. For example, the American Diabetes Association now recommends a BMI cut-off of ≥23 kg/m^2 for identifying Asian Americans at elevated risk for diabetes [18].

Worldwide obesity has almost tripled since 1975 [16]. In 2020, over 813 million adults globally were obese [19] (Fig. 1), and the World Obesity Federation estimates that by 2030 1.56 billion people globally aged five or older will be living with obesity [19]. In the U.S., the incidence of obesity has increased by 70% over the last 30 years for adults, such that in 2021 19 states had adult obesity rates at or above 35 percent, more than doubling since 2018 [20] (see Fig. 3). Non-Hispanic Blacks exhibited the highest age-adjusted prevalence of obesity, then Hispanics, non-Hispanic Whites and non-Hispanic Asians [20].

This increasing prevalence of obesity is also associated with increasing healthcare costs. The Organization for Economic Co-operation and Development (OECD) estimated that from 2020 to 2050, treating diseases resulting from overweight and obesity will cost a total of USD 425 billion per year, based on Purchasing Power Parity, in OECD, G20, EU28 and OECD accession and selected partner countries [21]. In the U.S., the aggregate medical costs of obesity are estimated at $260.6 billion [22], with costs for individuals with obesity an estimated $2,505 higher than those for individuals of normal weight [23, 24].

Obesity is associated with increased risks for conditions such as metabolic syndrome and T2D, heart disease, stroke, type 2 diabetes, osteoarthritis, and infertility in women [24–26]. Individuals with obesity also typically have chronic low-grade inflammation, which can lead to DNA damage over time, and eventually lead to increased risks for certain types of cancer including esophageal adenocarcinoma (gastroesophageal reflux disease and Barrett esophagus) and different types of liver cancer (chronic ulcerative colitis and hepatitis) [27]. Additionally, obesity is a risk factor for gallstones, which in turn increases the risks of gallbladder cancer [28]. Individuals with obesity also tend to have increased circulating levels of insulin, insulin-like growth factor 1

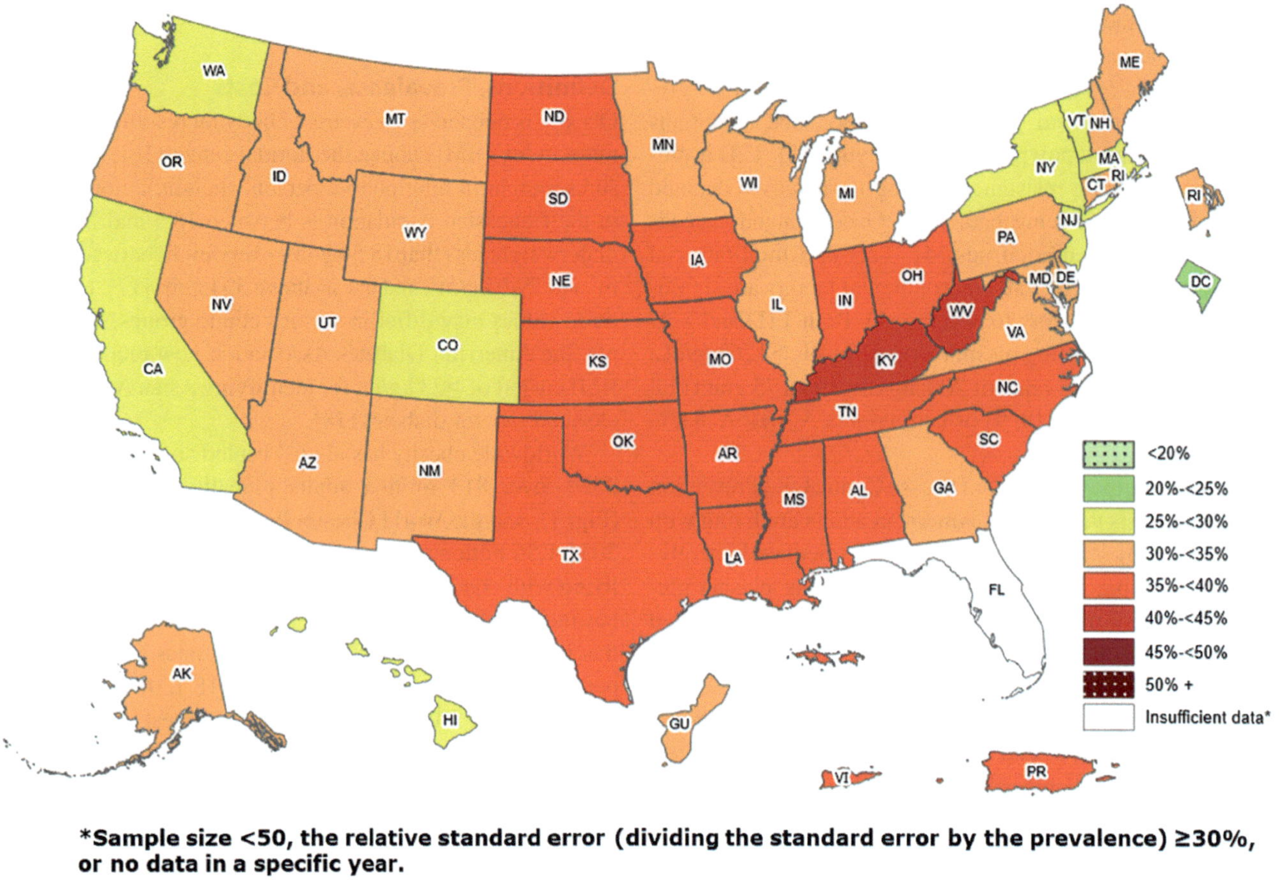

Fig. 3 Prevalence of self-reported obesity in US adults by state, 2021. (Source: Centers for Disease Control and Prevention (CDC))
Note: Use of this material, does not imply endorsement by the Centers for Disease Control and Prevention (CDC), the Agency for Toxic Substances and Disease Registry (ATSDR), United States Department of Health and Human Services (HHS), or the United States Government. This material is available on the CDC website at no charge

(IGF-1) and leptin [27]. Leptin appears to promote cell proliferation [29], and high levels of insulin and IGF-1 may contribute to the development of colon, kidney, prostate, and endometrial cancers [30].

2 The Gut Microbiome

The human body is host to approximately 3.8 × 10^{13} microorganisms, weighing about 0.2 kilograms [31], and existing in an approximate 1.3:1 ratio with human cells [32]. Their collective genomes contain over 150 times more genes than the human genome [33]. The gut microbiome includes microbes from all three domains of life: Bacteria, Eukarya (predominantly yeast) [34] and Archaea (predominantly methanogens) [35], as well as viruses such as Microviridae and crAss-like bacteriophages [36]. Bacteria are estimated to contribute 99.1% of the genes found in the gut, with the remainder contributed primarily by archaea, and only 0.1% are of eukaryotic and viral origins [37].

The predominant bacterial phyla in the gut are Firmicutes, Bacteroidetes, Actinobacteria, Proteobacteria, Fusobacteria, and Verrucomicrobia [38], with Firmicutes and Bacteroidetes together representing 90% of gut microbiota [39]. We note that many prokaryotic phylum names were changed in 2021—Firmicutes has now been renamed Bacillota, Proteobacteria is now Pseudomonadota, and Bacteroidetes is Bacteroidota—but for the purposes of this chapter we are using the previous names which appear in the publications discussed. Phylum Firmicutes alone has more than 200 different genera including *Lactobacillus*, *Bacillus*, *Clostridium*, *Enterococcus,* and *Ruminicoccus*, and Bacteroidetes includes genera such as *Bacteroides* and *Prevotella,* amongst others [38]. Functionally, the gut microbiome plays important roles in metabolism of lipids and carbohydrates metabolism, hormone modulation, protecting the host against pathogens, epithelial cell proliferation, and the synthesis of certain vitamins and amino acids (reviewed by Singer-Englar et al. [40]). The gut microbiome is integral to digestion, breaking down complex carbohydrates, proteins, and fats [41], and allows the host to extract calories from otherwise indigestible polysaccharides in the diet by utilizing a large array of glycoside hydrolases and polysaccharide lyases which are not encoded in the human genome [42]. Whilst important to host survival, these functions could also contribute to weight gain. In addition, microbial modulation of lipid accumulation, lipopolysaccharide content, and production of short chain fatty acids (SCFAs) can influence inflammation and/or or insulin signaling [43].

2.1 Dysbiosis of the Gut Microbiome

Research shows that the gut microbiome has co-evolved with the human host [42], and that changes in the balance of gut microbial populations can have significant consequences for human health. A healthy gut microbiome is characterized by microbial diversity and richness. Dysbiosis, which is defined as an imbalance in gut microbial populations, has been implicated in numerous conditions including inflammatory bowel disease (IBD), celiac disease, irritable bowel syndrome (IBS), as well as asthma, allergies, cardiovascular disease, metabolic syndrome, diabetes, and obesity [38, 44, 45]. Dysbiosis may also result in altered production of microbial antigens and metabolites [46, 47], which can alter intestinal barrier functions and disturb metabolic and signaling pathways in the host that are directly or indirectly associated with insulin resistance [47]. Alterations in the balances of gut microbial populations have also been demonstrated in obesity, and several recent studies have suggested that gut microbiota-derived endotoxemia may further contribute to the pathogenesis of obesity and insulin resistance [48]. Beyond the gut microbiome (bacteria), recent studies have shown that alterations in the gut mycobiome (fungi) may also be associated with T2D, including increases in *Candida albicans* which have been identified in human subjects [49] and which also contribute to insulin resistance in high fat diet (HFD)-fed mice, via a dectin-1 dependent pathway [49].

2.2 Microbial Dysbiosis and T2D

In one of the first seminal studies of the gut microbiome in T2D, performed using fecal samples from a cohort of adult Danish males, phylum Firmicutes and class Clostridia were found to be significantly reduced in subjects with T2D versus controls, whereas class Betaproteobacteria and genus *Lactobacillus* were highly enriched in subjects with T2D and positively correlated with plasma glucose levels [50]. In addition, the ratios of the phyla Bacteroidetes to Firmicutes and of the *Bacteroides-Prevotella* group to the *Clostridium coccoides-Eubacterium rectale* group exhibited significant positive correlations with plasma glucose concentrations, but not with BMI, in T2D subjects [50]. Subsequently, a metagenome-wide association study (MGWAS) of the fecal microbiome in elderly female European subjects found lower abundances of *Roseburia intestinalis* and *Faecalibacterium prausnitzii*, which are producers of the key SCFA butyrate, in T2D subjects [51]. Decreased abundance of butyrate-producing bacteria is associated with increased metabolic risks in human subjects, and studies indicate that butyrate can activate intestinal gluconeogenesis [52] and has anti-inflammatory properties that can alleviate obesity-related metabolic complications [53]. Subjects with T2D also exhibited increased levels of phylum Proteobacteria, class Clostridiales (phylum Firmicutes), and the species *Lactobacillus gasseri* and *Streptococcus mutans*, as well as increased expression of microbial genes involved in oxidative stress [51], which has been suggested to cause insulin resistance, β-cell dysfunction, impaired glucose tolerance and T2D [43, 54]. These results were consistent with several findings from a MGWAS of 345 Chinese subjects, which also found a moderate degree of dysbiosis in gut bacteria, increased *Lactobacillus* species and a decline in butyrate-producing bacteria in individuals with T2D, as well as an increase in microbial functions related to gut oxidative stress [55]. A later quantitative RT-PCR study of Japanese subjects with T2D and controls also reported some consistent findings, including significantly higher levels of total *Lactobacillus*, including the species *L. reuteri* and *L. plantarum*, and significantly lower levels of the genera *Atopobium* and *Prevotella* and the species *C. coccoides* in fecal samples of subjects with T2D versus controls [56]. More recently, Naderpoor et al. [57] explored the relationships between the fecal microbiome and insulin sensitivity and insulin secretion. They found that the abundances of the genera *Clostridium* and *Phascolarctobacterium* positively correlated with insulin sensitivity in overweight or obese adults who were otherwise healthy, whereas the abundance of genus *Dialister* was negatively correlated with insulin sensitivity. *Dialister* was positively correlated with fasting insulin and insulin AUC, whereas *Phascolarctobacterium* was negatively correlated with fasting insulin [57].

Of note, these studies included heterogenous subject populations, and in some cases lack gender balance and/or have incomplete data on medications used. While they demonstrate that alterations in the gut microbiome exist in subjects with T2D, it is important to account for the potential effects of these confounders on the gut microbiome. For example, metformin is the most commonly prescribed oral agent for T2D [58], and it has been suggested that metformin may alter gut microbiota composition by increasing the abundance of the mucin-degrading species *Akkermansia muciniphila* and also increasing circulating levels of the incretin glucagon-like peptide 1 (GLP-1) [59, 60]. A South American study showed that T2D subjects taking metformin had higher relative abundances of *A. muciniphila* and SCFA-producing bacteria including the genus *Megasphaera*, as well as *Bifidobacterium bifidum* and a *Prevotella* operational taxonomic unit (OTU) when compared to non-diabetic subjects [61]. T2D subjects taking metformin also had higher relative abundances of *A. muciniphila* and OTUs from

Prevotella and *Megasphaera* when compared to T2D subjects who were not taking metformin, suggesting that these were direct effects of metformin on the gut microbiome [61]. Consistent with this, metformin treatment has also been shown to increase the abundance of *A. muciniphila* and improve the glycemic profile of HFD-fed mice compared to non-metformin treated HFD-fed controls [62]. In another recent large-scale human study, Thingholm et al. examined gut microbiota in German subjects classed as lean non-diabetic (n = 633), obese non-diabetic (n = 494), and obese individuals with T2D (n = 153) [63]. While associations were identified between abundances of the genera *Akkermansia, Faecalibacterium, Oscillibacter,* and *Alistipes* and obesity, with more modest alterations for T2D, the authors noted that medications, including antihypertensives, oral antidiabetic agents, and dietary supplements, were significantly associated with microbiome variation [63], again illustrating the importance of controlling for medications when performing gut microbiome studies.

It is important to note that all of the studies described above use stool as a surrogate for the entire intestinal microbiome, despite the fact that the small bowel is more relevant for nutrient absorption and metabolism. Until recently, studying the small bowel was limited by challenges in collecting and analyzing samples, due to the difficulty of accessing the small bowel, and the low biomass and high viscosity of its contents. However, Leite et al. have now developed and validated novel techniques to study small bowel aspirates [64], which will enable studies to determine the role of small bowel microbial changes in both T2D and obesity (discussed in Sect. 5.2).

2.3 Microbial Dysbiosis and Obesity

Ley et al. first showed that the balances of the major gut phyla were altered in human subjects with obesity, with lower proportions of Bacteroidetes and higher proportions of Firmicutes in individuals with obesity compared to lean individuals [65]. These findings were consistent with findings from mouse models [66, 67]—an analysis of 5088 bacterial 16S rRNA gene sequences from the cecal microbiota of genetically obese *ob/ob* mice, lean and wild-type siblings, and their mothers (all of which were fed the same polysaccharide-rich diet) showed that obese animals had a 50% reduction in the abundance of phylum Bacteroidetes and a proportional increase in phylum Firmicutes when compared to lean mice, regardless of kinship [66]. Since then many, but not all, human studies have suggested that individuals with obesity have an increased Firmicutes: Bacteroidetes ratio. For example, in Japanese adults, this ratio was seen to be higher in subjects with obesity when compared to subjects without obesity [68]. However, Fernandes found no significant differences in this ratio between healthy, lean and overweight/obese groups [69], and Peters et al. also found no relationship between this ratio and BMI category in American adults [70]. Interestingly, studies of pregnant women found significantly higher levels of Bacteroides and *Staphylococcus aureus* in overweight pregnant women versus control normal-weight women. Mother's weight and BMI before pregnancy was shown to be correlated with higher concentrations of the genera *Bacteroides, Clostridium,* and *Staphylococcus*, and high *Bacteroides* concentrations were associated with excessive weight gain over pregnancy [71]. Lastly, a recent study of children with obesity found an association with decreased abundance of *Blautia* species, particularly *B. luti* and *B. wexlerae*, and this was even more pronounced in children with obesity and insulin resistance [72].

It should also be noted that an earlier characterization of fecal microbial communities in adult female monozygotic and dizygotic twin pairs and their mothers by Turnbaugh et al. [73] found that while each individual's gut microbial community varies with respect to the specific bacterial lineages present, bacterial diversity and functional redundancy results in a 'core microbiome' at the gene level, and that obesity was associated with a loss of bacterial diversity and changes in the representation of bacterial genes and metabolic pathways. This study is significant in that it illustrates the importance of bacterial diversity to a stable microbiome, and also demonstrates that while it is necessary to understand how the microbiome changes in different states (e.g. obesity, diabetes), identifying specific bacterial taxa may be less important than understanding how impacting shared bacterial functions affects the host. A later study of 123 non-obese and 169 Danish individuals with obesity further emphasized this point, demonstrating that individuals with low gut bacterial richness exhibited higher overall adiposity, insulin resistance, and dyslipidemia, coupled with a more pronounced inflammatory phenotype and a higher rate of weight gain over time compared to those with higher bacterial richness [67].

2.4 Role of Methanogenic Archaea in Metabolic Disorders

Archaea are another important component of the gut microbiome. These microbes are distinct from bacteria in that they that lack a nucleus, their cell walls lack peptidoglycan, and they have membranes that enclose lipids with hydrocarbons [35]. Methane-producing archaea (methanogens) are strictly anaerobic organisms that are found in diverse environments, including the gastrointestinal tract [74]. *Methanobrevibacter smithii* is the predominant methanogen in the human gut [75], followed by *Methanosphaera stadtmanae* [35]. *M.*

smithii generates methane using carbon dioxide and hydrogen from neighboring microbes [76]. This hydrogenotrophic methanogenesis causes hydrogen depletion, which increases the efficiency of bacterial fermentation and thus nutrient availability. Consistent with this, increased *M. smithii* levels have been suggested as a factor in weight gain and the development of obesity in rodent models [77, 78]. In humans, the number and proportion of *M. smithii* correlates strongly with the levels of methane on a breath test [75], and elevated breath methane levels have been demonstrated in human subjects with obesity [79, 80]. Further, methane-producing subjects exhibit impaired glucose tolerance compared to age- and BMI-matched controls [81], and methane production may also be linked to altered glycemic control in subjects with T1D [82].

In an interventional study, methane-producing prediabetic subjects with obesity who eradicated breath methane following treatment with targeted antibiotics exhibited reductions in stool *M. smithii* coupled with significant improvements in low-density lipoprotein, total cholesterol, and insulin levels on OGTT [83]. These findings suggest that eradication of gastrointestinal methanogens can lead to an improved metabolic state. Consistent with this, subjects with high breath methane and hydrogen exhibit decreased weight loss following bariatric surgery [84]. Mechanistically, co-colonization of gnotobiotic mice with *M. smithii* and *Bacteroides thetaiotaomicron* has been shown to alter the *B. thetaiotaomicron* transcriptome, increasing expression of three fructofuranosidases and facilitating polyfructose-containing glycan degradation [78]. This is accompanied by increased production of the SCFAs acetate and formate, which can serve as energy sources for the host. Increased *de novo* lipogenesis, epididymal fat pad weights and total body fat stores were also demonstrated in *B. thetaiotaomicron/M. smithii* biassociated animals [78]. Within the gut, methanogens localize in the same regions as L-cells which secrete GLP-1 [85], an incretin which influences appetite and glucose regulation. Moreover, mice fed a high-fat diet exhibit increases in body weight, GLP-1 secretion and fecal methanogen content, and GLP-1 response to oral glucose was also significantly elevated in the high-fat diet group [85]. Lastly, methane was shown to stimulate GLP-1 secretion in human and murine L cells [85].

M. smithii is also thought to play a role in the metabolic disorder anorexia nervosa (AN). Using real-time PCR, Armougom et al. showed that AN subjects had *M. smithii* concentrations that were much higher than in normal-weight controls ($p = 0.0171$) [86]. In a later study of 15 women with AN and 15 matched controls, women with AN were found to not only have lower energy intake, but significant differences at every taxonomic level when compared with controls. Specifically, next generation sequencing showed significant increases in *M. smithii* and the family Enterobacteriaceae in women with AN compared with healthy controls [87]. The SCFA-producing genera *Roseburia*, *Ruminococcus* and *Clostridium* were decreased, with concomitant reductions in butyrate and propionate levels. A possible explanation for these findings suggested by Armougom et al. is that increased levels of *M. smithii* in AN patients may represent an adaptation to increase nutrient harvest from very low calorie diets [87]. It is also possible that different strains of *M. smithii* are involved, although studies with e.g. shotgun sequencing allowing better resolution at the species and strain levels will be required to fully elucidate this.

3 Contributing Factors That Influence the Microbiome and Its Relationship with Diabetes and Obesity

3.1 Dietary Influences and the Microbiome

Diet is amongst the key factors which influence the etiology of diabetes and obesity. Human diets have evolved over time due to changes in food availability, increased prevalence of food processing, and food prices vs. population income [88]. Traditional, primarily plant-based diets are being supplanted by diets that are higher in sugars, animal fats, artificial ingredients and preservatives, and lower in dietary fiber, fruits and vegetables [89]. The Western diet is characterized by high levels of proteins derived from fatty, processed meats, saturated fats, refined grains, sugar, alcohol, salt, and corn-derived fructose syrup, and reduced consumption of fruits and vegetables [90]. Advances in food processing have led to the consumption of foods that are higher in calories, but lower in other nutrients (e.g., edible oils, refined grains) [91] and have increased the glycemic load of common meals [92]. Based on the U.S. National Health and Nutrition Examination Survey data, ultra-processed foods (which contain flavors, colors, sugars/sweeteners, emulsifiers and other additives) comprise 57.9% of energy intake, with 89.7% of this energy intake from added sugars [93]. The added sugars content in ultra-processed foods (21.1% of calories) is eight-fold higher than that of processed foods (2.4%) and five-fold higher than that of unprocessed or minimally processed foods and processed culinary ingredients grouped together (3.7%) [93].

Diet strongly influences gut microbiome composition, but differs with geography, lifestyle, and socioeconomic status in human populations [94]. Several studies have shown that the Western diet is associated with lower gut microbial diversity, including a landmark study by Yatsunenko et al. that analyzed fecal microbiomes from Malawian, Venezuelan Amerindian and US residents aged 0–17 and 18–70 years [95]. Across all ages, the Malawian and Amerindian populations were similar to each other but significantly different from the US population, which the authors suggested may be because the typical US diet is richer in protein, whereas Malawian and Amerindian diets are richer in corn and cassava, and hence contain more starch [95]. These differences were reflected in the associated microbial enzymatic profiles. Enzymatic commissions (EC) encoding glutamate synthase and α-amylase, which is involved in starch degradation, were more highly represented in Malawian and Amerindian adult microbiomes, while ECs involved in glutamine degradation and the catabolism of simple sugars, sugar substitutes and host glycans were more highly represented in US adults [95]. The importance of fermentable fibers is underscored in numerous studies showing that consumption of inulin, psyllium, oat beta glucan, and resistant starch can increase satiety (via decreases in ghrelin and increases in peptide YY (PYY)), improve glucose tolerance and glycemic control, and thus protect against diet-induced obesity [96–98]. Fibers are broken down by gut microbes, producing SCFAs such as butyrate and propionate, which influence GLP-1 and PYY production as well as ghrelin levels, and in a recent clinical study, a prebiotic fiber-enriched nutritional formula was shown to significantly reduce HbA_{1c} in subjects with T2D, with increases in the butyrate producers *Roseburia faecis* and *Anaerostipes hadrus* in the recipients' gut microbiomes [99].

There is also evidence that gut microbiota can respond rapidly to significant changes in diet. In a study of human volunteers placed on two different diets that varied according to their primary food source (a "plant-based diet" rich in grains, legumes, fruits, and vegetables; and an "animal-based diet" composed of meats, eggs, and cheeses), only the animal-based diet caused a shift in gut microbiome composition, with a significant increase in β-diversity within 1 day of starting the diet, which reverted to the original composition 2 days after the diet ended [100]. Subjects placed on the animal-based diet exhibited increased abundance of bile-tolerant microbes (*Alistipes, Bilophila,* and Bacteroides), and decreased levels of microbes which metabolize dietary plant polysaccharides (*Roseburia, E. rectale*, and *Ruminococcus bromii*) [100]. Metabolically, the animal-based diet resulted in significantly decreased levels of carbohydrate fermentation products, with significant positive relationships between clusters composed of putrefactive microbes (*Alistipes putredinis* and *Bacteroides* spp.) and those SCFAs which are the end products of amino acid fermentation (valerate, isovalerate and isobutyrate). There were also positive correlations between clusters comprised of saccharolytic microbes (*Roseburia, E. rectale,* and *F. prausnitzii*) and the products of carbohydrate fermentation [100, 101], and gene expression changes reflective of differences between the gut microbiomes of herbivorous and carnivorous mammals. The authors concluded that the human gut microbiome can rapidly switch between herbivorous and carnivorous functional profiles, likely due to seasonal food availabilities during human evolution. However, it should also be noted that Sonnenburg et al. have shown that while changes in the microbiome associated with consumption of a low-fiber diet are mostly reversible within a single-generation, consumption of the same diet over multiple generations results in a progressive and irreversible loss of microbial diversity [102].

Sugar-sweetened beverages (SSBs) such as soft drinks, energy drinks, fruit drinks, and vitamin water drinks are part of the Western diet and have particular impacts on the gut microbiome. They typically include naturally derived sweeteners including sucrose, high fructose corn syrup (HFCS), and fruit juice concentrates, and are the largest contributors to added sugar intake in the US diet [103]. Between 1970s and 2006, reported per capita consumption of SSBs rose from 64.4 kcal/day to 141.7 kcal/day, with adolescents and young adults consuming over 200 kcal/day [104], and positive associations have been reported between intake of SSBs and weight gain or risks for overweight and obesity in both children and adults [103]. In a meta-analysis of eight prospective cohort studies which included 310,819 participants and 15,043 cases of T2D, individuals in the highest SSB intake category (1–2 servings/day) had a 26% greater risk of developing T2D than those in the lowest category (none or less than 1 per month), and a one-serving-per-day increase in SSB intake was associated with a 15% increased risk for T2D [105]. In another study of over 50,000 women with 8 years of follow-up, Schulze et al. found that those who consumed 1 or more SSB per day had an 83% increased risk for T2D compared to those consuming less than 1 per month [106]. In an animal study, Noble et al. showed that the gut microbiome, but not body weight, adiposity index, or total caloric intake, was altered due to sugar consumption. At the genus level, *Prevotella* and *Lachnospiraceae incertae sedis* were decreased, whilst *Bacteroides, Alistipes, Lactobacillus, Clostridium sensu stricto, Bifidobacteriaceae*, and *Parasutterella* were increased by sugar consumption. Lastly, increased levels of bacterial lipopolysaccharide (LPS) may increase inflammatory responses and inhibit insulin receptor substrate phosphorylation through its actions on TLR4 [107], thus contributing to insulin resistance.

3.2 Food Additives and the Microbiome

Since the industrial revolution, the food technology industry has worked to maximize palatability, optimize process efficiency, increase shelf life, reduce costs, and improve food safety. Over 4,000 novel ingredients and additives have been entered into the food supply, including artificial sweeteners and emulsifiers [108, 109]. Gut microbiota alterations have been linked to both artificial sweeteners and emulsifiers, and their relationships to the development of T2D and obesity is an active area of research.

Artificial Sweeteners

Artificial sweeteners are non-nutritive sweeteners (NNS) which first entered consumer markets in the 1950s and gained popularity as a means of maintaining food palatability while significantly decreasing or eliminating caloric content. Artificial sweeteners may also trigger GLP-1 secretion by enteroendocrine cells in the gastrointestinal tract, thus curbing appetite and calorie intake [110]. However, it has also been suggested that NNS may actually increase appetite, and thus paradoxically lead to weight gain rather than weight loss [111, 112], although this is the subject of considerable debate [113]. The U.S. Food and Drug Administration has approved five NNS for human use, including aspartame and saccharin [114]. Aspartame is an amino-acid compound which is approximately 160 times sweeter than sugar, and which is absorbed from the gastrointestinal tract and metabolized in the liver to form phenylalanine, aspartic acid and methanol [115]. Aspartame contributes to weight gain, obesity, insulin resistance, and T2D [115]. In rodents, aspartame and its breakdown product phenylalanine inhibit intestinal alkaline phosphatase (IAP), an enzyme that prevents metabolic syndrome in mice. IAP activity was lower in mice consuming aspartame in their drinking water, and mice fed a high fat diet in addition to aspartame in their drinking water gained significantly more weight than mice fed a high fat diet without aspartame [116]. However, there is conflicting data as to the effects of aspartame intake on glucose tolerance. In a recent study of obese rats and their offspring, maternal aspartame intake with a high-fat high-sucrose diet was associated with increased body fat at weaning and increased body weight long term, with male offspring showing impaired glucose tolerance at 8 weeks [117]. In contrast, an earlier study of rats consuming either a standard diet or high-fat diet, both with or without low-dose aspartame in their drinking water, showed that animals given aspartame consumed fewer calories, gained less weight and had more favorable body compositions when on a high-fat diet compared to animals consuming a high fat diet and water alone [118]. In that study, aspartame consumption resulted in elevated fasting glucose levels and impaired insulin-stimulated glucose disposal both in rats on the standard diet and in those on the high-fat diet, independently of body composition [118]. With respect to the microbiome, aspartame increased total bacterial levels, as well as the abundances of family Enterobacteriaceae and the species *Clostridium leptum*, and *Roseburia* species were higher in the high-fat diet plus aspartame group than in the high-fat diet plus water group. Aspartame also attenuated the Firmicutes:Bacteroidetes ratio and serum metabolomics revealed an association between aspartame and elevations in the SCFA propionate, which is a substrate for gluconeogenesis [118].

A study of human subjects with and without obesity found that preloading with aspartame prior to lunch and dinner resulted in reduced glucose and insulin levels compared to sucrose at 20 min after the preload, with greater reductions following a stevia preload, but only stevia preloading resulted in lower postprandial insulin levels at 30 and 60 min after lunch [119]. A later study found that preloading with aspartame, monk fruit or stevia resulted in little differences in total energy intake or total AUC for glucose or insulin compared to preloading with sucrose [120]. Other studies have suggested that aspartame intake significantly influences the association between BMI and glucose tolerance, with subjects who reported aspartame intake exhibiting a stronger positive association between BMI and glucose tolerance than those reporting no aspartame intake [121]. These data suggest that aspartame consumption may be associated with greater obesity-related impairments in glucose tolerance, but the conflicting data highlight the importance of further research into the effects of aspartame on metabolic syndrome.

Saccharin, also known as ortho-sulfobenzoic acid, is an organic compound used as a non-nutritive sweetening agent, and is 200–700 times as sweet as table sugar. Suez et al. showed that mice consuming saccharin in their drinking water had a distinct microbiome composition that clustered separately from both their baseline microbiome compositions and from all control groups [122]. The microbiome of saccharin-consuming mice exhibited considerable dysbiosis, and more than 40 OTUs differing significantly from all control groups, and alterations in the relative abundances of order Clostridiales, genus *Bacteroides*, and species *L. reuteri*. Dysbiosis was also observed in mice consuming pure saccharin and a high-fat diet [122]. Interestingly, when microbiota from saccharin-consuming or glucose-consuming donor mice were transferred into germ-free recipients, the recipient microbiomes clustered separately, indicating that these were transferrable phenotypes [122]. Since this study, researchers have explored how saccharin alters gut microbiome diversity [114]. Labrecque et al. found that, depending on pregnancy status, the combination of ethanol and saccharin produced different effects than ethanol and water in mice, with reduced levels of *Clostridium* in mice consuming ethanol-saccharin vs ethanol-water [123]. Another study of male mice consuming saccharin in their drinking water for 6 months revealed elevated expression of pro-inflammatory markers such as iNOS and TNF-α in the liver, indicating that saccharin may induce inflammation [124]. LPS, flagella and fimbriae, and levels of bacterial genes involved in the production of multiple toxins were increased in saccharin-treated mice. In addition, gut microbiota in saccharin-treated mice exhibited a higher frequency and abundance of bacterial genes for multi-drug resistance [124].

A recently published human study comparing healthy adults taking one of several NNS (aspartame, saccharin, sucralose, and stevia) in combination with a glucose vehicle to those taking vehicle alone and those not taking any supplements, found that saccharin and sucralose impaired glucose tolerance in healthy adults, whereas aspartame, stevia, and the glucose vehicle alone did not [125]. All 4 NNS affected gut (stool) microbiome function, and both saccharin and sucralose also significantly affected gut microbiome composition [125]. Further, Hosseini et al. recently showed that consumption of non-aspartame sweeteners, but not aspartame alone, significantly affects the small bowel microbiome, with decreased microbial diversity and altered duodenal microbial metabolic functions involved in carbohydrate metabolism in subjects who consumed non-aspartame sweeteners such as sucralose [126].

Emulsifiers

Emulsifiers are added to most processed foods to improve food texture, prevent mixtures from separating, and extend shelf life. However, there is evidence that they can also alter gut microbial composition and encroachment on the mucus layer, which can induce intestinal inflammation and can promote the development of inflammatory bowel disease and metabolic syndrome [127]. In 2015, Chassaing et al. found that emulsifiers (specifically carboxymethylcellulose and polysorbate-80) induced low-grade inflammation, obesity and metabolic syndrome in wild-type mice [128]. The authors identified associations between the emulsifier-induced metabolic and microbial encroachment, altered microbial composition, and increased pro-inflammatory potential, suggesting that emulsifiers may be a contributing factor in the increased incidence of obesity, metabolic syndrome, and other chronic inflammatory diseases [128]. Supporting this, a 2017 study of the direct impact of carboxymethylcellulose and polysorbate-80 in an *in vitro* model found that both emulsifiers act directly on human microbiota, resulting in increased proinflammatory potential as indicated by increased levels of flagellin, which binds to TLR5 and triggers an immune response [129] Interestingly, while several gut commensal bacteria including species from the phyla Firmicutes and Proteobacteria, but not Bacteroidetes, can produce flagella, levels of flagellin are typically low in a healthy gut, suggesting that there are mechanisms in place to render commensal gut bacteria generally nonmotile [130].

3.3 Antibiotics and the Microbiome

The discovery of antibiotic drugs may represent one of the twentieth century's most important medical advances, and their use has saved millions of lives over the last 100 years [131]. In 2018, a total of 249.8 million outpatient antibiotic prescriptions were dispensed by U.S. community pharmacies [132]. However, there is evidence that antibiotic use, and particularly administration of antibiotics during the first 6 months of life [133] as well as repeated administration of antibiotics during childhood, may increase the risks for subsequent development of obesity [134] and that these effects are mediated through effects on the gut microbiome, which may not revert to pretreatment states as readily as was once thought [135]. As noted in a recent review by Fenneman et al. [135] the gut microbiome undergoes dramatic shifts in the first 2–3 years of life, correlating with the transitions from milk to solid foods and then to adult-like diets [136], and disruptions in the microbiome during these key early transitions may have significant and lasting consequences [135]. A study of 64,580 children in Pennsylvania found that 69% of children were exposed to antibiotics before age 24 months, but that the risks for early childhood obesity only reached significance in those with repeated (≥4) exposures to broad-spectrum antibiotics [137]. When further analyzed, the data indicated that these risks were greater in children who were younger at first antibiotic use [137]. Interestingly, comparing charts from the CDC reveals considerable overlap between the prevalence of obesity and the prevalence of antibiotic use in the US (see Figs. 3 and 4). A Finnish study of 12,062 children found similar results, but with more pronounced risks in boys [138]. Children exposed to antibiotics at any time in the first 24 months of life were heavier at ≥24 months, and BMIs were greater in boys who were first exposed at younger ages and who received multiple antibiotic treatments, particularly macrolides [138]. A 2017 retrospective cohort study of 21,714 children in the United Kingdom also found that exposure to antibiotics in the first 2 years of life was associated with increased risk of early childhood obesity [139]. The risks of obesity increased with antibiotic use in the first year of life, and with increasing number of prescriptions, particularly for those receiving 3 or more antibiotic prescriptions [139]. Another recent study from New Zealand found that exposure to antibiotics during the first 48 months of life was correlated with the development of obesity at 54 months, and number of antibiotic prescriptions during infancy was directly correlated with increased BMI during early childhood [140], and a study of 14,572 children in the US found that children who were exposed to antibiotics had increased risks for overweight or obesity coupled with a higher probability of multiple comorbidities, particularly those who received multiple prescriptions [141]. Supporting these findings, several studies have suggested that prenatal exposure to antibiotics taken by the mother may also predispose to overweight and obesity in early childhood [142–144].

Studies in adults have also suggested a link between antibiotic use and obesity. In a longitudinal study of adult women, chronic use of penicillin and bactericidals between ages 30 and 39 was found to be associated with obesity by the time of study enrollment (mean age 55) [145]. Bacteriostatic use in the 5 years before enrollment was linked to reductions in BMI after enrollment, but this inverse asso-

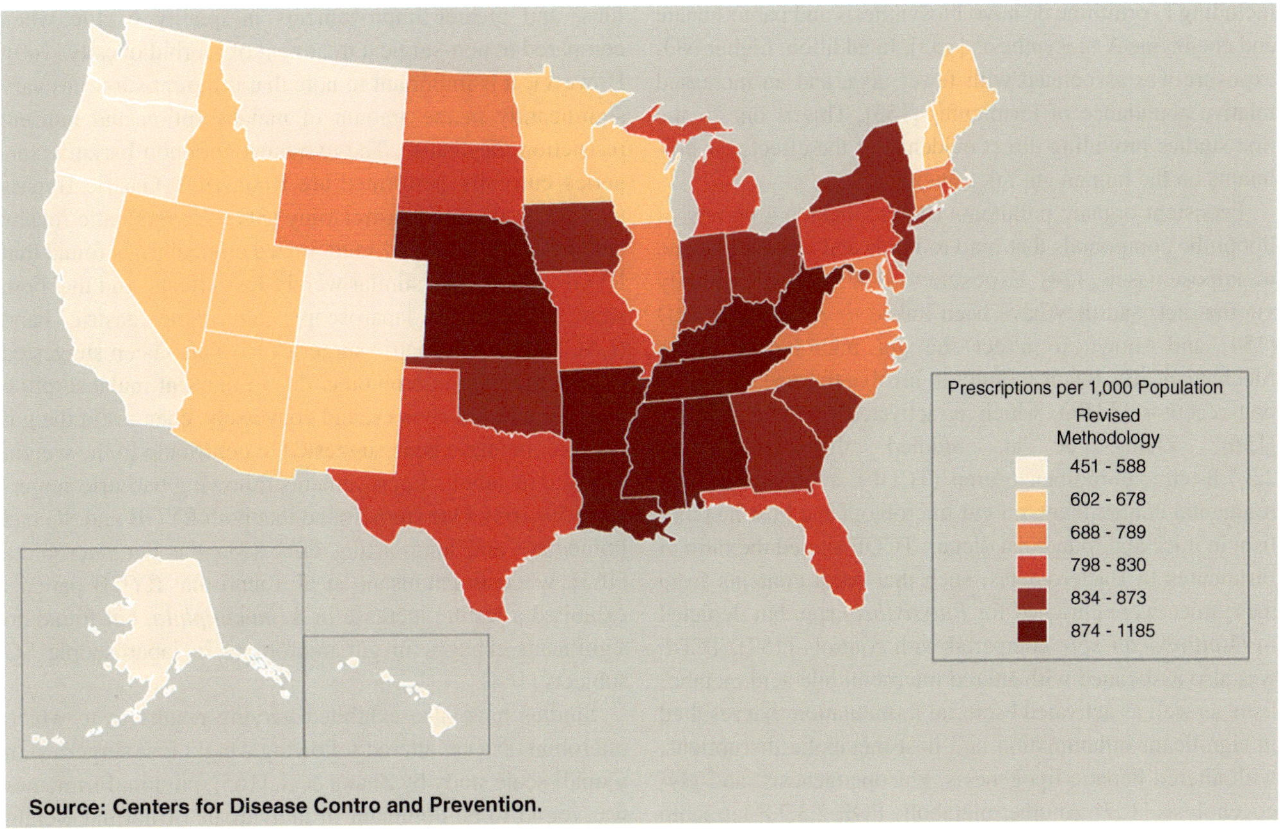

Fig. 4 Community Antibiotic Prescriptions per 1000 Population in the US by state, 2018. (Source: Centers for Disease Control and Prevention (CDC))
Note: Use of this material, does not imply endorsement by the Centers for Disease Control and Prevention (CDC), the Agency for Toxic Substances and Disease Registry (ATSDR), United States Department of Health and Human Services (HHS), or the United States Government. This material is available on the CDC website at no charge

ciation was only found in subjects taking antibiotics for skin conditions [145]. Another study of the use of vancomycin to treat infective endocarditis found that a major and significant weight gain occurred after intravenous treatment, particularly in males aged 65 and older that had not undergone cardiovascular surgery [146]. A later short-term study of prediabetic men with obesity who took amoxicillin, vancomycin, or placebo for 7 days showed that vancomycin (which targets Gram-positive bacteria), but not amoxicillin (broad-spectrum), decreased bacterial diversity and reduced the relative abundances of Firmicutes species involved in butyrate production and bile acid dihydroxylation, with concomitant increases in the relative abundances of Gram-negative Proteobacteria taxa such as genus *Enterococcus* and species *L. plantarum* [147]. No effects on various metabolic parameters including insulin sensitivity, energy harvest, gut permeability, systemic inflammation, and adipocyte size were seen. These results suggest that while short-term administration of these antibiotics can impact the gut microbiome, longer-term administration is required for effects on weight gain and development of obesity. While an overall relationship between antibiotic use and obesity appears to be established, many questions regarding cause and effect, particularly for the microbiome, remain unanswered.

3.4 Environmental Pollutants and the Microbiome

Air pollution is a known contributor to the development of multiple acute and chronic respiratory and cardiovascular diseases [148, 149] and has also been suggested to contribute to the development of T2D. Specifically, common air pollutants such as nitrogen dioxide (NO_2) and fine particulate matter <2.5 μm in diameter ($PM_{2.5}$) are suspected to contribute to T2D [150], by causing endothelial dysfunction and adipose inflammation, triggering insulin resistance [151]. The effects of environmental pollutants have been suggested to involve the gut microbiota [152]. This was supported by a recent study of the effects of air pollution on young adults in southern California, which showed that exposure to 24-h ozone (O_3) was associated with decreased Shannon diversity index, higher relative abundance of genus *Bacteroides*, specifically *B. caecimuris*, and effects on numerous gene pathways including L-ornithine de novo biosynthesis and pantothenate and coenzyme A biosynthesis [153]. In addition, higher NO_2 exposure was associated with fewer taxa, and an increased relative abundance of Firmicutes [153]. This is one of the first studies providing direct evidence of the effects air pollutants on the human gut microbiome.

Persistent organic pollutants (POPs) include a variety of lipophilic compounds that bind to lipids and can accumulate in adipose tissue [154]. Exposure to POPs occurs primarily via the diet, and they have been linked to obesity and T2D [154] and shown to affect the gut microbiome [155]. Mechanistically, this is thought to involve the aryl hydrocarbon receptor (AHR), which is activated by several POPs [156]. Zhang et al. studied the effects of 2,3,7,8-tetrachlorodibenzofuran (TCDF), a persistent environmental contaminant, on gut microbiota and host metabolism in mice and found that dietary TCDF altered the ratio of Firmicutes to Bacteroidetes, such that cecal contents from these mice were enriched for *Butyrivibrio* spp. but depleted in *Oscillobacter* spp. compared with controls [157]. TCDF was also associated with altered microbial bile acid metabolism, as well as activated bacterial fermentation that resulted in significant inflammation and host metabolic disruptions, with altered hepatic lipogenesis, gluconeogenesis, and glycogenolysis [157]. Another metabolic POP, 2,3,7,8-tetrachlo rodibenzo-p-dioxin (TCDD), did not appear to exacerbate streptozotocin-induced hyperglycemia in mice, but was associated with polydipsia and polyphagia, and significant effects on liver weight and pathology [158]. These were associated with significant alterations to the gut microbiome, including increased Firmicutes and decreased Bacteroidetes which negatively correlated with the changes in liver weight and pathology, as well as specific increases in families Lactobacillaceae and Desulfovibrionaceae, and decreases in Prevotellaceae [158]. Taken together, these studies demonstrate that environmental pollutants, both airborne and consumed through the diet, can alter the gut microbiome and affect metabolic parameters, through mechanisms which are being continually elucidated.

4 Effects of Interventions on the Gut Microbiome

4.1 Bariatric Surgeries and Their Effects on the Microbiome

Bariatric surgeries are designed to cause weight loss by restricting the amount of food the stomach can hold and impeding nutrient absorption [159]. Bariatric surgeries as a whole are associated with greater weight loss, higher remission rates of T2D and metabolic syndrome, better lipid profiles, and greater improvements in quality of life when compared to non-surgical treatment of morbid obesity [160]. However, it is important to note that different surgeries vary significantly in the amount of malabsorption and nutrient restriction they cause. The two most common bariatric surgeries currently performed are Roux-en-Y Gastric Bypass (RYGB) and sleeve gastrectomy (SG). A systematic review and meta-analysis of 11 randomized clinical trials found that RYGB and SG had similar weight-loss effects, and that both were superior to laparoscopic adjustable gastric band (LAGB) [161]. Bariatric surgeries have also been suggested to affect the gut microbiome, due to nutrient malabsorption and altered food choices, and conversely, changes in the gut microbiome have been suggested to contribute to the weight loss and metabolic improvements following bariatric surgeries [159, 162]. One study found that both RYGB and SG had immediate and long-lasting effects on the gut microbiota [163], whereas Palmisano et al. found that RYGB patients exhibited a lasting increase in *A. muciniphila*, but found no significant changes in gut microbiota in laparoscopic SG subjects [164].

Studies have also exhibited varying results as to which microbial taxa are altered following weight loss surgeries. In a small-scale study by Zhang et al. [165], phylum Firmicutes was found to be dominant in individuals of normal weight and those with obesity, and was significantly decreased in individuals after gastric bypass surgery. Individuals post-surgery also exhibited increases in class Gammaproteobacteria and decreases in Clostridia, and were enriched for families Enterobacteriaceae and Fusobacteriaceae and for specific *Akkermansia* spp. [165]. The first study to include subjects with T2D was a qPCR study by Furet et al. [166], which found lower levels of *Bacteroides/Prevotella* in subjects with obesity prior to RYGB that increased 3 months post-surgery, as well as increases in *Escherichia coli* that inversely correlated with fat mass and leptin levels independently of changes in food intake. *F. prausnitzii* was lower in subjects with diabetes and negatively associated with inflammatory markers throughout follow-up [166]. The same samples were reanalyzed 3 years later by pyrosequencing, which confirmed decreases in the Firmicutes genera *Lactobacillus*, *Dorea*, and *Blautia* after RYGB, and increases in the Bacteroidetes genera *Bacteroides* and *Alistipes*, as well as in *E. coli* [167]. This study also found associations between gut microbiota composition and white adipose tissue (WAT) gene expression that increased after RYGB [167]. Another study of subjects with obesity and T2D also found increases in Proteobacteria post-RYGB, but found that levels of *Faecalibacterium* species also decreased [168], in contrast to previous findings [166]. A recent study by Davies et al. suggested that abundances of both Firmicutes and Proteobacteria increased in T2D subjects with obesity following RYGB, whereas Bacteroidetes increased in those who underwent sleeve gastrectomy. In the same study, an increased abundance of Eubacteriaceae and *A. putredinis* prior to RYGB appeared to be predictive of T2D remission following surgery, and after surgery, subjects who achieved remission of T2D exhibited increases in Lachnospiraceae and *Roseburia* [169]. Changes in pH following bariatric surgeries may explain some of this variability, as pH increases as stomach volume decreases, and Firmicutes and Actinobacteria increase with pH increases, while Bacteroidetes decreases [170].

Mechanistically, recent studies have shown that bariatric surgeries alter the microbiome in favor of improved glucose metabolism [171, 172], with both duodenal-jejunal bypass surgery [171] and RYGB [172] increasing levels of microbes related to weight-loss, metabolic control (*Akkermansia*) and glycemic control (*Sutterella*). *A. muciniphila* has been shown in several studies to be decreased in severe obesity and to increase following bariatric surgeries [164, 165, 171], and it has been suggested that it mediates post-surgical metabolic changes through its effects on bile acids. Supporting this, Dao et al. recently showed that bile diversion to the ileum (a procedure similar to RYGB) induced weight loss, concomitant with increases in lymphatic GLP-1 and intestinal *A. muciniphila* [173]. Both RYGB and SG have previously been shown to alter SCFA levels [159], and Farup et al. recently showed that the total and relative amounts of the straight chain SCFA (acetic, propionic, and butyric acids) remained decreased, whereas the total and relative amounts of branched chain SCFA (isobutyric, isovaleric, and isocaproic acids) remained increased, in patients with obesity or severe obesity 6 months after either RYGB or laparoscopic SG [174].

Lastly, levels of methanogenic Archaea may also influence the outcomes of bariatric surgeries. Post-surgery breath methane and hydrogen levels were determined in subjects who underwent RYGB or SG (pre-surgery BMI ≥ 33), and percent total weight loss and percent change in BMI were prorated to 6 months after surgery [84]. Subjects who were positive for both methane and hydrogen on the breath exhibited lower prorated percent change in BMI when compared to all other subjects, and significantly lower prorated percent total weight loss ($p = 0.036$) [84]. These findings support a role for gut methanogens in host weight levels.

4.2 Duodenal Bypass Liners and the Microbiome

Duodenal-jejunal bypass liners (DJBLs) are devices which are implanted endoscopically, and comprise a nitinol metal anchor and an impermeable fluoropolymer sleeve that extends through the duodenum and into the jejunum. These devices were developed after endocrine and metabolic changes were noticed in patients following bariatric surgeries [175], and are designed to mimic their effects in inducing weight loss and improving glycemic control [176].

In a rat model of diabetes and obesity, the effects of bypass liners were studied in ZDF rats (a model of obesity and T2D) implanted with a duodenal endoluminal sleeve (DES) for 28 days [177]. At the phylum level, DES implantation was associated with decreased abundance of Bacteroidetes and increased Verrucomicrobia, with significant reductions in the genera *Lactobacillus, Streptococcus, Prevotella* and *Paraprevotella*, and increases in *Parabacteroides, Allobaculum* and *Bifidobacterium*. There was also a specific increase in *A. muciniphila* [177].

The effects of DJBL implantation on the duodenal microbiome and inflammatory responses was recently elucidated in a canine model [178]. Seven male mongrel dogs were implanted with EndoBarriers® (a specific brand of DJBL) for a 7-week period. This resulted in dramatic shifts in the duodenal microbiome compared to baseline, with a 54.76-fold decrease in the relative abundance of phylum Proteobacteria and concomitant increases in Fusobacteria, Firmicutes and Bacteroidetes [178]. At the family level, Enterobacteriaceae and Desulfovibrionaceae were significantly decreased in the duodenum at explantation, while Lactobacillaceae, Clostridiaceae, and Fusobacteriaceae were increased. Strikingly, these shifts were such that the duodenal microbiome at device removal was more similar to the fecal microbiome of unimplanted control dogs than their own duodenal microbiomes at baseline. Circulating cytokine levels were also significantly altered, with increases in IL2, IL6, IL18, GM-CSF, and TNFα at implantation week 3 and at explantation, and levels of IL2 and GM-CSF continued to be elevated 6 weeks after device removal [178]. In human subjects with obesity and T2D, analysis of fecal samples showed that DJBL implantation resulted in increased abundances of Proteobacteria, *Veillonella*, and *Lactobacillus* species, taxa which are typically found in the small bowel [179]. Of note, duodenal microbial resurfacing (DMR), a novel therapeutic strategy in which hydrothermal ablation of the duodenal mucosa is performed via endoscopy, has been proposed as a minimally invasive alternative to bariatric surgeries for the treatment of T2D, obesity, and NASH other metabolic conditions such as non-alcoholic fatty liver disease (NAFLD) and non-alcoholic steatohepatitis (NASH). DMR was recently shown to result in significant and durable improvements in HbA_{1c} and other metabolic parameters in subjects with T2D that persisted at 24 months post-procedure [180], effects which may be mediated through changes in the microbiome [181]. Although small bowel microbiome studies in DMR have yet to be performed, this is an interesting potential therapy for T2D.

4.3 Prebiotics and Probiotics

Prebiotics and probiotics have been suggested to restore homeostasis in gut microbiota, and ameliorate the dysbiosis and inflammation associated with T2D, obesity and other metabolic disorders. Prebiotics are compounds that promote the growth or action of beneficial gut microbes (e.g. lactic acid bacteria and Bifidobacteria) [182], and can be categorized into three major groups: fructooligosaccharides or galactooligosaccharides; lactulose; and non-digestible carbohydrates (e.g. inulin, cellulose, and pectin) [183]. They have been shown to increase the populations of protective gut microbes such as *Lactobacillus* and Bifidobacteria and decrease pathogens (reviewed in [182]), and also to decrease the production of pro-inflammatory cytokines such as IL-6, IL-1β and TNF-α, and to increase those of anti-inflammatory cytokines (transforming growth factor-β (TGF-β) and IL-10) [182]. In a recent clinical study, a prebiotic fiber-enriched nutritional formula was shown to significantly reduce HbA_{1c} in subjects with T2D, with increases in the butyrate producers *Roseburia faecis* and *Anaerostipes hadrus* in the recipients' gut microbiomes [99].

Probiotics are live microorganisms, most of which are bacterial species that exist in the human gut, such as *Lactobacillus, Bifidobacterium* and *Saccharomyces*, which have been shown to help maintain gut microbiota homeostasis and to regulate microbial metabolic activities [184]. The consumption of specific probiotic strains can help regulate gut membrane integrity and permeability, thus preventing gut leakiness, endotoxemia, and inflammation [185], and has also been shown to decrease pro-inflammatory and increase anti-inflammatory cytokines [186]. Further, Nagpal et al. showed that specific probiotic strains (*Lactobacillus* and *Enterococcus* strains derived from human infant fecal samples) can decrease growth of the pathogens *E. coli* and *Klebsiella pneumoniae* and increase SCFA production in culture [187]. When introduced into the mouse gut, these probiotics shifted gut microbiome composition and increased SCFA production, although the degree and duration of the effects were variable depending on whether a cocktail of 10 *Lactobacillus* and *Enterococcus* strains, *Lactobacillus* alone or *Enterococcus* alone was given, and on the number of probiotic doses given [187].

Other animal studies have also shown a wide range of results in terms of effects of probiotics on SCFA production, metabolic states, and microbiome composition. Sprague Dawley rats with streptozotocin-induced diabetes that were given *A. muciniphila* exhibited increased HDL and gut microbiome alpha diversity, as well as decreases in hepatic glycogen, serum plasminogen activator inhibitor, and TNF-α [188]. In mice with streptozotocin-induced diabetes, treatment with a *Saccharomyces boulardii* strain was shown to increase levels of phyla Bacteroidetes, Firmicutes, and Deferribacteres, with decreases in Proteobacteria and Verrucomicrobia, and to decrease serum triglycerides, increase hepatic triglycerides storage, modulate inflammatory profiles [189]. In an interesting recent study based on the observation that the prevalence of diabetes in Kazakh populations was lower than in other Chinese ethnic populations, thought to be due to the consumption of fermented camel's milk, diabetic *db/db* mice and controls were given low or high doses of a composite probiotic containing 10 different *Lactobacillus* strains and 4 yeast strains derived from camel's milk [190]. The results revealed decreased fasting blood glucose, glucose AUC and HbA_{1c}, and increased C-peptide levels in mice receiving probiotics, as well as decreased total cholesterol, triglycerides and LDL concentrations. Mice receiving probiotics also exhibited enhanced insulin secretion via glucose-triggered GLP-1 secretion, through upregulation of G protein-coupled receptor 43/41 [190]. Lastly, probiotics appeared to improve intestinal barrier function by increasing levels of SCFA-producing microbes *Lactobacillus, Bifidobacterium, C. leptum*, and *Roseburia*, with concomitant increases in propionate and butyrate. Levels of claudin-1 and mucin-2 were also upregulated, and circulating levels of bacterial LPS were decreased [190]. More recently, *Lactobacillus paracasei* IMC 502 was shown to significantly decrease blood glucose, HbA_{1c}, and lipid levels, and to improve insulin resistance and glucose intolerance in a T2D mouse model, with increases in the SCFAs acetate and butyrate as well as in GLP-1 and PYY [191]. In a small human study, direct duodenal infusion with the butyrate producer *Anaerobutyricum soehngenii* was shown to stimulate GLP-1 production and improve insulin sensitivity in subjects with metabolic syndrome [192]. Although many of these results need to be replicated in studies of humans subjects with T2D, they strongly suggest that specific probiotics can have beneficial effects on metabolic parameters in rodent models of diabetes, and that these effects are mediated through effects on the gut microbiome. Parallel to the recently-identified potential roles for the mycobiome in T2D, the edible fungus *Cordyceps militaris* and its active ingredient cordycepin [193], and culture medium from *Ceriporia lacerata* mycelium [194] have shown antidiabetic potentials in mouse [193] and human [194] studies, illustrating that potential therapeutics are not limited to bacteria.

However, it is also important to note that the contents of probiotics and what is listed on the label may differ, and consumers should be wary of this. For example, in a study of 25 different veterinary products, Weese and Martin found that only 22 product labels listed specific microorganisms, with one or more misspelled on 7 labels, and only 4 of 15 products that claimed to contain live organisms met or exceeded this claim when cultured on blood agar plates [195]. Another study evaluating 52 human dietary supplements found that only 30 of 52 products contained a correctly labeled classification, some of which included incorrect taxonomy, missing species, and un-labeled species, and that 17 product labels were below label viability claims [196]. These studies demonstrate the need for investigation into accurate taxonomy and relative amounts of mixed microbial populations contained in commercial probiotics.

5 What's Next

5.1 Fecal Microbiota Transplantation (FMT)

FMT delivers stool from a healthy human donor to a recipient via colonoscopy, enema, nasogastric (NG) tube, or as capsules, and has been used to treat *Clostridioides difficile* colitis and other serious gastrointestinal conditions [197] (see Chap. "The Role of the Microbiome in Inflammatory Bowel Disease"). It has also gained attention as a potential means of improving insulin sensitivity [198], and studies into the use FMT as a therapeutic for obesity are currently underway.

In 2019, a pilot study by Allegretti et al. evaluated 22 individuals with obesity, (BMI $\geq$5 kg/m^2) without a diagnosis of diabetes, nonalcoholic steatohepatitis, or metabolic syndrome [199]. Subjects were given FMT capsules (containing fecal matter from a single lean donor) or placebo capsules, with subject follow-up through 26 weeks post-FMT. Patients who received FMT had sustained shifts in their microbiomes toward those of the donor, but there were no significant differences in AUC for GLP-1 between the groups at 12 weeks, nor were there any significant changes in BMI [199].

Another study evaluating the effects of FMT on improvements in insulin activity in subjects with obesity also failed to find differences between subject receiving FMT (capsule form, derived from 4 metabolically healthy lean donors) and those receiving placebo [200]. Although there were measurable improvements in insulin sensitivity in the FMT group, the changes did not reach statistical significance and there were no differences between groups for other measured metabolic outcomes (HbA_{1c} and body composition). Similar to previous findings [199], 16S sequencing showed the microbiomes of FMT recipients exhibited more similar compositions to donor material, and less similar compositions to their own baseline samples [200]. Taken together, the results of these studies may suggest that while FMT does alter the recipient microbiome, the metabolic benefits are not significant.

5.2 Analysis of the Small Bowel Microbiome

The small bowel is the primary site of nutrient absorption and metabolism [201]. However, the majority of studies of the 'gut' microbiome in diabetes and obesity use fecal samples. While fecal sampling has obvious advantages for ease of collection, the small bowel differs significantly from the colon in terms of pH and transit time [202, 203]. Moreover, small bowel luminal contents are lower in volume, have significantly lower microbial biomass, and are much more viscous than colonic contents, such that techniques developed for microbial isolation and sequencing from stool samples may not work for small intestinal samples. Recently, Leite et al. described the development of methods to overcome these challenges, including the use of a mucolytic agent to improve bacterial recovery and DNA isolation [64], and also described the successful use of these methods to characterize microbial populations in different small bowel segments, all of which differed significantly from stool [204]. Given the relevance of the small bowel to metabolic conditions such as T2D and obesity, characterizing the small intestinal microbiome could play a key role in identifying specific therapeutic targets [205].

5.3 Effects of the Circadian Clock on Metabolism

It is becoming increasingly understood that host metabolism is strongly influenced by the circadian clock, and that the circadian clock is in turn influenced by the gut microbiota [206–208]. Mammalian circadian clocks are regulated at the transcriptional and posttranscriptional levels [208, 209], and synchronize host metabolic pathways with 24-h light-dark cycles, allowing for optimal energy utilization, and disruption of sleep cycles has long been known to increase risks for obesity and metabolic disorders [210]—for example, obstructive sleep apnea is a significant risk factor for metabolic syndrome, insulin resistance, and cardiovascular disease [211]. Although a comprehensive review of this fascinating topic is beyond the scope of this chapter, Thaiss et al. showed that the composition of the gut microbiota undergoes diurnal oscillations, including oscillations in genera from the orders Clostridiales, Lactobacillales, and Bacteroidales [212], and these were later shown to be influenced by host diet and nutrient load as well as feeding cycles [209]. Studies of ileostomy samples have also shown diurnal differences in the ileal microbiome [213]. The proliferation and regeneration of intestinal epithelial cells (IECs), and their functions in nutrient absorption and metabolism, are regulated by the circadian clock [214]. IECs express the circadian transcription factor NFIL3, which oscillates diurnally. Wang et al. found that NFIL3 expression is significantly reduced in germ-free mice, and that the amplitude of NFIL3 transcription is controlled by gut microbiota [206]. Specifically, NFIL3 expression is repressed by the circadian repressor REV-ERBα, through a mechanism involving sub-epithelial dendritic and group 3 innate lymphoid cells that is mediated via IL-23 and IL-22 following epithelial sensing of bacterial flagellin and LPS

[206]. NFIL3 in turn controls lipid uptake and metabolism for storage in adipose tissue [206]. This is just one example of how the gut microbiota control host metabolism via regulation of the circadian clock. Separately, it has been suggested that modern eating practices (3 daily meals plus snacks) disrupts the circadian clock and gut microbiome composition, and that time-restricted eating (TRE) may restore circadian rhythms, thus explaining the improvements in insulin resistance, cognitive function, and glycemic control that are seen with TRE [215]. It will be interesting to see if this exciting area of research translates into clinical treatments.

6 Concluding Remarks

The dual pandemics of obesity and diabetes continue to be at the forefront of human disease. The downstream effects are seen manifesting in heart disease, liver disease, immune changes, neurocognition and others. While significant effort has been utilized into trying to defeat these disease states, we are failing miserably. Why? Perhaps it's because each of these diseases is complex, and likely there are many subtypes—based on etiology, genetics, and other factors. Treatment and prevention will not be a "one step" process. In fact, looking into novel areas, such as the gastrointestinal microbiome, allows the opportunity to explore areas of causation that have yet to be tapped. When looking at the microbiome of the GI tract, it is most sensical to focus on the area where confluence of hormonal, nutritive and digestive functions meet: the small intestine. While the potential of the microbiome is enormous, narrowing down the overwhelming possibilities of contributing factors is daunting—not just for diabetes and obesity, but for other diseases as well. It is only with perseverance (and some trial and error) that we can start to hone down microbial therapeutic targets. Moving forward, this untapped potential should be a strategic target for research in human health and disease.

References

1. World Health Organization. Fact sheet—diabetes. Geneva: World Health Organization; 2020.
2. International Diabetes Federation. IDF diabetes atlas. 10th ed. Brussels: International Diabetes Federation; 2021.
3. National Diabetes Statistics Report, 2022—estimates of diabetes and its burden in the United States. Centers for Disease Control and Prevention, US Department of Health and Human Services, Atlanta, GA; 2022.
4. American Diabetes Association. 2. classification and diagnosis of diabetes: standards of medical care in diabetes—2021. Diabetes Care. 2020;44:S15–33. https://doi.org/10.2337/dc21-S002.
5. Atkinson MA, Skyler JS. In: Skyler JS, editor. Atlas of diabetes. 4th ed. Springer; 2012. p. 65–94.
6. Kao K, Sabin M. Type 2 diabetes mellitus in children and adolescents. Aust Fam Physician. 2016;45:401–6.
7. Sirdah MM, Reading NS. Genetic predisposition in type 2 diabetes: a promising approach toward a personalized management of diabetes. Clin Genet. 2020;98(6):525. https://doi.org/10.1111/cge.13772.
8. Vijan S. In the clinic. Type 2 diabetes. Ann Intern Med. 2015;162:ITC1–16. https://doi.org/10.7326/aitc201503030.
9. American Diabetes Association. Diagnosis and classification of diabetes mellitus. Diabetes Care. 2014;37:S81. https://doi.org/10.2337/dc14-S081.
10. Iglay K, et al. Prevalence and co-prevalence of comorbidities among patients with type 2 diabetes mellitus. Curr Med Res Opin. 2016;32:1243–52. https://doi.org/10.1185/03007995.2016.1168291.
11. Henning RJ. Type-2 diabetes mellitus and cardiovascular disease. Futur Cardiol. 2018;14:491–509. https://doi.org/10.2217/fca-2018-0045.
12. Singh RB, Mengi SA, Xu YJ, Arneja AS, Dhalla NS. Pathogenesis of atherosclerosis: a multifactorial process. Exp Clin Cardiol. 2002;7:40–53.
13. Kahn BB, Flier JS. Obesity and insulin resistance. J Clin Invest. 2000;106:473–81. https://doi.org/10.1172/JCI10842.
14. Insulin resistance & prediabetes, 2018. National Institute of Diabetes and Digestive and Kidney Diseases (NIDDK). https://www.niddk.nih.gov/health-information/diabetes/overview/what-is-diabetes/prediabetes-insulin-resistance.
15. Caricilli AM, Saad MJ. The role of gut microbiota on insulin resistance. Nutrients. 2013;5:829–51. https://doi.org/10.3390/nu5030829.
16. World Health Organization. Fact sheet—obesity and overweight. Geneva: World Health Organization; 2020.
17. American Heart Association. American Heart Association annual report 2018–19. American Heart Association; 2019.
18. Hsu WC, Araneta MR, Kanaya AM, Chiang JL, Fujimoto W. BMI cut points to identify at-risk Asian Americans for type 2 diabetes screening. Diabetes Care. 2015;38:150–8. https://doi.org/10.2337/dc14-2391.
19. World Obesity Atlas 2023. World Obesity Federation, 2023.
20. Overweight and obesity. Centers for Disease Control and Prevention, Atlanta, GA; 2022.
21. The heavy burden of obesity. OECD; 2019.
22. Trogdon JG, Finkelstein EA, Hylands T, Dellea PS, Kamal-Bahl SJ. Indirect costs of obesity: a review of the current literature. Obes Rev. 2008;9:489–500. https://doi.org/10.1111/j.1467-789X.2008.00472.x.
23. Cawley J, et al. Direct medical costs of obesity in the United States and the most populous states. J Manag Care Spec Pharm. 2021;27:354–66. https://doi.org/10.18553/jmcp.2021.20410.
24. Adult Obesity Facts. Centers for Disease Control and Prevention, Atlanta, GA, 2020.
25. Bener A, et al. Obesity index that better predict metabolic syndrome: body mass index, waist circumference, waist hip ratio, or waist height ratio. J Obes. 2013;2013:269038. https://doi.org/10.1155/2013/269038.
26. Ahmad N, Adam SI, Nawi AM, Hassan MR, Ghazi HF. Abdominal obesity indicators: waist circumference or waist-to-hip ratio in Malaysian adults population. Int J Prev Med. 2016;7:82. https://doi.org/10.4103/2008-7802.183654.
27. Obesity and Cancer. National Cancer Institute (NCI); 2017. https://www.cancer.gov/about-cancer/causesprevention/risk/obesity/obesity-fact-sheet.
28. Randi G, Franceschi S, La Vecchia C. Gallbladder cancer worldwide: geographical distribution and risk factors. Int J Cancer. 2006;118:1591–602. https://doi.org/10.1002/ijc.21683.
29. Ptak A, Kolaczkowska E, Gregoraszczuk EL. Leptin stimulation of cell cycle and inhibition of apoptosis gene and protein expression in OVCAR-3 ovarian cancer cells. Endocrine. 2013;43:394–403. https://doi.org/10.1007/s12020-012-9788-7.

30. Gallagher EJ, LeRoith D. Obesity and diabetes: the increased risk of cancer and cancer-related mortality. Physiol Rev. 2015;95:727–48. https://doi.org/10.1152/physrev.00030.2014.
31. Sender R, Fuchs S, Milo R. Revised estimates for the number of human and bacteria cells in the body. PLoS Biol. 2016;14:e1002533. https://doi.org/10.1371/journal.pbio.1002533.
32. Gilbert JA, et al. Current understanding of the human microbiome. Nat Med. 2018;24:392–400. https://doi.org/10.1038/nm.4517.
33. Qin J, et al. A human gut microbial gene catalogue established by metagenomic sequencing. Nature. 2010;464:59–65.
34. Lozupone CA, Stombaugh JI, Gordon JI, Jansson JK, Knight R. Diversity, stability and resilience of the human gut microbiota. Nature. 2012;489:220–30. https://doi.org/10.1038/nature11550.
35. Gaci N, Borrel G, Tottey W, O'Toole PW, Brugere JF. Archaea and the human gut: new beginning of an old story. World J Gastroenterol. 2014;20:16062–78. https://doi.org/10.3748/wjg.v20.i43.16062.
36. Shkoporov AN, et al. The human gut virome is highly diverse, stable, and individual specific. Cell Host Microbe. 2019;26:527–541.e525. https://doi.org/10.1016/j.chom.2019.09.009.
37. Eckburg PB, et al. Diversity of the human intestinal microbial flora. Science. 2005;308:1635–8. https://doi.org/10.1126/science.1110591.
38. Rinninella E, et al. What is the healthy gut microbiota composition? A changing ecosystem across age, environment, diet, and diseases. Microorganisms. 2019;7:14. https://doi.org/10.3390/microorganisms7010014.
39. Arumugam M, et al. Enterotypes of the human gut microbiome. Nature. 2011;473:174–80. https://doi.org/10.1038/nature09944.
40. Singer-Englar T, Barlow G, Mathur R. Obesity, diabetes, and the gut microbiome: an updated review. Expert Rev Gastroenterol Hepatol. 2018;13:3. https://doi.org/10.1080/17474124.2019.1543023.
41. Han H, et al. Gut microbiota and type 1 diabetes. Int J Mol Sci. 2018;19:995. https://doi.org/10.3390/ijms19040995.
42. Ley RE, et al. Evolution of mammals and their gut microbes. Science. 2008;320:1647–51. https://doi.org/10.1126/science.1155725.
43. Tilg H, Moschen AR. Microbiota and diabetes: an evolving relationship. Gut. 2014;63:1513–21. https://doi.org/10.1136/gutjnl-2014-306928.
44. Lloyd-Price J, Abu-Ali G, Huttenhower C. The healthy human microbiome. Genome Med. 2016;8:51. https://doi.org/10.1186/s13073-016-0307-y.
45. Barlow GM, Mathur R. Type 2 diabetes and the microbiome. J Endocr Soc. 2022;7:bvac184. https://doi.org/10.1210/jendso/bvac184.
46. Belizario JE, Faintuch J, Garay-Malpartida M. Gut microbiome dysbiosis and immunometabolism: new frontiers for treatment of metabolic diseases. Mediat Inflamm. 2018;2018:2037838. https://doi.org/10.1155/2018/2037838.
47. Sharma S, Tripathi P. Gut microbiome and type 2 diabetes: where we are and where to go? J Nutr Biochem. 2019;63:101–8. https://doi.org/10.1016/j.jnutbio.2018.10.003.
48. Hur KY, Lee MS. Gut microbiota and metabolic disorders. Diabetes Metab J. 2015;39:198–203. https://doi.org/10.4093/dmj.2015.39.3.198.
49. Bao L, Zhang Y, Zhang G, Jiang D, Yan D. Abnormal proliferation of gut mycobiota contributes to the aggravation of type 2 diabetes. Commun Biol. 2023;6:226. https://doi.org/10.1038/s42003-023-04591-x.
50. Larsen N, et al. Gut microbiota in human adults with type 2 diabetes differs from non-diabetic adults. PLoS One. 2010;5:e9085. https://doi.org/10.1371/journal.pone.0009085.
51. Karlsson FH, et al. Gut metagenome in European women with normal, impaired and diabetic glucose control. Nature. 2013;498:99–103. https://doi.org/10.1038/nature12198.
52. De Vadder F, et al. Microbiota-generated metabolites promote metabolic benefits via gut-brain neural circuits. Cell. 2014;156:84–96. https://doi.org/10.1016/j.cell.2013.12.016.
53. Bach Knudsen KE, et al. Impact of diet-modulated butyrate production on intestinal barrier function and inflammation. Nutrients. 2018;10:1499. https://doi.org/10.3390/nu10101499.
54. Wright E Jr, Scism-Bacon JL, Glass LC. Oxidative stress in type 2 diabetes: the role of fasting and postprandial glycaemia. Int J Clin Pract. 2006;60:308–14. https://doi.org/10.1111/j.1368-5031.2006.00825.x.
55. Qin J, et al. A metagenome-wide association study of gut microbiota in type 2 diabetes. Nature. 2012;490:55–60. https://doi.org/10.1038/nature11450.
56. Sato J, et al. Gut dysbiosis and detection of "live gut bacteria" in blood of Japanese patients with type 2 diabetes. Diabetes Care. 2014;37:2343–50. https://doi.org/10.2337/dc13-2817.
57. Naderpoor N, et al. Faecal microbiota are related to insulin sensitivity and secretion in overweight or obese adults. J Clin Med. 2019;8:452. https://doi.org/10.3390/jcm8040452.
58. Wu T, Horowitz M, Rayner CK. New insights into the antidiabetic actions of metformin: from the liver to the gut. Expert Rev Gastroenterol Hepatol. 2017;11:157–66. https://doi.org/10.1080/17474124.2017.1273769.
59. Derrien M, et al. Modulation of mucosal immune response, tolerance, and proliferation in mice colonized by the Mucin-degrader akkermansia muciniphila. Front Microbiol. 2011;2:166. https://doi.org/10.3389/fmicb.2011.00166.
60. Everard A, et al. Cross-talk between Akkermansia muciniphila and intestinal epithelium controls diet-induced obesity. Proc Natl Acad Sci U S A. 2013;110:9066–71. https://doi.org/10.1073/pnas.1219451110.
61. de la Cuesta-Zuluaga J, et al. Metformin is associated with higher relative abundance of mucin-degrading *Akkermansia muciniphila* and several short-chain fatty acid–producing microbiota in the gut. Diabetes Care. 2017;40:54–62. https://doi.org/10.2337/dc16-1324.
62. Shin NR, et al. An increase in the Akkermansia spp. population induced by metformin treatment improves glucose homeostasis in diet-induced obese mice. Gut. 2014;63:727–35. https://doi.org/10.1136/gutjnl-2012-303839.
63. Thingholm LB, et al. Obese individuals with and without type 2 diabetes show different gut microbial functional capacity and composition. Cell Host Microbe. 2019;26:252–264.e210. https://doi.org/10.1016/j.chom.2019.07.004.
64. Leite GGS, et al. Optimizing microbiome sequencing for small intestinal aspirates: validation of novel techniques through the REIMAGINE study. BMC Microbiol. 2019;19:239. https://doi.org/10.1186/s12866-019-1617-1.
65. Ley RE, Turnbaugh PJ, Klein S, Gordon JI. Microbial ecology: human gut microbes associated with obesity. Nature. 2006;444:1022–3.
66. Ley RE, et al. Obesity alters gut microbial ecology. Proc Natl Acad Sci U S A. 2005;102:11070–5.
67. Le Chatelier E, et al. Richness of human gut microbiome correlates with metabolic markers. Nature. 2013;500:541–6. https://doi.org/10.1038/nature12506.
68. Kasai C, et al. Comparison of the gut microbiota composition between obese and non-obese individuals in a Japanese population, as analyzed by terminal restriction fragment length polymorphism and next-generation sequencing. BMC Gastroenterol. 2015;15:100. https://doi.org/10.1186/s12876-015-0330-2.
69. Fernandes J, Su W, Rahat-Rozenbloom S, Wolever TM, Comelli EM. Adiposity, gut microbiota and faecal short chain fatty acids

are linked in adult humans. Nutr Diabetes. 2014;4:e121. https://doi.org/10.1038/nutd.2014.23.
70. Peters BA, et al. A taxonomic signature of obesity in a large study of American adults. Sci Rep. 2018;8:9749. https://doi.org/10.1038/s41598-018-28126-1.
71. Collado MC, Rautava S, Aakko J, Isolauri E, Salminen S. Human gut colonisation may be initiated in utero by distinct microbial communities in the placenta and amniotic fluid. Sci Rep. 2016;6:23129. https://doi.org/10.1038/srep23129.
72. Benitez-Paez A, et al. Depletion of Blautia species in the microbiota of obese children relates to intestinal inflammation and metabolic phenotype worsening. mSystems. 2020;5:e00857. https://doi.org/10.1128/mSystems.00857-19.
73. Turnbaugh PJ, et al. A core gut microbiome in obese and lean twins. Nature. 2009;457:480–4. https://doi.org/10.1038/nature07540.
74. Hook SE, Wright AD, McBride BW. Methanogens: methane producers of the rumen and mitigation strategies. Archaea. 2010;2010:945785. https://doi.org/10.1155/2010/945785.
75. Kim G, et al. Methanobrevibacter smithii is the predominant methanogen in patients with constipation-predominant IBS and methane on breath. Dig Dis Sci. 2012;57:3213–8. https://doi.org/10.1007/s10620-012-2197-1.
76. Sahakian AB, Jee SR, Pimentel M. Methane and the gastrointestinal tract. Dig Dis Sci. 2010;55:2135–43.
77. Turnbaugh PJ, et al. An obesity-associated gut microbiome with increased capacity for energy harvest. Nature. 2006;444:1027–31.
78. Samuel BS, Gordon JI. A humanized gnotobiotic mouse model of host-archaeal-bacterial mutualism. Proc Natl Acad Sci U S A. 2006;103:10011–6.
79. Basseri RJ, et al. Intestinal methane production in obese individuals is associated with a higher body mass index. Gastroenterol Hepatol (N Y). 2012;8:22–8.
80. Mathur R, et al. Methane and hydrogen positivity on breath test is associated with greater body mass index and body fat. J Clin Endocrinol Metab. 2013;98:E698–702. https://doi.org/10.1210/jc.2012-3144.
81. Mathur R, et al. Methane-producing human subjects have higher serum glucose levels during oral glucose challenge than non-methane producers: a pilot study of the effects of enteric methanogens on glycemic regulation. Res J Endocrinol Metabol. 2014;2:2. https://doi.org/10.7243/2053-3640-2-2.
82. Cesario V, et al. Methane intestinal production and poor metabolic control in type I diabetes complicated by autonomic neuropathy. Minerva Endocrinol. 2014;39:201–7.
83. Mathur R, et al. Metabolic effects of eradicating breath methane using antibiotics in prediabetic subjects with obesity. Obesity (Silver Spring). 2016;24:576–82. https://doi.org/10.1002/oby.21385.
84. Mathur R, et al. Intestinal methane production is associated with decreased weight loss following bariatric surgery. Obes Res Clin Pract. 2016;10(6):728. https://doi.org/10.1016/j.orcp.2016.06.006.
85. Laverdure R, Mezouari A, Carson MA, Basiliko N, Gagnon J. A role for methanogens and methane in the regulation of GLP-1. Endocrinol Diabetes Metab. 2018;1:e00006. https://doi.org/10.1002/edm2.6.
86. Armougom F, Henry M, Vialettes B, Raccah D, Raoult D. Monitoring bacterial community of human gut microbiota reveals an increase in Lactobacillus in obese patients and Methanogens in anorexic patients. PLoS One. 2009;4:e7125. https://doi.org/10.1371/journal.pone.0007125.
87. Borgo F, et al. Microbiota in anorexia nervosa: the triangle between bacterial species, metabolites and psychological tests. PLoS One. 2017;12:e0179739. https://doi.org/10.1371/journal.pone.0179739.
88. Kearney J. Food consumption trends and drivers. Philos Trans R Soc Lond Ser B Biol Sci. 2010;365:2793–807. https://doi.org/10.1098/rstb.2010.0149.
89. World Health Organization. Fact sheet—healthy diet. Geneva: World Health Organization; 2020.
90. Statovci D, Aguilera M, MacSharry J, Melgar S. The impact of western diet and nutrients on the microbiota and immune response at mucosal interfaces. Front Immunol. 2017;8:838. https://doi.org/10.3389/fimmu.2017.00838.
91. Popkin BM, Adair LS, Ng SW. Global nutrition transition and the pandemic of obesity in developing countries. Nutr Rev. 2012;70:3–21. https://doi.org/10.1111/j.1753-4887.2011.00456.x.
92. Ludwig DS, et al. High glycemic index foods, overeating, and obesity. Pediatrics. 1999;103:E26. https://doi.org/10.1542/peds.103.3.e26.
93. Martinez Steele E, et al. Ultra-processed foods and added sugars in the US diet: evidence from a nationally representative cross-sectional study. BMJ Open. 2016;6:e009892. https://doi.org/10.1136/bmjopen-2015-009892.
94. De Filippo C, et al. Impact of diet in shaping gut microbiota revealed by a comparative study in children from Europe and rural Africa. Proc Natl Acad Sci U S A. 2010;107:14691–6. https://doi.org/10.1073/pnas.1005963107.
95. Yatsunenko T, et al. Human gut microbiome viewed across age and geography. Nature. 2012;486:222–7. https://doi.org/10.1038/nature11053.
96. Brooks L, et al. Fermentable carbohydrate stimulates FFAR2-dependent colonic PYY cell expansion to increase satiety. Mol Metab. 2017;6:48–60. https://doi.org/10.1016/j.molmet.2016.10.011.
97. Abutair AS, Naser IA, Hamed AT. Soluble fibers from psyllium improve glycemic response and body weight among diabetes type 2 patients (randomized control trial). Nutr J. 2016;15:86. https://doi.org/10.1186/s12937-016-0207-4.
98. Pino JL, Mujica V, Arredondo M. Effect of dietary supplementation with oat β-glucan for 3 months in subjects with type 2 diabetes: a randomized, double-blind, controlled clinical trial. J Funct Foods. 2021;77:104311. https://doi.org/10.1016/j.jff.2020.104311.
99. Frias JP, et al. A microbiome-targeting fibre-enriched nutritional formula is well tolerated and improves quality of life and haemoglobin A1c in type 2 diabetes: a double-blind, randomized, placebo-controlled trial. Diabetes Obes Metab. 2023;25:1203–12. https://doi.org/10.1111/dom.14967.
100. David LA, et al. Diet rapidly and reproducibly alters the human gut microbiome. Nature. 2014;505:559–63. https://doi.org/10.1038/nature12820.
101. Sonnenburg JL, et al. Glycan foraging in vivo by an intestine-adapted bacterial symbiont. Science. 2005;307:1955–9. https://doi.org/10.1126/science.1109051.
102. Sonnenburg ED, et al. Diet-induced extinctions in the gut microbiota compound over generations. Nature. 2016;529:212–5. https://doi.org/10.1038/nature16504.
103. Malik VS, Hu FB. Sweeteners and risk of obesity and type 2 diabetes: the role of sugar-sweetened beverages. Curr Diab Rep. 2012; https://doi.org/10.1007/s11892-012-0259-6.
104. Popkin BM. Patterns of beverage use across the lifecycle. Physiol Behav. 2010;100:4–9. https://doi.org/10.1016/j.physbeh.2009.12.022.
105. Malik VS, et al. Sugar-sweetened beverages and risk of metabolic syndrome and type 2 diabetes: a meta-analysis. Diabetes Care. 2010;33:2477–83. https://doi.org/10.2337/dc10-1079.
106. Schulze MB, et al. Sugar-sweetened beverages, weight gain, and incidence of type 2 diabetes in young and middle-aged women. JAMA. 2004;292:927–34. https://doi.org/10.1001/jama.292.8.927.
107. Kim JJ, Sears DD. TLR4 and insulin resistance. Gastroenterol Res Pract. 2010;2010:212563. https://doi.org/10.1155/2010/212563.
108. Substances added to food. U.S. Food and Drug Administration; 2018. https://www.usc.es/caa/EdulcWeb/EAFUS.pdf.

109. Simmons AL, Schlezinger JJ, Corkey BE. What are we putting in our food that is making us fat? Food additives, contaminants, and other putative contributors to obesity. Curr Obes Rep. 2014;3:273–85. https://doi.org/10.1007/s13679-014-0094-y.
110. Brown RJ, Walter M, Rother KI. Ingestion of diet soda before a glucose load augments glucagon-like peptide-1 secretion. Diabetes Care. 2009;32:2184–6. https://doi.org/10.2337/dc09-1185.
111. Yang Q. Gain weight by "going diet?" Artificial sweeteners and the neurobiology of sugar cravings: neuroscience 2010. Yale J Biol Med. 2010;83:101–8.
112. Barrios-Correa AA, et al. Chronic intake of commercial sweeteners induces changes in feeding behavior and signaling pathways related to the control of appetite in BALB/c mice. Biomed Res Int. 2018;2018:3628121. https://doi.org/10.1155/2018/3628121.
113. Lohner S, Toews I, Meerpohl JJ. Health outcomes of non-nutritive sweeteners: analysis of the research landscape. Nutr J. 2017;16:55. https://doi.org/10.1186/s12937-017-0278-x.
114. Ruiz-Ojeda FJ, Plaza-Diaz J, Saez-Lara MJ, Gil A. Effects of sweeteners on the gut microbiota: a review of experimental studies and clinical trials. Adv Nutr. 2019;10:S31–48. https://doi.org/10.1093/advances/nmy037.
115. Ferland A, Brassard P, Poirier P. Is aspartame really safer in reducing the risk of hypoglycemia during exercise in patients with type 2 diabetes? Diabetes Care. 2007;30:e59. https://doi.org/10.2337/dc06-1888.
116. Gul SS, et al. Inhibition of the gut enzyme intestinal alkaline phosphatase may explain how aspartame promotes glucose intolerance and obesity in mice. Appl Physiol Nutr Metab. 2017;42:77–83. https://doi.org/10.1139/apnm-2016-0346.
117. Nettleton JE, et al. Maternal low-dose aspartame and stevia consumption with an obesogenic diet alters metabolism, gut microbiota and mesolimbic reward system in rat dams and their offspring. Gut. 2020;69(10):1807. https://doi.org/10.1136/gutjnl-2018-317505.
118. Palmnas MS, et al. Low-dose aspartame consumption differentially affects gut microbiota-host metabolic interactions in the diet-induced obese rat. PLoS One. 2014;9:e109841. https://doi.org/10.1371/journal.pone.0109841.
119. Anton SD, et al. Effects of stevia, aspartame, and sucrose on food intake, satiety, and postprandial glucose and insulin levels. Appetite. 2010;55:37–43. https://doi.org/10.1016/j.appet.2010.03.009.
120. Tey SL, Salleh NB, Henry J, Forde CG. Effects of aspartame-, monk fruit-, stevia- and sucrose-sweetened beverages on postprandial glucose, insulin and energy intake. Int J Obes. 2017;41:450–7. https://doi.org/10.1038/ijo.2016.225.
121. Kuk JL, Brown RE. Aspartame intake is associated with greater glucose intolerance in individuals with obesity. Appl Physiol Nutr Metab. 2016;41:795–8. https://doi.org/10.1139/apnm-2015-0675.
122. Suez J, et al. Artificial sweeteners induce glucose intolerance by altering the gut microbiota. Nature. 2014;514:181–6. https://doi.org/10.1038/nature13793.
123. Labrecque MT, Malone D, Caldwell KE, Allan AM. Impact of ethanol and saccharin on fecal microbiome in pregnant and non-pregnant mice. J Pregnancy Child Health. 2015;2:1000193. https://doi.org/10.4172/2376-127X.1000193.
124. Bian X, et al. Saccharin induced liver inflammation in mice by altering the gut microbiota and its metabolic functions. Food Chem Toxicol. 2017;107:530–9. https://doi.org/10.1016/j.fct.2017.04.045.
125. Suez J, et al. Personalized microbiome-driven effects of non-nutritive sweeteners on human glucose tolerance. Cell. 2022;185:3307–3328.e3319. https://doi.org/10.1016/j.cell.2022.07.016.
126. Hosseini A, Barlow GM, Leite G, et al. Consuming Artificial Sweeteners May Alter the Structure and Function of Duodenal Microbial Communities. iScience 2023:108530. https://doi.org/10.1016/j.isci.2023.108530.
127. Viennois E, Gewirtz AT, Chassaing B. Chronic inflammatory diseases: are we ready for microbiota-based dietary intervention? Cell Mol Gastroenterol Hepatol. 2019;8:61–71. https://doi.org/10.1016/j.jcmgh.2019.02.008.
128. Chassaing B, et al. Dietary emulsifiers impact the mouse gut microbiota promoting colitis and metabolic syndrome. Nature. 2015;519:92–6. https://doi.org/10.1038/nature14232.
129. Chassaing B, Van de Wiele T, De Bodt J, Marzorati M, Gewirtz AT. Dietary emulsifiers directly alter human microbiota composition and gene expression ex vivo potentiating intestinal inflammation. Gut. 2017;66:1414–27. https://doi.org/10.1136/gutjnl-2016-313099.
130. Cullender TC, et al. Innate and adaptive immunity interact to quench microbiome flagellar motility in the gut. Cell Host Microbe. 2013;14:571–81. https://doi.org/10.1016/j.chom.2013.10.009.
131. Moser C, et al. Antibiotic therapy as personalized medicine—general considerations and complicating factors. APMIS. 2019;127:361–71. https://doi.org/10.1111/apm.12951.
132. Outpatient antibiotic prescriptions—United States, 2018. Ctr Dis Control Prev; 2020. https://www.cdc.gov/antibioticuse/data/report-2018.html.
133. Trasande L, et al. Infant antibiotic exposures and early-life body mass. Int J Obes. 2013;37:16–23. https://doi.org/10.1038/ijo.2012.132.
134. Turta O, Rautava S. Antibiotics, obesity and the link to microbes—what are we doing to our children? BMC Med. 2016;14:57. https://doi.org/10.1186/s12916-016-0605-7.
135. Fenneman AC, Weidner M, Chen LA, Nieuwdorp M, Blaser MJ. Antibiotics in the pathogenesis of diabetes and inflammatory diseases of the gastrointestinal tract. Nat Rev Gastroenterol Hepatol. 2023;20:81–100. https://doi.org/10.1038/s41575-022-00685-9.
136. Koenig JE, et al. Succession of microbial consortia in the developing infant gut microbiome. Proc Natl Acad Sci U S A. 2011;108(Suppl 1):4578–85. https://doi.org/10.1073/pnas.1000081107.
137. Bailey LC, et al. Association of antibiotics in infancy with early childhood obesity. JAMA Pediatr. 2014;168:1063–9. https://doi.org/10.1001/jamapediatrics.2014.1539.
138. Saari A, Virta LJ, Sankilampi U, Dunkel L, Saxen H. Antibiotic exposure in infancy and risk of being overweight in the first 24 months of life. Pediatrics. 2015;135:617–26. https://doi.org/10.1542/peds.2014-3407.
139. Scott FI, et al. Administration of antibiotics to children before age 2 years increases risk for childhood obesity. Gastroenterology. 2016;151:120–129.e125. https://doi.org/10.1053/j.gastro.2016.03.006.
140. Chelimo C, Camargo CA Jr, Morton SMB, Grant CC. Association of repeated antibiotic exposure up to age 4 years with body mass at age 4.5 years. JAMA Netw Open. 2020;3:e1917577. https://doi.org/10.1001/jamanetworkopen.2019.17577.
141. Aversa Z, et al. Association of infant antibiotic exposure with childhood health outcomes. Mayo Clin Proc. 2020;96(1):66–77. https://doi.org/10.1016/j.mayocp.2020.07.019.
142. Zhang M, et al. Association of prenatal antibiotics with measures of infant adiposity and the gut microbiome. Ann Clin Microbiol Antimicrob. 2019;18:18. https://doi.org/10.1186/s12941-019-0318-9.
143. Jess T, et al. Antibiotic use during pregnancy and childhood overweight: a population-based nationwide cohort study. Sci Rep. 2019;9:11528. https://doi.org/10.1038/s41598-019-48065-9.
144. Isaevska E, et al. Maternal antibiotic use and vaginal infections in the third trimester of pregnancy and the risk of obesity in preschool children. Pediatr Obes. 2020;15:e12632. https://doi.org/10.1111/ijpo.12632.

145. Furlong M, Deming-Halverson S, Sandler DP. Chronic antibiotic use during adulthood and weight change in the sister study. PLoS One. 2019;14:e0216959. https://doi.org/10.1371/journal.pone.0216959.
146. Thuny F, et al. Vancomycin treatment of infective endocarditis is linked with recently acquired obesity. PLoS One. 2010;5:e9074. https://doi.org/10.1371/journal.pone.0009074.
147. Reijnders D, et al. Effects of gut microbiota manipulation by antibiotics on host metabolism in obese humans: a randomized double-blind placebo-controlled trial. Cell Metab. 2016;24:63–74. https://doi.org/10.1016/j.cmet.2016.06.016.
148. Chen H, Goldberg MS. The effects of outdoor air pollution on chronic illnesses. Mcgill J Med. 2009;12:58–64.
149. Cohen AJ, et al. Estimates and 25-year trends of the global burden of disease attributable to ambient air pollution: an analysis of data from the Global Burden of Diseases Study 2015. Lancet. 2017;389:1907–18. https://doi.org/10.1016/S0140-6736(17)30505-6.
150. Dendup T, Feng X, Clingan S, Astell-Burt T. Environmental risk factors for developing type 2 diabetes mellitus: a systematic review. Int J Environ Res Public Health. 2018;15:78. https://doi.org/10.3390/ijerph15010078.
151. Rajagopalan S, Brook RD. Air pollution and type 2 diabetes: mechanistic insights. Diabetes. 2012;61:3037–45. https://doi.org/10.2337/db12-0190.
152. Jin Y, Wu S, Zeng Z, Fu Z. Effects of environmental pollutants on gut microbiota. Environ Pollut. 2017;222:1–9. https://doi.org/10.1016/j.envpol.2016.11.045.
153. Fouladi F, et al. Air pollution exposure is associated with the gut microbiome as revealed by shotgun metagenomic sequencing. Environ Int. 2020;138:105604. https://doi.org/10.1016/j.envint.2020.105604.
154. Lee DH, Porta M, Jacobs DR Jr, Vandenberg LN. Chlorinated persistent organic pollutants, obesity, and type 2 diabetes. Endocr Rev. 2014;35:557–601. https://doi.org/10.1210/er.2013-1084.
155. Jin UH, et al. Microbiome-derived tryptophan metabolites and their aryl hydrocarbon receptor-dependent agonist and antagonist activities. Mol Pharmacol. 2014;85:777–88. https://doi.org/10.1124/mol.113.091165.
156. Howell G 3rd, Mangum L. Exposure to bioaccumulative organochlorine compounds alters adipogenesis, fatty acid uptake, and adipokine production in NIH3T3-L1 cells. Toxicol In Vitro. 2011;25:394–402. https://doi.org/10.1016/j.tiv.2010.10.015.
157. Zhang L, et al. Persistent organic pollutants modify gut microbiota-host metabolic homeostasis in mice through aryl hydrocarbon receptor activation. Environ Health Perspect. 2015;123:679–88. https://doi.org/10.1289/ehp.1409055.
158. Lefever DE, et al. TCDD modulation of gut microbiome correlated with liver and immune toxicity in streptozotocin (STZ)-induced hyperglycemic mice. Toxicol Appl Pharmacol. 2016;304:48–58. https://doi.org/10.1016/j.taap.2016.05.016.
159. Peat CM, Kleiman SC, Bulik CM, Carroll IM. The intestinal microbiome in bariatric surgery patients. Eur Eat Disord Rev. 2015;23:496–503. https://doi.org/10.1002/erv.2400.
160. Gloy VL, et al. Bariatric surgery versus non-surgical treatment for obesity: a systematic review and meta-analysis of randomised controlled trials. BMJ. 2013;347:f5934. https://doi.org/10.1136/bmj.f5934.
161. Kang JH, Le QA. Effectiveness of bariatric surgical procedures: a systematic review and network meta-analysis of randomized controlled trials. Medicine (Baltimore). 2017;96:e8632. https://doi.org/10.1097/MD.0000000000008632.
162. Crommen S, Mattes A, Simon MC. Microbial adaptation due to gastric bypass surgery: the nutritional impact. Nutrients. 2020;12:1199. https://doi.org/10.3390/nu12041199.
163. Paganelli FL, et al. Roux-Y gastric bypass and sleeve gastrectomy directly change gut microbiota composition independent of surgery type. Sci Rep. 2019;9:10979. https://doi.org/10.1038/s41598-019-47332-z.
164. Palmisano S, et al. Changes in gut microbiota composition after bariatric surgery: a new balance to decode. J Gastrointest Surg. 2019;24(8):1736–46. https://doi.org/10.1007/s11605-019-04321-x.
165. Zhang H, et al. Human gut microbiota in obesity and after gastric bypass. Proc Natl Acad Sci U S A. 2009;106:2365–70. https://doi.org/10.1073/pnas.0812600106.
166. Furet JP, et al. Differential adaptation of human gut microbiota to bariatric surgery-induced weight loss: links with metabolic and low-grade inflammation markers. Diabetes. 2010;59:3049–57. https://doi.org/10.2337/db10-0253.
167. Kong L-C, et al. Gut microbiota after gastric bypass in human obesity: increased richness and associations of bacterial genera with adipose tissue genes. Am J Clin Nutr. 2013;98:16–24. https://doi.org/10.3945/ajcn.113.058743.
168. Graessler J, et al. Metagenomic sequencing of the human gut microbiome before and after bariatric surgery in obese patients with type 2 diabetes: correlation with inflammatory and metabolic parameters. Pharmacogenomics J. 2013;13:514–22. https://doi.org/10.1038/tpj.2012.43.
169. Davies N, O'Sullivan JM, Plank LD, Murphy R. Gut microbial predictors of type 2 diabetes remission following bariatric surgery. Obes Surg. 2020;30(9):3536. https://doi.org/10.1007/s11695-020-04684-0.
170. Murphy R, et al. Differential changes in gut microbiota after gastric bypass and sleeve gastrectomy bariatric surgery vary according to diabetes remission. Obes Surg. 2017;27:917–25. https://doi.org/10.1007/s11695-016-2399-2.
171. Cortez RV, et al. Shifts in intestinal microbiota after duodenal exclusion favor glycemic control and weight loss: a randomized controlled trial. Surg Obes Relat Dis. 2018;14:1748–54. https://doi.org/10.1016/j.soard.2018.07.021.
172. Wang C, et al. The genus Sutterella is a potential contributor to glucose metabolism improvement after Roux-en-Y gastric bypass surgery in T2D. Diabetes Res Clin Pract. 2020;162:108116. https://doi.org/10.1016/j.diabres.2020.108116.
173. Dao MC, et al. Akkermansia muciniphila abundance is lower in severe obesity, but its increased level after bariatric surgery is not associated with metabolic health improvement. Am J Physiol Endocrinol Metab. 2019;317:E446–59. https://doi.org/10.1152/ajpendo.00140.2019.
174. Farup PG, Valeur J. Changes in faecal short-chain fatty acids after weight-loss interventions in subjects with morbid obesity. Nutrients. 2020;12:802. https://doi.org/10.3390/nu12030802.
175. Glaysher MA, et al. A randomised controlled trial of a duodenal-jejunal bypass sleeve device (EndoBarrier) compared with standard medical therapy for the management of obese subjects with type 2 diabetes mellitus. BMJ Open. 2017;7:e018598. https://doi.org/10.1136/bmjopen-2017-018598.
176. Ruban A, Ashrafian H, Teare JP. The EndoBarrier: duodenal-jejunal bypass liner for diabetes and weight loss. Gastroenterol Res Pract. 2018;2018:7823182. https://doi.org/10.1155/2018/7823182.
177. Kim T, Holleman CL, Ptacek T, Morrow CD, Habegger KM. Duodenal endoluminal barrier sleeve alters gut microbiota of ZDF rats. Int J Obes. 2017;41:381–9. https://doi.org/10.1038/ijo.2016.224.
178. Paszkiewicz RL, et al. EndoBarrier® implantation results in fecalization of the small bowel microbiome and systemic inflammation in a canine model. Curr Res Clin Diab Obes. 2020;1:102.
179. de Jonge C, et al. Metabolic improvement in obese patients after duodenal-jejunal exclusion is associated with intestinal microbiota composition changes. Int J Obes. 2019;43:2509–17. https://doi.org/10.1038/s41366-019-0336-x.
180. van Baar ACG, et al. Durable metabolic improvements 2 years after duodenal mucosal resurfacing (DMR) in patients with

type 2 diabetes (REVITA-1 Study). Diabetes Res Clin Pract. 2022;184:109194. https://doi.org/10.1016/j.diabres.2022.109194.
181. Meiring S, et al. A changed gut microbiota diversity is associated with metabolic improvements after duodenal mucosal resurfacing with glucagon-like-peptide-1 receptor agonist in type 2 diabetes in a pilot study. Front Clin Diabetes Healthc. 2022;3:856661. https://doi.org/10.3389/fcdhc.2022.856661.
182. Shokryazdan P, Faseleh Jahromi M, Navidshad B, Liang JB. Effects of prebiotics on immune system and cytokine expression. Med Microbiol Immunol. 2017;206:1–9. https://doi.org/10.1007/s00430-016-0481-y.
183. Yoo JY, Kim SS. Probiotics and prebiotics: present status and future perspectives on metabolic disorders. Nutrients. 2016;8:173. https://doi.org/10.3390/nu8030173.
184. Wilkins T, Sequoia J. Probiotics for gastrointestinal conditions: a summary of the evidence. Am Fam Physician. 2017;96:170–8.
185. de Oliveira GLV, Leite AZ, Higuchi BS, Gonzaga MI, Mariano VS. Intestinal dysbiosis and probiotic applications in autoimmune diseases. Immunology. 2017;152:1–12. https://doi.org/10.1111/imm.12765.
186. Marino E, et al. Gut microbial metabolites limit the frequency of autoimmune T cells and protect against type 1 diabetes. Nat Immunol. 2017;18:552–62. https://doi.org/10.1038/ni.3713.
187. Nagpal R, et al. Human-origin probiotic cocktail increases short-chain fatty acid production via modulation of mice and human gut microbiome. Sci Rep. 2018;8:12649. https://doi.org/10.1038/s41598-018-30114-4.
188. Zhang L, et al. Akkermansia muciniphila can reduce the damage of gluco/lipotoxicity, oxidative stress and inflammation, and normalize intestine microbiota in streptozotocin-induced diabetic rats. Pathog Dis. 2018;76:4. https://doi.org/10.1093/femspd/fty028.
189. Albuquerque R, et al. Saccharomyces boulardii Tht 500101 changes gut microbiota and ameliorates hyperglycaemia, dyslipidaemia, and liver inflammation in streptozotocin-diabetic mice. Benef Microbes. 2019;10:901–12. https://doi.org/10.3920/BM2019.0056.
190. Wang Y, et al. Composite probiotics alleviate type 2 diabetes by regulating intestinal microbiota and inducing GLP-1 secretion in db/db mice. Biomed Pharmacother. 2020;125:109914. https://doi.org/10.1016/j.biopha.2020.109914.
191. Gu Y, et al. Lactobacillus paracasei IMC 502 ameliorates type 2 diabetes by mediating gut microbiota-SCFA-hormone/inflammation pathway in mice. J Sci Food Agric. 2023;103:2949–59. https://doi.org/10.1002/jsfa.12267.
192. Koopen A, et al. Duodenal Anaerobutyricum soehngenii infusion stimulates GLP-1 production, ameliorates glycaemic control and beneficially shapes the duodenal transcriptome in metabolic syndrome subjects: a randomised double-blind placebo-controlled cross-over study. Gut. 2022;71:1577–87. https://doi.org/10.1136/gutjnl-2020-323297.
193. Liu X, et al. Cordyceps militaris extracts and cordycepin ameliorate type 2 diabetes mellitus by modulating the gut microbiota and metabolites. Front Pharmacol. 2023;14:1134429. https://doi.org/10.3389/fphar.2023.1134429.
194. Kim BH, Yim SV, Hwang SD, Kim YS, Kim JH. A clinical trial on anti-diabetic efficacy of submerged culture medium of Ceriporia lacerata mycelium. BMC Complement Med Ther. 2023;23:83. https://doi.org/10.1186/s12906-023-03895-z.
195. Weese JS, Martin H. Assessment of commercial probiotic bacterial contents and label accuracy. Can Vet J. 2011;52:43–6.
196. Morovic W, Hibberd AA, Zabel B, Barrangou R, Stahl B. Genotyping by PCR and high-throughput sequencing of commercial probiotic products reveals composition biases. Front Microbiol. 2016;7:1747. https://doi.org/10.3389/fmicb.2016.01747.
197. Tian Y, et al. Fecal microbiota transplantation for ulcerative colitis: a prospective clinical study. BMC Gastroenterol. 2019;19:116. https://doi.org/10.1186/s12876-019-1010-4.
198. Khan MT, Nieuwdorp M, Backhed F. Microbial modulation of insulin sensitivity. Cell Metab. 2014;20:753–60. https://doi.org/10.1016/j.cmet.2014.07.006.
199. Allegretti JR, et al. Effects of fecal microbiota transplantation with oral capsules in obese patients. Clin Gastroenterol Hepatol. 2020;18:855–863.e852. https://doi.org/10.1016/j.cgh.2019.07.006.
200. Yu EW, et al. 622—Double-blind randomized placebo-controlled trial of weekly fecal microbiota transplantation (FMT) capsules in obese adults: evaluating microbiota engraftment and improvements in insulin sensitivity. Gastroenterology. 2019;156:S-129–30.
201. Leser TD, Molbak L. Better living through microbial action: the benefits of the mammalian gastrointestinal microbiota on the host. Environ Microbiol. 2009;11:2194–206. https://doi.org/10.1111/j.1462-2920.2009.01941.x.
202. Costello EK, et al. Bacterial community variation in human body habitats across space and time. Science. 2009;326:1694–7. https://doi.org/10.1126/science.1177486.
203. Hillman ET, Lu H, Yao T, Nakatsu CH. Microbial ecology along the gastrointestinal tract. Microbes Environ. 2017;32:300–13. https://doi.org/10.1264/jsme2.ME17017.
204. Leite GGS, et al. Mapping the segmental microbiomes in the human small bowel in comparison with stool: a REIMAGINE study. Dig Dis Sci. 2020;65:2595–604. https://doi.org/10.1007/s10620-020-06173-x.
205. Aydin O, Nieuwdorp M, Gerdes V. The gut microbiome as a target for the treatment of type 2 diabetes. Curr Diab Rep. 2018;18:55. https://doi.org/10.1007/s11892-018-1020-6.
206. Wang Y, et al. The intestinal microbiota regulates body composition through NFIL3 and the circadian clock. Science. 2017;357:912–6. https://doi.org/10.1126/science.aan0677.
207. Kubo M. Diurnal rhythmicity programs of microbiota and transcriptional oscillation of circadian regulator, NFIL3. Front Immunol. 2020;11:552188. https://doi.org/10.3389/fimmu.2020.552188.
208. de Assis LVM, Oster H. The circadian clock and metabolic homeostasis: entangled networks. Cell Mol Life Sci. 2021;78:4563–87. https://doi.org/10.1007/s00018-021-03800-2.
209. Gutierrez Lopez DE, Lashinger LM, Weinstock GM, Bray MS. Circadian rhythms and the gut microbiome synchronize the host's metabolic response to diet. Cell Metab. 2021;33:873–87. https://doi.org/10.1016/j.cmet.2021.03.015.
210. Beccuti G, Pannain S. Sleep and obesity. Curr Opin Clin Nutr Metab Care. 2011;14:402–12. https://doi.org/10.1097/MCO.0b013e3283479109.
211. Allaband C, et al. Intermittent hypoxia and hypercapnia alter diurnal rhythms of luminal gut microbiome and metabolome. mSystems. 2021;6:e0011621. https://doi.org/10.1128/mSystems.00116-21.
212. Thaiss CA, et al. Transkingdom control of microbiota diurnal oscillations promotes metabolic homeostasis. Cell. 2014;159:514–29. https://doi.org/10.1016/j.cell.2014.09.048.
213. Booijink CC, et al. High temporal and inter-individual variation detected in the human ileal microbiota. Environ Microbiol. 2010;12:3213–27. https://doi.org/10.1111/j.1462-2920.2010.02294.x.
214. Stokes K, et al. The circadian clock gene BMAL1 coordinates intestinal regeneration. Cell Mol Gastroenterol Hepatol. 2017;4:95–114. https://doi.org/10.1016/j.jcmgh.2017.03.011.
215. Harris C, Czaja K. Can circadian eating pattern adjustments reduce risk or prevent development of T2D? Nutrients. 2023;15:1762. https://doi.org/10.3390/nu15071762.

The Role of Gut Bacteria in Functional Gastrointestinal Disorders

John A. Damianos and Eric D. Shah

Highlights

- Disorders of gut-brain interactions (DGBIs) were formerly termed functional gastrointestinal disorders (FGIDs). The brain-gut axis comprises interactions between the central and enteric nervous systems and effects on gastrointestinal function (motility, absorption, etc.).
- Gut microbes that affect motility include *Lactobacillus acidophilus*, *Bifidobacterium bifidum*, *Micrococcus luteus, Escherichia coli*, and the methanogen *Methanobrevibacter smithii*. Their effects may be mediated via effects of metabolites (e.g. short chain fatty acids), neurotransmitters (e.g. serotonin), and cell components (e.g. lipopolysaccharide) on the enteric nervous system.
- Gut microbes and their metabolites may also attenuate visceral hypersensitivity. For example, *Lactobacillus rhamnosus* prevents dorsal root ganglia excitability following balloon rectal distension in rodents, and *Bifidobacterium dentium* can produce the neurotransmitter GABA, which ameliorates hypersensitivity.
- Disruption of the epithelial barrier and increased intestinal permeability may contribute to inflammatory bowel disease, and impaired duodenal mucosal integrity has been linked to functional dyspepsia (FD). Decreased *Prevotella* and increased *Bifidobacteria* and *Clostridioides* have been identified in FD patients, and improved following 12 weeks treatment with *Lactobacillus gasseri* OLL2716.
- A low FODMAP (fermentable oligosaccharides, disaccharides, monosaccharides, and polyols) diet can relieve distension and pain in irritable bowel syndrome subjects, but may also reduce beneficial butyrate-producing and prebiotic bacteria such as *Bifidobacteria*, *F. prausnitzii*, and *Akkermansia muciniphila.* In contrast, probiotic bacteria such as *Lactobacillus plantarum* DSM 9843 and *Bifidobacterium infantis* 35,624 may be beneficial.

1 Introduction

Disorders of gut-brain interactions (DGBIs), formerly termed functional gastrointestinal disorders (FGIDs), contribute a significant burden to primary care and gastroenterology practice. Up to one-third of the population is estimated to meet criteria for a DGBI, and DGBIs comprise up to 60% of gastroenterology outpatient appointments [1, 2]. The financial impact of DGBIs to patients and the healthcare system is also substantial; a 2002 report estimated the annual cost of irritable bowel syndrome (IBS) to be $15 million, a figure which has certainly risen since then [3]. IBS itself accounts for 25–50% of gastroenterology referrals [4].

DGBIs are definedby the Rome Foundation as "disorders of gut–brain interaction." They comprise a group of disorders classified by "GI symptoms related to any combination of the following: motility disturbance, visceral hypersensitivity, altered mucosal and immune function, altered gut microbiota, and altered central nervous system (CNS) processing" [5]. These include 33 adult and 20 pediatric disorders defined by symptoms as delineated in the Rome criteria. Crucial to understanding the pathophysiology of DGBIs is an appreciation of the gut-brain axis (see Fig. 1). This axis comprises the intricate interactions between the central nervous and enteric nervous systems and their effect on gastrointestinal function (i.e., motility, absorption, etc.). Each of the proposed pathophysiologic mechanisms underlying DGBIs may exist independently or result from alterations in the gut microbiota.

J. A. Damianos
Department of Gastroenterology and Hepatology, Mayo Clinic, Rochester, MN, USA

E. D. Shah (✉)
Department of Internal Medicine, University of Michigan, Ann Arbor, MI, USA
e-mail: ershah@med.umich.edu

M. Pimentel et al. (eds.), *Clinical Understanding of the Human Gut Microbiome*, https://doi.org/10.1007/978-3-031-46712-7_6

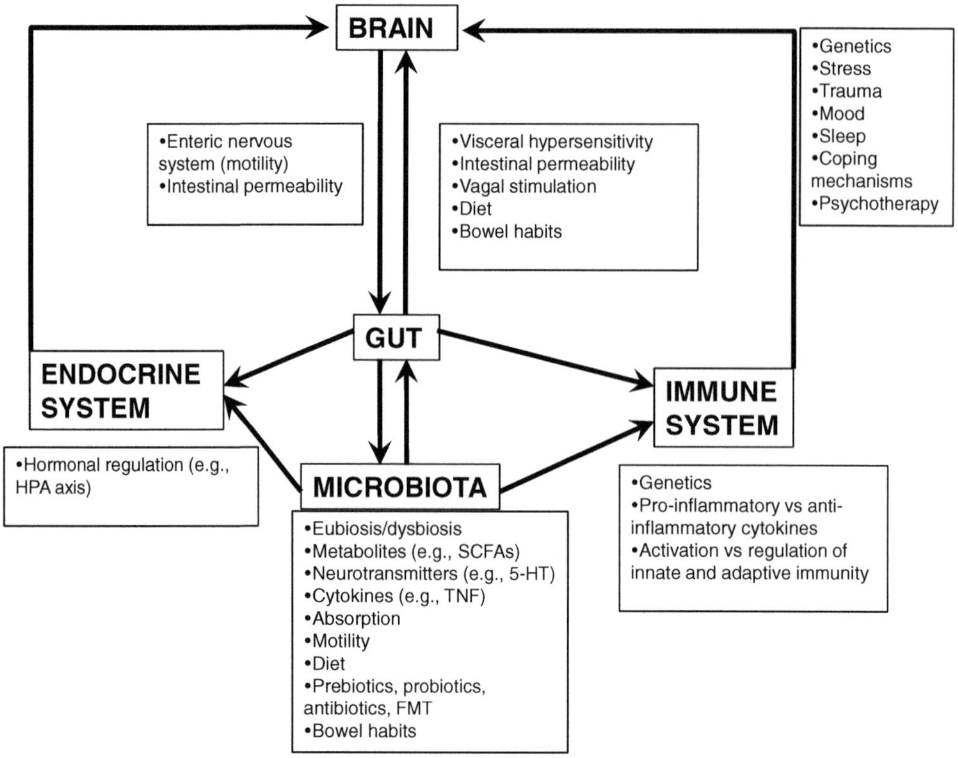

Fig. 1 Components and determinants of the gut-brain axis

1.1 Motility Disturbance

Gastrointestinal motility results from a complex interplay between the enteric nervous system (ENS), the gut wall musculature, and the gut microbiota. Neurons in the myenteric plexus and submucosal plexus express a variety of ion channels, G-protein coupled receptors, toll-like receptors, and other signaling channels that are responsive to ligands present in the lumen. Substrates from the microbiota include microbial metabolites (e.g., short chain fatty acids), neurotransmitters (e.g., serotonin), and microbial cell components which can directly or indirectly affect motility via interactions with the ENS. Studies in germ-free rats have elucidated the multifactorial interaction between specific bacterial taxa and motility [6]. For example, *Lactobacillus acidophilus* and *Bifidobacterium bifidum* have been shown to promote small intestinal transit, whereas *Micrococcus luteus* and *Escherichia coli* slow transit. Bacterial endotoxin lipopolysaccharide appears to delay gastric emptying and induce sphincter of Oddi dysfunction. Mice that have received fecal microbiota from human IBS-D patients exhibit accelerated gastrointestinal transit times [7]. Human studies have confirmed the significance of the microbiota on motility. There is evidence, for example, that an abundance of methane-producing bacteria such as *Methanobrevibacter smithii* can lead to constipation, while hydrogen sulfide-producing bacteria such as *Fusobacterium* and *Desulfovibrio* species cause diarrhea [8, 9]. Additionally, multiple human studies have examined the effects of probiotics on motility. For example, *Bacillus animalis, Bacillus lactis,* and *Lactobacillus casei* have all been found to decrease transit time and increase stooling frequency [10, 11]. Together, these findings support the hypothesis that dysbiosis breeds dysmotility, raising the possibility of treating motility disorders and DGBIs that feature dysmotility by targeting the microbiota.

1.2 Visceral Hypersensitivity

Visceral hypersensitivity represents "a decreased pain threshold following nociceptor activation, or to an exaggerated response to the painful stimulus" [12]. It is a product of both peripheral and central sensing and interpretation of visceral nociceptor stimulation. Visceral hypersensitivity can result from dysfunction at any step in the nociceptive pathway including: "(1) sensitization of primary sensory afferents innervating the viscera, (2) hyperexcitability of spinal ascending neurons (central sensitization) receiving synaptic input from the viscera, (3) dysregulation of descending pathways that modulate spinal nociceptive transmission, and (4) changes in the central perception of painful stimuli" [12]. Individuals with visceral hypersensitivity exhibit exaggerated responses to painful stimuli, such as rectal balloon distension in studies or to physiologic gas production by the microbiota in response to a meal, and these abnormal responses can be mapped to increased activation of pain areas in the brain using fMRI [13]. The microbiota directly influence visceral hypersensitivity through production of metabolites that interact with the visceral sensory neurons. Dysbiosis in the microbiome and mycobiome have been documented in patients with visceral hypersensitivity [14, 15]. Studies in both rodents and humans have identified particular taxa that attenuate visceral hypersensitivity. For example, probiotic *Lactobacillus rhamnosus* prevents dorsal root ganglia excitability in response to balloon rectal distension in rodents, and *Lactobacilli* have been shown to induce visceral analgesia via upregulation of cannabinoid receptors [12]. *Bifidobacterium dentium* has demonstrated ability to produce the neurotransmitter GABA which acts via central and peripheral mechanisms to ameliorate visceral hypersensitivity [16]. Additionally, the abundance of *Bifidobacteria* in the human microbiome is inversely associated with abdominal pain [17]. Since the microbiota contribute directly to the development of visceral hypersensitivity, targeting the microbiome presents an opportunity to treat DGBIs due to abnormal pain responses.

1.3 Altered CNS Processing

Visceral hypersensitivity is closely related to altered central pain sensitization. This aspect of the bidirectional gut-brain axis may partly explain the relationship between DGBIs and psychiatric disorders. There is particular interest in early-life trauma and its effect on pain processing. Indeed, there is a well-established connection between PTSD and IBS [18], and adverse childhood experiences correlate with IBS severity [19]. Psychopharmacologic interventions for IBS likely target this aspect of pathogenesis (tricyclic antidepressants NNT 4.5, selective serotonin reuptake inhibitors NNT 5). Similarly, psychological therapies such as cognitive behavioral therapy, relaxation therapy, and hypnosis have been found to be effective (NNT 4–6) [20]. The efficacy of psychological therapies suggests top-down modulation of the enteric nervous system is possible. While the role of gut microbiota with respect to such top-down processing is still unclear, there is evidence that composition of the microbiome may predict response to cognitive behavioral therapy [21]. Bottom-up modulation is suggested by the ability of probiotics to improve psychiatric symptoms (so-called *psychobiotics*), such as *Bifidobacterium longum* NCC3001 which has been demonstrated to improve depression symptoms in IBS and alter brain activity on fMRI [22], as well as the ability of fecal microbiota transplantation (FMT) to improve psychiatric symptoms in patients with IBS, functional diarrhea, and functional constipation [23].

1.4 Altered Mucosal and Immune Function

The intestinal epithelial barrier comprises the intricate interaction between gut mucosal epithelial cells, tight junctions, and the luminal microbiota that separates the body from the outside world. Integrity of this barrier is crucial for digestive and systemic health, as dysfunction of the epithelial barrier has been implicated in a host of gastrointestinal and non-gastrointestinal illnesses including inflammatory bowel disease, irritable bowel syndrome, enteric infections, obesity, celiac disease, and food allergy [24], as well as even depression [25] and Parkinson's disease [26]. Disruption of the epithelial barrier allows for increased intestinal permeability, permitting endogenous and pathogenic bacteria and fungi to access the lamina propria and bloodstream. This phenomenon (often referred to as "leaky gut" physiology) has been demonstrated to incite inflammation in the gut and systemically, and may be a key contributor in the pathogenesis of inflammatory bowel diseases such as Crohn's disease and ulcerative colitis [24]. Metabolic endotoxemia refers to the phenomenon whereby lipopolysaccaride (a component of the cell wall in Gram negative bacteria) translocates via the perturbed epithelial barrier into systemic circulation and incites a host inflammatory response through several pathways including nuclear factor-κB, CD14, and toll-like receptors [27]. Post-infectious irritable bowel syndrome (PI-IBS) patients have been shown to have increased intestinal permeability [28] and there is evidence that antibodies to vinculin, a cytoskeletal protein component of the epithelial barrier, can be used as a biomarker of the disease [29]. Impaired duodenal mucosal integrity may also contribute to the pathogenesis of functional dyspepsia [30].

The gut microbiota play a central role in the entrainment of the immune system from birth and continue to influence its functioning and regulation throughout life. Indeed, the immune system is most concentrated in the gut [31]. The architecture of the gut epithelium includes components of the innate and adaptive immune systems that interface with the systemic immune response. Gut epithelial cells themselves express a number of receptors that trigger gene transcription, interleukin secretion, and other immune signaling molecules. Composition of the microbiome also modulates the number and function of immune cells such as tissue-resident macrophages (microglia) and basophils. In addition to directly regulating specific components of the immune system, the microbiota also utilize substrate from the diet to serve as additional modulators of immune function. Dietary fiber, for example, is digested by the microbiota into various metabolites including short-chain fatty acids (SCFAs), such as butyrate, which acts as a powerful anti-inflammatory agent. Deficiency of butyrate-producing bacteria has been found in patients with IBD [32] and IBS [33].

The remainder of this chapter will highlight specific DGBIs with special attention to approaches to modulate the microbiome, including diet, prebiotics, probiotics, antibiotics, and FMT.

2 Functional Dyspepsia

The Rome IV criteria for functional dyspepsia are as follows:

1. One or more of the following:
   (a) Bothersome postprandial fullness
   (b) Bothersome early satiation
   (c) Bothersome epigastric pain
   (d) Bothersome epigastric burning
   AND
2. No evidence of structural disease (including at upper endoscopy) that is likely to explain the symptoms.

Patients are further divided into having either *postprandial distress syndrome* (PDS) or *epigastric pain syndrome* (EPS). The criteria for each are as follows:

Postprandial distress syndrome.

Must include one or both of the following at least 3 days per week:

1. Bothersome postprandial fullness (i.e., severe enough to impact on usual activities)
2. Bothersome early satiation (i.e., severe enough to prevent finishing a regular-size meal). No evidence of organic, systemic, or metabolic disease that is likely to explain the symptoms on routine investigations (including at upper endoscopy).

Epigastric pain syndrome.

Must include at least 1 of the following symptoms at least 1 day a week:

1. Bothersome epigastric pain (i.e., severe enough to impact on usual activities) AND/OR
2. Bothersome epigastric burning (i.e., severe enough to impact on usual activities). No evidence of organic, systemic, or metabolic disease that is likely to explain the symptoms on routine investigations (including at upper endoscopy) [34].

The pathophysiology of FD is multifactorial and may include abnormal gastric emptying, impaired gastric accommodation, gastric and duodenal hypersensitivity, *Helicobacter pylori* infection, low-grade duodenal inflammation (evidenced by duodenal eosinophilia), abnormal duodenal mucosal permeability, interaction with food antigens, enteric infection, and psychosocial factors [34]. Dysbiosis may be a cause or an effect of these proposed mechanisms. Indeed, *H. pylori* eradication can result in symptom resolution in a subset of patients [35, 36], and it has been proposed that resultant dysbiosis may underlie symptom development [37]. Additionally, there is evidence that the duodenal microbiome composition differs from that of healthy controls. A small study characterized the bacterial composition of the duodenum and found that compared to controls, FD patients had a greater abundance of *Streptococci* and reduced anaerobic *Prevotella*, *Veillonella*, and *Actinomyces* [38]. There was a negative correlation between mucosal bacterial load and quality of life, and decreased bacterial load was associated with reduced diversity. Another study found FD patients to have reduced *Prevotella* and increased *Bifidobacteria* and *Clostridioides* as well as decreased overall diversity. Composition of the duodenal microbiome in FD is also associated with gastric emptying and symptoms [39]. Treatment of FD patients for 12 weeks with yogurt containing probiotic *Lactobacillus gasseri* OLL2716 significantly increased *Prevotella* and reduced *Bifidobacteria* and *Clostridioides* such that the bacterial composition closely resembled that of healthy controls [40]. Another study confirmed these results by treating 106 individuals with *H. pylori* negative FD for 12 weeks with *L. gasseri* OLL2716 and found a significant improvement of symptoms, with 35.3% reporting elimination of symptoms compared to 17.3% of the placebo group. Spore-forming *Bacillus* probiotics have also been shown to improve FD symptoms, with concomitant improvement in the microbial and immune parameters [41]. One randomized control trial of 86 Chinese patients found a 2-week course of the non-absorbable antibiotic rifaximin to be superior to placebo in resolving global FD symptoms, especially among women [42]. Rifaximin is effective irrespective of comorbid IBS, and the effect is strongest in those with small intestinal bacterial overgrowth [43]. Diet is also a known trigger of FD symptoms. One review on the role of diet in the generation of FD symptoms found that fatty foods, wheat (especially gluten), dairy, processed foods, alcohol, caffeine, spicy food, and fruit may exacerbate FD [44]. While specific mechanisms were proposed for each group (e.g., caffeine affecting motility, gluten inducing cytokine release), surprisingly, there was no mention of the microbiota as a potential mediator of symptom generation. While the role of FODMAP ingestion is unclear in the pathogenesis of FD, several of the problem foods are themselves FODMAPs (e.g., wheat, fruits). It is possible that restricting FODMAPs or at least specific food triggers alleviates symptoms in part via modulation of microbiome composition [45].

In summary, research on the role of the microbiome in FD is mounting, with growing interest in the duodenal microenviroment, specifically surrounding inflammation and dysbiosis. The duodenal microbiome of FD patients appears to differ from that of healthy controls, and, in addition to eradication of *H. pylori* and elimination of trigger foods, it is possible that treatment with certain probiotics or rifaximin can resolve the dysbiosis and improve or even eliminate symptoms. More research is warranted to better characterize the dysbiosis of FD for better targeting of treatment.

3 Irritable Bowel Syndrome

Irritable bowel syndrome (IBS) is a heterogenous disorder defined by the Rome IV as follows:

Recurrent abdominal pain on average at least 1 day/week in the last 3 months, associated with two or more of the following criteria:

1. Related to defecation
2. Associated with a change in the frequency of stool
3. Associated with a change in the form (appearance) of stool [46].

Patients are further classified as IBS with predominant constipation (IBS-C), IBS with predominant diarrhea (IBS-D), IBS with mixed bowel habits (IBS-M) or IBS, unsubtyped (IBS-U). IBS is often a term invoked broadly to classify chronic abdominal pain with diarrhea and/or constipation not due to disorders such as IBD or celiac disease, which have clearly defined serum markers (e.g., anti-TTG antibodies) or histopathologic features (e.g., crypt abscesses, granulomas). Furthermore, IBS is a syndrome, characterized by symptoms (i.e., pain and disordered defecation) and not by etiology. As such, various other conditions can cause IBS-like symptoms, such as lactose intolerance, disaccharidase deficiencies, small intestine bacterial overgrowth, bile acid malabsorption, non-celiac gluten sensitivity, fructose intolerance, histamine intolerance, and small-fiber polyneuropathy, among others [47]. It is crucial to investigate the underlying pathophysiology of IBS-like symptoms to appropriately diagnose the patient and accurately target treatment; for example, lactose intolerance is easily cured by restricting ingestion of lactose-containing foods; bile acid diarrhea is effective treated with bile acid sequestrants. The heterogeneity within the IBS literature including investigation of pathophysiologic markers and response to treatment is at least partly due to the misattribution of IBS-like symptoms to non-IBS conditions. IBS also exists along a spectrum of sensory and motor symptoms, overlapping with disorders such as centrally mediated abdominal pain syndrome (predominantly pain), functional diarrhea (predominantly diarrhea), and functional constipation (predominantly constipation), as well as often overlapping with other DGBIs such as FD [48, 49].

The composition of the microbiome of IBS patients appears to differ from that of healthy controls. One systematic review and meta-analysis of 13 studies analyzing the composition of the IBS microbiome found decreased *Lactobacillus*, *Bifidobacterium*, and *F. prausnitzii*, especially in IBS-D [50]. A more recent systematic review of 24 articles identified increased Enterobacteriaceae, Lactobacillaceae, and *Bacteroides*, and decreased uncultured Clostridiales, *Faecalibacterium*, and *Bifidobacterium* in IBS patients compared to controls. Additionally, many of these studies found decreased microbial diversity in IBS [51]. Machine learning algorithms can reliably distinguish IBS from healthy individuals via the microbiota, and can also discriminate between patients with severe versus mild/moderate symptoms. In one analysis, symptom severity was found to be negatively associated with microbial richness, exhaled methane (CH4), methanogen species, and enterotypes enriched with Clostridiales or *Prevotella* species [52]. IBS is also able to be reliably distinguished from both IBD and healthy individuals by the microbiota [53]. One study was able to identify a bacterial signature that correlated with structural analysis of brain regions including sensory- and salience-related areas, as well as with early life trauma [54]. Another fascinating study revealed patients with IBS-D to have a similar microbiome composition to patients with depression, marked by decreased diversity and increased Bacterioides and *Prevotella* [55]. The specific subtypes of IBS may be driven by the particular dysbiosis. Overabundance of stool methanogens such as *Methanobrevibacter smithii* correlates with constipation in IBS patients, while an overabundance of hydrogen sulfide-producing bacteria such as *Fusobacterium* and *Desulfovibrio* species correlates with diarrhea [9]. There is also growing interest in the role of the *mycobiome*, referring to fungal species, in the pathogenesis of IBS [56]. In one study, the diversity and stability of fungal species significantly differed between IBS patients and healthy controls, and between patients with visceral hypersensitivity and those without [14]. Abnormalities of the mycobiome may also directly contribute to abdominal pain in IBS [56]. Clearly, dysbiosis is a feature of IBS, and its extent may predict symptom severity and/or response to treatment. Furthermore, specific microbiome signatures may be able to distinguish subtypes of IBS [57]. Employing either or both top-down and bottom-up processes in FGIDs can modulate the gut-brain axis, possibly in part mediated via the microbiota.

4 Post-Infectious IBS

The strongest risk factor for IBS is enteric infection. It is estimated that 10% of individuals who acquire an acute gastroenteritis will go on to develop what is termed post-infection IBS (PI-IBS). The probability of acquiring PI-IBS varies by pathogen, but may reach 45% for certain pathogens, with protozoa conferring the highest risk [28]. Studies on PI-IBS have implicated increased intestinal permeability, altered serotonin metabolism, and increased density of enterochromaffin cells and T-lymphocytes in the lamina propria. Other risk factors include female sex, younger age, severity of infection, and psychological factors such as anxiety, depression, somatization, and neuroticism. Dysbiosis is prominent in PI-IBS, and its microbial signature is distinct from that of conventional IBS [28]. There is some evidence that PI-IBS arises via an autoimmune etiology. Antibodies to vinculin and to cytolethal distending toxin B (CdtB) are associated with IBS-D and can reliably discriminate IBS-D from IBD [58]. CtdB is a protein produced by all enteric pathogens. Through the process of molecular mimicry, following an enteric infection antibodies to CdtB cross-react with vinculin, an actin-binding protein found in the gut pacemaker apparatus including interstitial cells of Cajal (ICC cells) and myenteric ganglia [59]. The American Gastroenterological Association (AGA) notes that while anti-CdtB and anti-vinculin antibodies have good specificity and high positive predictive value, their sensitivity is less than 50%, meaning that a negative study does not rule out IBS-D. However, a recent study suggests that with a pre-test probability for IBS-D of 57%, testing positive for both antibodies results in a post-test probability of >98% [58]. Currently, the AGA advises more research on the clinical utility of these tests before recommending them for diagnostic purposes [60]. The efficacy of specific treatments for PI-IBS is still being evaluated. It is unclear what proportion of IBS represents PI-IBS, but the antibiotic rifaximin may be beneficial as it is in nonconstipated IBS in general [61]. One study of patients with PI-IBS undergoing FMT for recurrent CDI found persistence of IBS symptoms 6 months after FMT in almost two-thirds of patients [62]. A small randomized control trial assessed glutamine supplementation for PI-IBS and found a 14-fold reduction in IBS symptom severity, decreased bowel movement frequency, increased Bristol scale stool consistency, and decreased intestinal permeability [63]. Due to findings of low-grade inflammation in PI-IBS, anti-inflammatory therapies have been tested but without success. A study of prednisolone in PI-IBS revealed no change in number of enterochromaffin cells on rectal biopsy or symptomatic improvement [64]. Similarly, a trial of mesalamine showed no benefit [65]. As a result, the current management of PI-IBS is extrapolated from that of conventional IBS. Better characterization of the pathophysiology and dysbiosis of PI-IBS should lead to more targeted treatment.

5 SIBO

Small intestine bacterial overgrowth (SIBO) refers to an overabundance of coliform bacteria in the small bowel. SIBO was initially recognized as a cause of malabsorption in patients who had undergone Billroth procedures or those with small bowel dysmotility (e.g., systemic sclerosis). More recently, SIBO has been implicated in a wide range of diseases to varying degrees of pathophysiologic significance, including IBD, immune deficiencies, chronic pancreatitis, liver disease, obesity, type II diabetes, gallstones, and even deep vein thrombosis [66]. There is tremendous interest in the role of SIBO in IBS. Before analyzing the relationship between these two entities, it is imperative to first understand the diagnostic challenges and controversies surrounding SIBO.

Historically, the gold standard for diagnosis of SIBO has been $\geq 10^5$ colony forming units (CFUs) per milliliter via jejunal aspirate. This method of detecting SIBO has several limitations including its invasiveness and cost. Additionally, there is concern for false negative values given that most endoscopes cannot reach the distal small bowel, where overgrowth is more likely to be located. Furthermore, aspirates are likely contaminated with oropharyngeal flora, further confounding interpretation of the aspirate and potentially resulting in false positives. In the context of these challenges, breath testing has emerged as an alternative diagnostic modality. The theory behind breath testing stems from the fact that while intestinal gas is composed of gases from swallowed air and food, gut metabolism, diffusion of bloodstream gases, and microbial metabolism, the gases hydrogen (H_2) and methane (CH_4) are produced exclusively via gut microbial fermentation [67]. Specific bacteria break down dietary carbohydrates and release these gases which diffuse into the abdominal venous system and are exhaled by the lungs. Breath testing utilizes a carbohydrate substrate (glucose or lactulose for SIBO, or others for specific malabsorption syndromes) for patients to ingest, after which breath H_2 and CH_4 (and sometimes hydrogen sulfide (H_2S)) are measured via gas chromatography [68]. According to the 2017 North American consensus, indications for breath testing include diagnosis of SIBO, evaluation of excessive methane excretion in patients with constipation, diagnosis of carbohydrate maldigestion syndromes (e.g., lactose intolerance, fructose intolerance), and assessment of conditions with bloating [67]. The consensus recommends the following for dosing and diagnosis:

- Dosing: Lactulose (10 g), glucose (75 g), fructose (25 g), and lactose (25 g).
- A $\geq$ 20 p.p.m. rise in hydrogen by 90 min following glucose or lactulose breath testing is diagnostic for SIBO.
- Methane levels of $\geq$10 p.p.m. are considered methane-positive (termed intestinal methanogen overgrowth, IMO).
- A $\geq$ 20 p.p.m. rise in hydrogen above baseline is considered diagnostic for carbohydrate maldigestion [67].

There is also evidence that a single fasting methane level $\geq$ 10 ppm is diagnostic of IMO [69]. IMO has distinct clinical features compared with hydrogen SIBO, including a higher tendency for constipation, older age, less likely to have B12 deficiency, and unlikely to be associated with prior abdominal surgery [70].

It is worth noting that there is also growing interest in a third type of SIBO caused by overgrowth of H_2S-producing bacteria, as this has been shown to correlate with diarrhea [9, 71].

It has historically been difficult to assess the sensitivity and specificity of breath testing for a number of reasons including the limitations of employing jejunal aspirate as the gold standard, the comparison to an accurate control group (given that SIBO is present in many GI and non-GI diseases), and inconsistencies in breath testing methodology and interpretation. For example, one systematic review found the sensitivity of glucose breath testing across 13 studies varied from 20% to 93% and specificity from 30% to 86%; similarly, among 11 studies, the lactulose breath test sensitivity ranged 31–68% and specificity 44–100% [72]. A more recent study comparing breath testing to jejunal aspirate found glucose breath testing to have a sensitivity of 27% and a specificity of 100%, whereas lactulose breath testing had a sensitivity of 0% and a specificity of 98% [73]. Indeed, a study comparing lactulose and hydrogen breath testing found that 88.9% of patients had a falsely positive lactulose test [74]. Another study comparing IBS patients and healthy controls found that lactulose breath testing was unable to distinguish between the two groups [75]. The low sensitivity of the lactulose breath test has led some to reject it as part of SIBO diagnosis, favoring instead glucose breath testing in spite of its relatively low sensitivity [76]. Additionally, studies have employed varied substrate doses and diagnostic cut-offs to identify SIBO, further confounding interpretation. The overall pooled sensitivity and specificity for glucose breath testing is 54% and 83%, and for lactulose is 42% and 71% [68]. Another consideration is the natural variation of orocecal transit times; fast transit times can lead to false positive readings, especially in the context of lactulose which hastens intestinal transit [77]. Indeed, there is scintigraphic evidence that almost 50% of positive breath tests are false positives from fast transit [78]. Conversely, the 90-min cutoff of breath testing does not account for those with prolonged orocecal transit time such as those with systemic sclerosis, leading to false negatives [68]. A summary of sensitivity and specificity of breath testing based on substrate is found in Table 1. With the pitfalls of jejunal aspirate as gold standard and the controversy surrounding breath testing, there is great interest in alternative methods of detecting SIBO, and there is ongoing research into technologies such as ingested capsules, 16S RNA sequencing, and metabolomics. This research will prove particularly important since incorrect identification of SIBO may lead to inappropriate exposure to antibiotics, potentially adversely disrupting the microbiome and contributing to antibiotic resistance. A final consideration in the SIBO controversy is its significance. While the presence of bacterial overgrowth is associated with many GI and non-GI diseases, it may not be the cause of symptoms. Furthermore, up to 13% of healthy individuals may test positive for SIBO [80]. For example, one study found that the presence of SIBO does not underlie chronic functional GI symptoms, but that small intestinal dysbiosis (i.e., altered microbial composition) does. The study additionally found that individuals following a high-fiber diet may test positive for SIBO attributable to their diet [81]. These issues have led some to propose SIBO to mean “a condition of gut bacteria excess that clinically responds to antibiotic therapy” [59, 72]. Indeed, there is evidence that symptom improvement after antibiotic therapy correlates with reduction of hydrogen breath testing, suggesting that SIBO itself is veritably the cause of the symptoms [82]. Similarly, FD patients with SIBO are the most likely to benefit from rifaximin [43]. A more recent study confirmed that lactulose breath testing predicts response to rifaximin; the odds of symptom improvement were 4.3 times higher in patients with a positive baseline breath test [83]. Overall, it appears that the absolute number of microbes is less important than the type of microbes. Duodenal microbiome sequencing of patients with SIBO has identified certain “disruptor” taxa (such as *Enterobacteriaceae*) which dominate the microbiota and correlate with gastrointestinal symptoms [84].

Table 1 Sensitivity and specificity of breath testing to identify IBS according to substrate used. (Adapted from Shah et al. [79])

Carbohydrate (number of studies)	Sensitivity (%)	Specificity (%)
Fructose (1)	70	35
Glucose (3)	15	97
Lactulose (5)	72	66
Sucrose (1)	32	82
Xylose (2)	17	100

SIBO has long been hypothesized to be associated with IBS. Due to the aforementioned challenges inherent with the diagnosis of SIBO, it has been difficult to characterize the true prevalence of SIBO in IBS; indeed, studies have reported SIBO prevalence anywhere from 4% to 84%. The literature has been conflicting, with some studies showing no difference between IBS patients and healthy control with respect to the presence of SIBO [75, 85]. A 2018 systematic review and meta-analysis of 50 studies, however, found that one-third of IBS patients tested positive for SIBO, and the odds of SIBO in IBS compared to healthy controls were approximately fivefold, with a pooled prevalence of 38%. Notably, female sex and old age were positively associated with SIBO among IBS patients [86]. A more recent systematic review and meta-analysis of 25 studies found an odds ratio of 3.7 for SIBO in IBS, which increased to 4.9 in studies using only healthy controls. Breath testing identified SIBO in 35.5% of IBS patients compared to 29.7% in controls, whereas culture identified SIBO in 13.9% of IBS patients compared to 5% of controls [87]. The overall quality of evidence was found to be low, with high heterogeneity in the data. Clearly though, SIBO is associated with IBS in at least a subset of patients. One significant outstanding question, however, is what the significance of SIBO in IBS actually is. As previously mentioned, SIBO has been found in many gastrointestinal disorders from IBD to cancer, and it is not known if SIBO is a significant contributor to pathogenesis, a consequence of the underlying disease process, an innocent bystander phenomenon, or anywhere in between. In IBS, this question is particularly challenging to answer given the heterogeneity of IBS pathogenesis and phenotypes. It is easily conceivable that underlying dysmotility causing slow transit could lead to bacterial overgrowth. Another theory advanced by Pimentel et al. link post-infectious IBS to SIBO via an autoimmune mechanism [88]. Specifically, the antibodies anti-CdtB and anti-vinculin commonly found in PI-IBS have been demonstrated to target, via molecular mimicry, the neuromuscular apparatus of gut motility including interstitial cells of Cajal and myenteric ganglia cells. This enteric neuropathy may lead to intestinal stasis, setting the stage for bacterial overgrowth and the development of IBS symptomatology (see Fig. 2). This model does not capture all aspects of the pathophysiology of PI-IBS however [28].

One major consideration that lends credence to the IBS-SIBO hypothesis is the success of the antibiotic rifaximin in treating nonconstipated IBS. The three TARGET trials demonstrated a modest improvement of IBS symptoms with a 14 day course of thrice daily rifaximin (550 g) that is sustained at 4 and 10 weeks. The number needed to treat (NNT) was 11. TARGET 3 found that 59.2% of responders relapsed

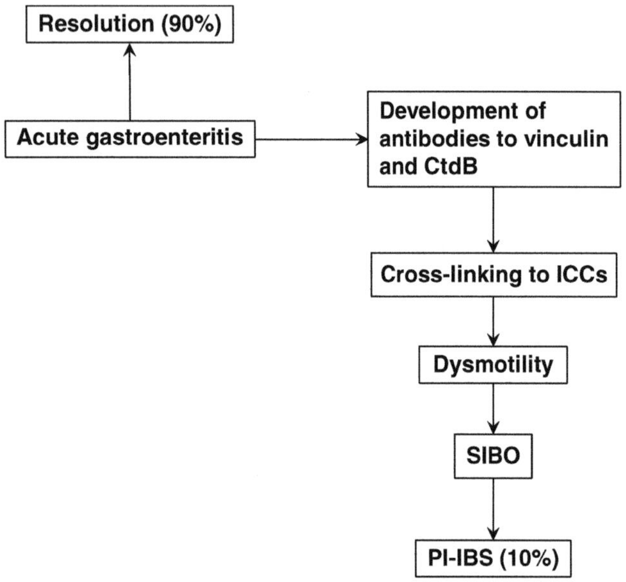

Fig. 2 Pathophysiologic sequence linking acute gastroenteritis to post-infectious irritable bowel syndrome

at 18 weeks, and 38.1% of those responded to re-treatment. Unfortunately, participants in the TARGET trials were never tested for SIBO, so the possibility still exists that the benefit of rifaximin is independent of SIBO. However, one study did find that patients with a positive baseline lactulose breath test have a 4.3 times increased odds in responding to rifaximin, suggesting that rifaximin does indeed modulate the microbiota. While the effects of rifaximin have not completely been elucidated, evidence points to a reduction in gut microbial richness and an increase in intestinal transit time as well as anti-inflammatory actions and strengthening the intestinal epithelial barrier [89, 90]. However, the NNT of 11 for rifaximin is higher than that of antidepressants, antispasmodics, and placebo without deception, which range from 2.5 to 5 [77]. The favorable effects of these interventions point to other hypotheses of IBS pathogenesis (e.g., antidepressants for a psychosocial hypothesis, antispasmodics for visceral hypersensitivity). Taken together, it is likely that SIBO is implicated in a subset of IBS patients, possibly those with PI-IBS; furthermore, this subset of patients may receive the highest benefit from treatment with rifaximin. One study supported this hypothesis by showing that norfloxacin improved IBS symptoms in patients who tested positive for SIBO compared to IBS patients without SIBO [8].

5.1 Modulation of the Gut Microbiome in IBS

Given that the microbiome is a key player in the pathogenesis of IBS, modulation of the microbiota may be able to be harnessed in treatment. One of the quickest and most reliable ways to alter the microbiome is through diet [91]. Most IBS patients cite diet as relating to their symptoms [92], so there is great interest in dietary therapy. Increasing soluble fiber, for example, has been shown to improve symptoms and abdominal pain [93]. It is widely accepted that intake of fermentable oligosaccharides, disaccharides, monosaccharides, and polyols (FODMAPs) exacerbate symptoms of IBS. FODMAPs are poorly absorbed in the small intestine and exert an osmotic effect, increasing luminal water content and increasing intestinal transit time. Furthermore, fermentation by the microbiota leads to generation of H_2 and CH_4, increasing luminal tension causing distension and pain in those with visceral hypersensitivity. Multiple randomized trials have confirmed that reducing dietary FODMAPs (the low-FODMAP diet) improves IBS symptoms [94], with some studies reporting up to 70% of patients benefiting [95]. Because many FODMAPs are prebiotics, restriction of FODMAPs leads to changes in the microbiota. Specifically, the diet has been found to reduce total bacterial abundance, butyrate-producing bacteria, and prebiotic bacteria, specifically *Bifidobacteria*, *Faecalibacterium prausnitzii*,

Akkermansia muciniphila, *Ruminococcus gnavus*, and *Clostridioides* [96]. This reduction of beneficial bacteria is potentially concerning, especially given that reduced bacterial diversity is associated with IBS symptoms and that *Bifidobacteria* abundance is known to be inversely correlated with abdominal pain [17]. One randomized clinical trial demonstrated the efficacy of probiotic supplementation to restore *Bifidobacteria* levels in patients adhering to the low-FODMAP diet [2]. With the theoretically concerning changes in the microbiota induced by the low-FODMAP diet and its highly restrictive nature, it is recommended that implementation of the diet be performed in close conjunction with a registered dietician to follow a sequence of FODMAP restriction, FODMAP reintroduction, and FODMAP personalization [97].

Prebiotics refer to dietary substrates that preferentially support the metabolism and proliferation of beneficial gut bacteria. Dietary fiber, for example, increases numbers of beneficial *Lactobacilli* and *Bifidobacteria* and may decrease numbers of pathogenic taxa [98]. One systematic review and meta-analysis identified three studies of prebiotic supplementation in IBS but consistently found no benefit compared to placebo [99]. Dietary prebiotics may be different, however, as FODMAPS tend to worsen symptoms and soluble fiber tends to improve symptoms [100].

Probiotics are "live microorganisms that confer a health benefit when consumed in adequate amounts" [101]. Similar to the endogenous microbiota, probiotics exert various mechanisms of action including enhancement of the gut epithelial barrier, enriched digestion, production of bioactive metabolites, interaction with the innate and adaptive immune system, and protection against enteric pathogens. There has been tremendous interest in probiotics in IBS. A 2018 systematic review and meta-analysis found an overall benefit for probiotics as a class and benefit for specific strains and combinations of strains in IBS with respect to global IBS symptoms and abdominal pain. Specifically, *Lactobacillus plantarum* DSM 9843, *E. coli* DSM1752 and *Streptococcus faecium* were found to be beneficial. *Bifidobacteria* may also be beneficial, particularly *Bifidobacterium infantis* 35,624. The NNT of combination probiotics was found to be 7 [99]. There appears to be greater benefit in combination probiotics compared to both placebo and single-strain [102]. Probiotic research in IBS has been confounded by marked methodological heterogeneity among studies, species and specific strains studied, and IBS subtype. As such, while the data appear to support probiotic supplementation and particularly combination probiotics, more research is needed to clearly delineate which specific strains and combinations of strains are beneficial for each IBS subtype. Relatedly, several challenges exist in the current probiotic landscape including lack of regulation of commercially available formulations, questions regarding the ability of certain strains to colonize the intestinal epithelium, uncertainty surrounding duration of supplementation and durability of effect, and lack of large randomized control trials. Furthermore, the future of probiotic research should focus on clarifying underlying pathophysiology of disease to better target probiotic treatment. Akin to antibiotic selection based on pathogen and susceptibility, this disease-specific and strain-specific approach would allow for more accurate selection of the appropriate probiotic for the appropriate condition.

Synbiotics refer to combination prebiotics and probiotics, usually taken as a supplement. There is a paucity of trials evaluating synbiotics in IBS, but those undertaken thus far have not demonstrated benefit [99].

Antibiotics have also been found to be beneficial in IBS (which is paradoxical, since antibiotic use is also a risk factor for the development of IBS [90]). In addition to the prior discussion of the TARGET trials of rifaximin (NNT 11), further studies have confirmed a benefit of rifaximin in nonconstipated IBS with NNT 9 [99]. Neomycin (NNT 5) and norfloxacin (NNT 3) have also been studied, though not as extensively as rifaximin [99]. As discussed earlier, the benefit of rifaximin may be greatest in those with positive lactulose breath testing for SIBO [83], and those with confirmed eradication of SIBO are more likely to report improvement of symptoms [90]. There is of course concern in utilizing antibiotics for IBS, most specifically regarding antibacterial resistance and further disturbance of endogenous microbiota.

Finally, there has been growing interest in FMT in IBS. FMT will be discussed at length in Chap. "Fecal Transplant: The Benefits and Harms of Fecal Microbiota Transplantation". Initial encouraging case reports and prospective studies have prompted larger randomized control trials. Multiple systematic reviews and meta-analyses of randomized control trials for IBS each found no benefit compared to placebo [103–105]. However, significant methodological heterogeneity was present among the trials, potentially confounding results. For example, method of administration (oral capsule, nasojejunal tube, colonoscopy) may be important, as the reviews consistently find more significant benefit in studies endoscopically delivered FMT [106]. Additionally, the selection of stool appears to be crucial. Donor FMT is more effective than autologous FMT. Donor selection is another contributor to heterogeneity and confounding among studies. Indeed, most studies published have not examined the microbial composition of donor stool, which puts quality into question. The inclusion of all subtypes may also be of significance. Duration of treatment and usage of fresh versus frozen stool also differed among trials. Subgroup analysis found that single-dose FMT using colonoscopy and nasojejunal tubes in comparison with autologous FMT for placebo treatment with an NNT of 5; number needed to harm in multidose capsule FMT was 3

[105]. A more recent randomized control trial found an 89.1% response rate compared to 23.6% placebo [107]. Subsequent research has begun to address these confounders to better understand the mechanism and potential benefit or harm of FMT in IBS. One study found that FMT delivered endoscopically into the jejeunum improved symptoms and quality of life and also corrected abnormalities in the microbiota and short-chain fatty acids in patients with IBS-D [108]. Another study found that among 10 IBS patients who did not initially respond to FMT delivered endoscopically into the jejeunum, 7 of them did respond to repeat high-dose FMT [109]. A further unresolved question centers on the origin and quality of donor stool. For example, there is debate over what constitutes a "healthy" donor. Some have proposed the idea of a "superdonor," characterized by a lifetime of health combined with a healthy lifestyle. In one study, a "super donor" (a 36-year-old male born vaginally and breast fed, with a normal BMI, who exercises regularly, was not taking any medications, and had only taken antibiotics three times in his life) stool sample yielded significant response and symptom remission in IBS patients [109]. Others have considered combining multiple donors' stool to increase the microbiological diversity of the sample. Clearly, many unanswered questions remain with respect to FMT for IBS and more research is warranted. While FMT may eventually prove beneficial for IBS, it is not currently recommended.

6 Functional Diarrhea, Functional Constipation, and Centrally Mediated Abdominal Pain Syndrome

The Rome criteria for DGBIs exist primarily for research purposes such as designing clinical trials. In clinical practice, however, the actual presentations of DGBIs may vary considerably. Indeed, the Rome Foundation acknowledges that DGBIs exist along a spectrum of symptomatology [5]. For example, a recent analysis found that IBS and functional diarrhea overlap significantly in terms of clinical and psychological features, suggesting that they exist along a continuum [110]. As such, it is helpful to conceptualize patient symptoms along a symptom axis consisting of sensory and motor symptoms. Some patients may have predominantly sensory symptoms, consistent with a diagnosis of centrally mediated abdominal pain syndrome (CAPS), predominantly motor symptoms, consistent with the diagnoses of either functional diarrhea or functional constipation, or both sensory and motor symptoms, consistent with irritable bowel syndrome. Determining a patient's most distressing symptom or symptoms (sensory, motor, or both) can help clarify diagnosis and lead to better symptom-based management. Like in IBS, there is growing interest in characterizing microbiome abnormalities in CAPS, functional diarrhea, and functional constipation. For example, the inverse association between *Bifidobacteria* levels and abdominal pain has garnered significant interest in probiotic treatment to restore levels and improve abdominal pain. Patients with functional constipation have demonstrable microbiome disturbances and there is early promising data for the usage of prebiotics, probiotics (including *B. lactis*, *L. casei* Shirota, and *L. reuteri* DSM 17938), and even FMT to correct the dysbiosis and increase stooling frequency [11, 111]. The emerging understanding of methanogens' (e.g. *Methanobrevibacter smithii*) contribution to dysmotility and constipation especially in the context of SIBO has also led to the investigation of antibiotics such as rifaximin and neomycin to treat slow-transit constipation [8, 112]. The importance of methanogens in the pathophysiology of constipation supports the role of breath testing in diagnosis, since methanogen-predominant SIBO responds better to dual therapy of neomycin and rifaximin compared to neomycin monotherapy [113].

7 Conclusion

Functional gastrointestinal disorders represent a heterogenous group of disordered gut-brain interactions. Pathophysiology is incompletely elucidated but involves dysfunction along the bidirectional gut-brain access with contributions from psychosocial stress and abuse, visceral hypersensitivity, immune dysregulation, disrupted mucosal epithelial barrier function, and dysbiosis. The role of the microbiota is implicated in each of these aspects, so there is great interest in characterizing the nature of dysbiosis and correcting it via combinations of diet, prebiotics, probiotics, antibiotics, and fecal microbiota transplant. Due to the relative novelty of microbiome research and the significant heterogeneity of DGBIs, this line of research is nascent and presents a host of unanswered questions. While many commercial direct-to-consumer products purport to characterize the composition of the microbiome and provide tailored recommendations based on the results, these products have not been validated scientifically, and there is not yet enough research to accurately characterize a distinct microbial signature for each DGBI. With continued investigation, however, it may be possible to identify such signatures and thereby be able to create and utilize validated non-invasive methods of detecting dysbiosis with the goal of correcting it for symptom improvement. Ideally, this would take the form of disease-specific and strain-specific tailored treatment. Finally, there is growing appreciation of the role of fungi and viruses in host health and disease. Future research should take into account these players to fully understand the structure and function of the intestinal microbiome in health, pathogenesis of disease, and prevention and treatment of both gastrointestinal and non-gastrointestinal illness.

References

1. Corazziari E. Definition and epidemiology of functional gastrointestinal disorders. Best Pract Res Clin Gastroenterol. 2004;18(4):613–31. https://doi.org/10.1016/j.bpg.2004.04.012.
2. Staudacher H, Lomer M, Farquharson F, Louis P, Fava F, Franciosi E, Scholz M, et al. A diet low in FODMAPs reduces symptoms in patients with irritable bowel syndrome and a probiotic restores Bifidobacterium species: a randomized controlled trial. Gastroenterology. 2017;153(4):936–47. https://doi.org/10.1053/j.gastro.2017.06.010.
3. Sandler R, Everhart J, Donowitz J, Adams E, Cronin K, Goodman C, Gemmen E, Shah S, Avdic A, Rubin R. The burden of selected digestive diseases in the United States. Gastroenterology. 2002;122(5):1500–11. https://doi.org/10.1053/gast.2002.32978.
4. Everhart J, Renault P. Irritable bowel syndrome in office-based practice in the United States. Gastroenterology. 1991;100(4):998–1005. https://doi.org/10.1016/0016-5085(91)90275-p.
5. Drossman D. Functional gastrointestinal disorders: history, pathophysiology, clinical features, and Rome IV. Gastroenterology. 2016;150(6):1262–79. https://doi.org/10.1053/j.gastro.2016.02.032.
6. Hyland N, Cryan J. Microbe-host interactions: influence of the gut microbiota on the enteric nervous system. Dev Biol. 2016;417(2):182–7. https://doi.org/10.1016/j.ydbio.2016.06.027.
7. Palma D, Giada ML, Jun L, Dang V, Deng Y, Jury J, Umeh G, et al. Transplantation of fecal microbiota from patients with irritable bowel syndrome alters gut function and behavior in recipient mice. Sci Transl Med. 2017;9(379):eaaf6397. https://doi.org/10.1126/scitranslmed.aaf6397.
8. Ghoshal U, Shukla R, Srivastava D, Ghoshal U. Irritable bowel syndrome, particularly the constipation-predominant form, involves an increase in *Methanobrevibacter smithii*, which is associated with higher methane production. Gut Liver. 2016;10(6):932–8. https://doi.org/10.5009/gnl15588.
9. Villanueva-Millan MJ, Leite G, Wang J, Morales W, Parodi G, Pimentel ML, Barlow GM, et al. Methanogens and hydrogen sulfide producing bacteria guide distinct gut microbe profiles and irritable bowel syndrome subtypes. Am J Gastroenterol. 2022;117(12):2055. https://doi.org/10.14309/ajg.0000000000001997.
10. Dimidi E, Hristodoulides S, Scott SM, Whelan K. Mechanisms of action of probiotics and the gastrointestinal microbiota on gut motility and constipation. Adv Nutr. 2017;8(3):484–94. https://doi.org/10.3945/an.116.014407.
11. Ohkusa T, Koido S, Nishikawa Y, Sato N. Gut microbiota and chronic constipation: a review and update. Front Med. 2019;6:19. https://doi.org/10.3389/fmed.2019.00019.
12. Chichlowski M, Colin R. Visceral pain and gastrointestinal microbiome. J Neurogastroenterol Motil. 2015;21(2):172–81. https://doi.org/10.5056/jnm15025.
13. Guleria A, Karyampudi A, Singh R, Khetrapal CL, Verma A, Ghoshal UC, Kumar D. Mapping of brain activations to rectal balloon distension stimuli in male patients with irritable bowel syndrome using functional magnetic resonance imaging. J Neurogastroenterol Motil. 2017;23(3):415–27. https://doi.org/10.5056/jnm16148.
14. Botschuijver S, Roeselers G, Levin E, Jonkers DM, Welting O, Heinsbroek SEM, de Weerd HH, et al. Intestinal fungal dysbiosis is associated with visceral hypersensitivity in patients with irritable bowel syndrome and rats. Gastroenterology. 2017;153(4):1026–39. https://doi.org/10.1053/j.gastro.2017.06.004.
15. Grabauskas G, Xiaoyin W, Kao KD, Owyang C. Gut dysbiosis contributes to visceral hypersensitivity in irritable bowel syndrome via regulation of the levels of resolvin D1 in the colonic mucosa. Gastroenterology. 2019;156(6):S-165.
16. Pokusaeva K, Johnson C, Luk B, Uribe G, Fu Y, Oeuzguen N, Matsunami RK, et al. GABA-producing bifidobacterium dentium modulates visceral sensitivity in the intestine. J Neurogastroenterol Motil. 2017;29(1):e12904. https://doi.org/10.1111/nmo.12904.
17. Parkes GC, Rayment NB, Hudspith BN, Petrovska L, Lomer MC, Brostoff J, Whelan K, Sanderson JD. Distinct microbial populations exist in the mucosa-associated microbiota of sub-groups of irritable bowel syndrome. J Neurogastroenterol Motil. 2012;24(1):31–9. https://doi.org/10.1111/j.1365-2982.2011.01803.x.
18. Ng QX, Sen AY, Soh WL, Venkatanarayanan N, Lim DY, Yeo W-S. Systematic review with meta-analysis: the association between post-traumatic stress disorder and irritable bowel syndrome. J Gastroenterol Hepatol. 2019;34(1):68–73. https://doi.org/10.1111/jgh.14446.
19. Park SH, Videlock EJ, Shih W, Presson AP, Mayer EA, Chang L. Adverse childhood experiences are associated with irritable bowel syndrome and gastrointestinal symptom severity. J Neurogastroenterol Motil. 2016;28(8):1252–60. https://doi.org/10.1111/nmo.12826.
20. Ford A, Lacy B, Harris L, Quigley E, Moayyedi P. Effect of antidepressants and psychological therapies in irritable bowel syndrome: an updated systematic review and meta-analysis. Am J Gastroenterol. 2019;114(1):21–39. https://doi.org/10.1038/s41395-018-0222-5.
21. Jacobs JP, Lackner JM, Lagishetty V, Gudleski GD, Firth RS, Tillisch K, Naliboff BD, Labus JS, Mayer EA. Intestinal Microbiota Predict Response to Cognitive Behavioral Therapy for Irritable Bowel Syndrome. Gastroenterology. 2018;154(6):S–181. https://doi.org/10.1016/S0016-5085(18)31017-5.
22. Pinto-Sanchez MI, Hall GB, Ghajar K, Nardelli A, Bolino C, Lau JT, Martin F-P, et al. Probiotic Bifidobacterium longum NCC3001 reduces depression scores and alters brain activity: a pilot study in patients with irritable bowel syndrome. Gastroenterology. 2017;153(2):448–59. https://doi.org/10.1053/j.gastro.2017.05.003.
23. Kurokawa S, Kishimoto T, Mizuno S, Masaoka T, Naganuma M, Liang K-c, Kitazawa M, et al. The effect of fecal microbiota transplantation on psychiatric symptoms among patients with irritable bowel syndrome, functional diarrhea and functional constipation: an open-label observational study. J Affect Disord. 2018;235:506–12. https://doi.org/10.1016/j.jad.2018.04.038.
24. König J, Wells J, Cani P, García-Ródenas C, MacDonald T, Mercenier A, Whyte J, Troost F, Brummer R-J. Human intestinal barrier function in health and disease. Clin Transl Gastroenterol. 2016;7(10):196. https://doi.org/10.1038/ctg.2016.54.
25. Calarge CA, Devaraj S, Shulman RJ. Gut permeability and depressive symptom severity in unmedicated adolescents. J Affect Disord. 2019;246:586–94. https://doi.org/10.1016/j.jad.2018.12.077.
26. Schwiertz AS, Spiegel J, Dillmann U, Grundmann D, Bürmann J, Faßbender K, Schäfer K-H, Unger MM. Fecal markers of intestinal inflammation and intestinal permeability are elevated in Parkinson's disease. Parkinsonism Relat Disord. 2018;50:104–7. https://doi.org/10.1016/j.parkreldis.2018.02.022.
27. Cox AJ, West NP, Cripps AW. Obesity, inflammation, and the gut microbiota. Lancet Diabetes Endocrinol. 2015;3(3):207–15. https://doi.org/10.1016/S2213-8587(14)70134-2.
28. Barbara G, Grover M, Bercik P, Corsetti M, Ghoshal UC, Ohman L, Rajilić-Stojanović M. Rome foundation working team report on post-infection irritable bowel syndrome. Gastroenterology. 2019;156(1):46–58. https://doi.org/10.1053/j.gastro.2018.07.011.
29. Barlow G, Rezaie A, Lin E, Pimentel M. A definitive blood test for post-infectious irritable bowel syndrome? Expert Rev Gastroenterol Hepatol. 2016;10(11):1197–9. https://doi.org/10.1080/17474124.2016.1242411.

30. Vanheel H, Vicario M, Vanuytsel T, Van Oudenhove L, Martinez C, Keita ÅV, Pardon N, et al. Impaired duodenal mucosal integrity and low-grade inflammation in functional dyspepsia. Gut. 2014;63(2):262–71. https://doi.org/10.1136/gutjnl-2012-303857.
31. Obata Y, Pachnis V. The effect of microbiota and the immune system on the development and organization of the enteric nervous system. Gastroenterology. 2016;151(5):836–44. https://doi.org/10.1053/j.gastro.2016.07.044.
32. Sun M, Wei W, Liu Z, Cong Y. Microbiota metabolite short chain fatty acids, GPCR, and inflammatory bowel diseases. J Gastroenterol. 2017;52(1):1–8.
33. Pozuelo M, Panda S, Santiago A, Mendez S, Accarino A, Santos J, Guarner F, Azpiroz F, Manichanh C. Reduction of butyrate- and methane-producing microorganisms in patients with irritable bowel syndrome. Nat Sci Rep. 2015;5:12693.
34. Stanghellini V, Chan FKL, Hasler WL, Malagelada JR, Suzuki H, Tack J, Talley NJ. Gastroduodenal disorders. Gastroenterology. 2016;150(6):1380–92. https://doi.org/10.1053/j.gastro.2016.02.011.
35. Du L-J, Chen B-R, Kim JJ, Kim S, Shen J-H, Dai N. Helicobacter pylori eradication therapy for functional dyspepsia: systematic review and meta-analysis. World J Gastroenterol. 2016;22(12):3486–95. https://doi.org/10.3748/wjg.v22.i12.3486.
36. Kang SJ, Park B, Shin CM. Helicobacter pylori eradication therapy for functional dyspepsia: a meta-analysis by region and H. pylori prevalence. J Clin Med. 2019;8(9):1324. https://doi.org/10.3390/jcm8091324.
37. Kim Y-J, Chung WC, Kim BW, Kim SS, Kim JI, Kim NJ, Yoo J, Kim SH. Is helicobacter pylori associated functional dyspepsia correlated with dysbiosis. J Neurogastroenterol Motil. 2017;23(4):504–16. https://doi.org/10.5056/jnm17066.
38. Zhong L, Shanahan ER, Raj A, Koloski NA, Fletcher L, Morrison M, Walker MM, Talley NJ, Holtmann G. Dyspepsia and the microbiome: time to focus on the small intestine. Gut. 2017;66:1168–9.
39. Shanahan ER, Kang S, Staudacher H, Shah A, Do A, Burns G, Chachay VS, et al. Alterations to the duodenal microbiota are linked to gastric emptying and symptoms in functional dyspepsia. Gut. 2022;72:929–38. https://doi.org/10.1136/gutjnl-2021-326158.
40. Nakae H, Tsuda A, Matsuoka T, Mine T, Koga Y. Gastric microbiota in the functional dyspepsia patients treated with probiotic yogurt. BMJ Open Gastroenterol. 2016;3(1):e000109. https://doi.org/10.1136/bmjgast-2016-000109.
41. Wauters L, Slaets H, De Paepe K, Ceulemans M, Wetzels S, Geboers K, Toth J, et al. Efficacy and safety of spore-forming probiotics in the treatment of functional dyspepsia: a pilot randomised, double-blind, placebo-controlled trial. Lancet Gastroenterol Hepatol. 2021;6(10):784–92. https://doi.org/10.1016/S2468-1253(21)00226-0.
42. Tan VPY, Liu KSH, Lam FYF, Hung IFN, Yuen MF, Leung WK. Randomised clinical trial: Rifaximin versus placebo for the treatment of functional dyspepsia. Aliment Pharmacol Ther. 2017;45(6):767–76. https://doi.org/10.1111/apt.13945.
43. Shah A, Gurusamy SR, Hansen T, Callaghan G, Talley NJ, Koloski N, Walker MM, Jones MP, Morrison M, Holtmann GJ. Concomitant irritable bowel syndrome does not influence the response to antimicrobial therapy in patients with functional dyspepsia. Dig Dis Sci. 2022;67(6):2299–309. https://doi.org/10.1007/s10620-021-07149-1.
44. Duncanson KR, Talley NJ, Walker MM, Burrows TL. Food and functional dyspepsia: a systematic review. J Hum Nutr Diet. 2018;31(3):390–407. https://doi.org/10.1111/jhn.12506.
45. Tack J, Tornblom H, Tan V, Carbone F. Evidence-based and emerging dietary approaches to upper disorders of gut–brain interaction. Am J Gastroenterol. 2022;117(6):965–72. https://doi.org/10.14309/ajg.0000000000001780.
46. Lacy B, Patel N. Rome criteria and a diagnostic approach to irritable bowel syndrome. J Clin Med. 2017;6(11):99. https://doi.org/10.3390/jcm6110099.
47. Camilleri M, Boeckxstaens G. Irritable bowel syndrome: treatment based on pathophysiology and biomarkers. Gut. 2023;72(3):590–9. https://doi.org/10.1136/gutjnl-2022-328515.
48. Barberio B, Yiannakou Y, Houghton LA, Black CJ, Savarino EV, Ford AC. Overlap of Rome IV irritable bowel syndrome and functional dyspepsia and effect on natural history: a longitudinal follow-up study. Clin Gastroenterol Hepatol. 2022a;20(2):e89–101. https://doi.org/10.1016/j.cgh.2021.04.011.
49. Barberio B, Yiannakou Y, Houghton LA, Black CJ, Savarino EV, Ford AC. Overlap of rome IV irritable bowel syndrome and functional dyspepsia and effect on natural history: a longitudinal follow-up study. Clin Gastroenterol Hepatol. 2022b;20(2):e89; https://www.cghjournal.org/article/S1542-3565(21)00445-6/fulltext#.YHMoFo3NwsQ.twitter. Accessed 21 Mar 2023.
50. Liu H-N, Hao W, Chen Y-Z, Chen Y-J, Shen X-Z, Liu T-T. Altered molecular signature of intestinal microbiota in irritable bowel syndrome patients compared with healthy controls: a systematic review and meta-analysis. Dig Liver Dis. 2017;49(4):331–7. https://doi.org/10.1016/j.dld.2017.01.142.
51. Pittayanon R, Lau JT, Yuan Y, Leontiadis GI, Tse F, Surette M, Moayyedi P. Gut microbiota in patients with irritable bowel syndrome-a systematic review. Gastroenterology. 2019;157(1):97–108. https://doi.org/10.1053/j.gastro.2019.03.049.
52. Tap J, Derrien M, Törnblom H, Brazeilles R, Cools-Portier S, Doré J, Störsrud S, Le Nevé B, Öhman L, Simrén M. Identification of an intestinal microbiota signature associated with severity of irritable bowel syndrome. Gastroenterology. 2017;152(1):111–23. https://doi.org/10.1053/j.gastro.2016.09.049.
53. Vila AV, Imhann F, Collij V, Jankipersadsing SA, Gurry T, Mujagic Z, Kurilshikov A, et al. Gut microbiota composition and functional changes in inflammatory bowel disease and irritable bowel syndrome. Sci Transl Med. 2018;10(472):eaap8914.
54. Labus JS, Hollister EB, Jacobs J, Kirbach K, Oezguen N, Gupta A, Acosta J, et al. Differences in gut microbial composition correlate with regional brain volumes in irritable bowel syndrome. Microbiome. 2017;5:49. https://doi.org/10.1186/s40168-017-0260-z.
55. Liu Y, Zhang L, Wang X, Wang Z, Zhang J, Jiang R, Wang X, et al. Similar fecal microbiota signatures in patients with diarrhea-predominant irritable bowel syndrome and patients with depression. Clin Gastroenterol Hepatol. 2016;14(11):1602–11. https://doi.org/10.1016/j.cgh.2016.05.033.
56. van Thiel I, de Jonge W, van den Wijngaard R. Fungal feelings in the irritable bowel syndrome: the intestinal mycobiome and abdominal pain. Gut Microbes. 2023;15(1):2168992. https://doi.org/10.1080/19490976.2023.2168992.
57. Jalanka J, Salonen A, Fuentes S, de Vos WM. Microbial signatures in post-infectious irritable bowel syndrome—toward patient stratification for improved diagnostics and treatment. Gut Microbes. 2015;6(6):364–9. https://doi.org/10.1080/19490976.2015.1096486.
58. Morales W, Rezaie A, Barlow G, Pimentel M. Second-generation biomarker testing for irritable bowel syndrome using plasma anti-CdtB and anti-vinculin levels. Dig Dis Sci. 2019;64(11):3115–21. https://doi.org/10.1007/s10620-019-05684-6.
59. Pimentel M. Breath testing for small intestinal bacterial overgrowth: should we bother? Am J Gastroenterol. 2016;111(3):307–8. https://doi.org/10.1038/ajg.2016.30.
60. Carrasco-Labra A, Lytvyn L, Falck-Ytter Y, Surawicz CM, Chey WD. AGA technical review on the evaluation of functional diarrhea and diarrhea-predominant irritable bowel syndrome in adults (IBS-D). Gastroenterology. 2019;157(3):859–80. https://doi.org/10.1053/j.gastro.2019.06.014.

61. Pimentel M, Lembo A, Chey WD, Zakko S, Ringel Y, Jing Y, Mareya SM, Shaw AL, Bortey E, Forbes WP. Rifaximin therapy for patients with irritable bowel syndrome without constipation. N Engl J Med. 2011;364(1):22–3. https://doi.org/10.1056/NEJMoa1004409.
62. Kassim OB, Smith JG, Schora D, Adams N, David E, Yen E. Post-infectious irritable bowel syndrome persists after fecal microbiota transplant for C. difficile infection. Am J Gastroenterol. 2018;113:S93–4.
63. Zhou QQ, Verne ML, Fields JZ, Lefante JJ, Basra S, Salameh H, Nicholas G, Verne. Randomised placebo-controlled trial of dietary glutamine supplements for postinfectious irritable bowel syndrome. Gut. 2019;68(8):996–1002. https://doi.org/10.1136/gutjnl-2017-315136.
64. Dunlop SP, Jenkins D, Neal KR, Naesdal J, Borgaonker M, Collins SM, Spiller RC. Randomized, double-blind, placebo-controlled trial of prednisolone in post-infectious irritable bowel syndrome. Aliment Pharmacol Ther. 2003;18(1):77–84. https://doi.org/10.1046/j.1365-2036.2003.01640.x.
65. Lam C, Tan W, Leighton M, Hastings M, Lingaya M, Falcone Y, Zhou X, et al. A mechanistic multicentre, parallel group, randomised placebo-controlled trial of mesalazine for the treatment of IBS with diarrhoea (IBS-D). Gut. 2016;65(1):91–9. https://doi.org/10.1136/gutjnl-2015-309122.
66. Quigley E. The spectrum of small intestinal bacterial overgrowth (SIBO). Curr Gastroenterol Rep. 2019;21(1):3. https://doi.org/10.1007/s11894-019-0671-z.
67. Rezaie A, Buresi M, Lembo A, Lin H, McCallum R, Rao S, Schmulson M, Valdovinos M, Zakko S, Pimentel M. Hydrogen and methane-based breath testing in gastrointestinal disorders: the North American consensus. Am J Gastroenterol. 2017;112(5):775–84. https://doi.org/10.1038/ajg.2017.46.
68. Tansel A, Levinthal DJ. Understanding our tests: hydrogen-methane breath testing to diagnose small intestinal bacterial overgrowth. Clin Transl Gastroenterol. 2022;14(4):e00567. https://doi.org/10.14309/ctg.0000000000000567.
69. Takakura W, Pimentel M, Rao S, Villanueva-Millan MJ, Chang C, Morales W, Sanchez M, et al. A single fasting exhaled methane level correlates with fecal methanogen load, clinical symptoms and accurately detects intestinal methanogen overgrowth. Am J Gastroenterol. 2022;117(3):470–7. https://doi.org/10.14309/ajg.0000000000001607.
70. Madigan KE, Bundy R, Weinberg RB. Distinctive clinical correlates of small intestinal bacterial overgrowth with methanogens. Clin Gastroenterol Hepatol. 2022;20(7):1598–1605.e2. https://doi.org/10.1016/j.cgh.2021.09.035.
71. Liu JJ, Brenner DM. Updates and caveats to breath testing for intestinal overgrowth. Am J Gastroenterol. 2022;117(9):1390–3; https://journals.lww.com/ajg/Fulltext/2022/09000/Updates_and_Caveats_to_Breath_Testing_for.2.aspx.
72. Khoshini R, Dai S-C, Lezcano S, Pimentel M. A systematic review of diagnostic tests for small intestinal bacterial overgrowth. Dig Dis Sci. 2008;53(6):1443–54.
73. Ghoshal U, Srivastava D, Ghoshal U, Misra A. Breath tests in the diagnosis of small intestinal bacterial overgrowth in patients with irritable bowel syndrome in comparison with quantitative upper gut aspirate culture. Eur J Gastroenterol Hepatol. 2014;26(7):753–60. https://doi.org/10.1097/MEG.0000000000000122.
74. Kirton E, Anderson-Leary L, Gallagher J, Hewitt S, Burke J, Jackson W. PWE-081 hydrogen breath testing: high incidence of small intestinal bacterial overgrowth diagnosis using lactulose versus glucose. Gut. 2019;68(Suppl 2):A210. https://doi.org/10.1136/gutjnl-2019-BSGAbstracts.401.
75. Bratten JR, Spanier J, Jones MP. Lactulose breath testing does not discriminate patients with irritable bowel syndrome from healthy controls. Am J Gastroenterol. 2008;103(4):958–63. https://doi.org/10.1111/j.1572-0241.2008.01785.x.
76. Usai-Satta P, Giannetti C, Oppia F, Cabras F. The North American consensus on breath testing: the controversial diagnostic role of lactulose in SIBO. Am J Gastroenterol. 2018;113(3):440. https://doi.org/10.1038/ajg.2017.392.
77. Aziz I, Tornblom H, Simren M. Small intestinal bacterial overgrowth as a cause for irritable bowel syndrome: guilty or not guilty? Curr Opin Gastroenterol. 2017;33(3):196–202. https://doi.org/10.1097/MOG.0000000000000348.
78. Lin EC, Massey BT. Scintigraphy demonstrates high rate of false-positive results from glucose breath tests for small bowel bacterial overgrowth. Clin Gastroenterol Hepatol. 2016;14(2):203–8. https://doi.org/10.1016/j.cgh.2015.07.032.
79. Shah ED, Basseri RJ, Chong K, et al. Abnormal breath testing in IBS: a meta-analysis. Dig Dis Sci. 2010;55:2441–9.
80. Rao SSC, Bhagatwala J. Small intestinal bacterial overgrowth: clinical features and therapeutic management. Clin Transl Gastroenterol. 2019;10(10):e00078. https://doi.org/10.14309/ctg.0000000000000078.
81. Saffouri GB, Shields-Cutler RR, Chen J, Yang Y, Lekatz HR, Hale VL, Cho JM, et al. Small intestinal microbial dysbiosis underlies symptoms associated with functional gastrointestinal disorders. Nat Commun. 2019;10(1):2012. https://doi.org/10.1038/s41467-019-09964-7.
82. Sharara AI, Aoun E, Abdul-Baki H, Mounzer R, Sidani S, Elhajj I. A randomized double-blind placebo-controlled trial of Rifaximin in patients with abdominal bloating and flatulence. Am J Gastroenterol. 2006;101(2):326–33. https://doi.org/10.1111/j.1572-0241.2006.00458.x.
83. Rezaie A, Heimanson Z, McCallum R, Pimentel M. Lactulose breath testing as a predictor of response to rifaximin in patients with irritable bowel syndrome with diarrhea. Am J Gastroenterol. 2019;114(12):1886. https://doi.org/10.14309/ajg.0000000000000444.
84. Barlow JT, Leite G, Romano AE, Sedighi R, Chang C, Celly S, Rezaie A, Mathur R, Pimentel M, Ismagilov RF. Quantitative sequencing clarifies the role of disruptor taxa, oral microbiota, and strict anaerobes in the human small-intestine microbiome. Microbiome. 2021;9(1):214. https://doi.org/10.1186/s40168-021-01162-2.
85. Posserud I, Stotzer P-O, Björnsson ES, Abrahamsson H, Simrén M. Small intestinal bacterial overgrowth in patients with irritable bowel syndrome. Gut. 2007;56(6):802–8. https://doi.org/10.1136/gut.2006.108712.
86. Chen B, Kim JJ-W, Zhang Y, Lijun D, Dai N. Prevalence and predictors of small intestinal bacterial overgrowth in irritable bowel syndrome: a systematic review and meta-analysis. J Gastroenterol. 2018;53(7):807–18. https://doi.org/10.1007/s00535-018-1476-9.
87. Shah A, Talley NJ, Jones M, Kendall BJ, Koloski N, Walker MM, Morrison M, Holtmann GJ. Small intestinal bacterial overgrowth in irritable bowel syndrome: a systematic review and meta-analysis of case-control studies. Am J Gastroenterol. 2020;115:190–201. https://doi.org/10.14309/ajg.0000000000000504.
88. Pimentel M, Morales W, Pokkunuri V, Brikos C, Kim SM, Kim SE, Triantafyllou K, et al. Autoimmunity links vinculin to the pathophysiology of chronic functional bowel changes following campylobacter Jejuni infection in a rat model. Dig Dis Sci. 2015;60(5):1195–205. https://doi.org/10.1007/s10620-014-3435-5.
89. Acosta A, Camilleri M, Shin A, Nord SL, O'Neill J, Gray AV, Lueke AJ, et al. Effects of Rifaximin on transit, permeability, fecal microbiome, and organic acid excretion in irritable bowel syndrome. Clin Transl Gastroenterol. 2016;7:e173. https://doi.org/10.1038/ctg.2016.32.

90. Chey WD, Shah ED, DuPont HL. Mechanism of action and therapeutic benefit of Rifaximin in patients with irritable bowel syndrome: a narrative review. Ther Adv Gastroenterol. 2020;13:1756284819897531. https://doi.org/10.1177/1756284819897531.
91. David LA, Maurice CF, Carmody RN, Gootenberg DB, Button JE, Wolfe BE, Ling AV, et al. Diet rapidly and reproducibly alters the human gut microbiome. Nature. 2014;505(7484):559–63. https://doi.org/10.1038/nature12820.
92. Simren M, Mansson A, Langkilde AM, Svedlund J, Abrahamsson H, Bengtsson U, Bjornsson ES. Food-related gastrointestinal symptoms in the irritable bowel syndrome. Digestion. 2001;63(2):108–15. https://doi.org/10.1159/000051878.
93. Nagarajan N, Morden A, Bischof D, King EA, Kosztowski M, Wick EC, Stein EM. The role of fiber supplementation in the treatment of irritable bowel syndrome: a systematic review and meta-analysis. Eur J Gastroenterol Hepatol. 2015;27(9):1002–10. https://doi.org/10.1097/MEG.0000000000000425.
94. Halmos EP, Power VA, Shepherd SJ, Gibson PR, Muir JG. A diet low in FODMAPs reduces symptoms of irritable bowel syndrome. Gastroenterology. 2014;146(1):67–75.e5. https://doi.org/10.1053/j.gastro.2013.09.046.
95. Staudacher HM, Irving PM, Lomer MCE, Whelan K. Mechanisms and efficacy of dietary FODMAP restriction in IBS. Nat Rev Gastroenterol Hepatol. 2014;11(4):256–66. https://doi.org/10.1038/nrgastro.2013.259.
96. Halmos EP, Christophersen CT, Bird AR, Shepherd SJ, Gibson PR, Muir JG. Diets that differ in their FODMAP content alter the colonic luminal microenvironment. Gut. 2015;64(1):93–100. https://doi.org/10.1136/gutjnl-2014-307264.
97. Whelan K, Martin LD, Staudacher HM, Lomer MCE. The low FODMAP diet in the management of irritable bowel syndrome: an evidence-based review of FODMAP restriction, reintroduction and personalisation in clinical practice. J Hum Nutr Diet. 2018;31(2):239–55. https://doi.org/10.1111/jhn.12530.
98. Carlson JL, Erickson JM, Lloyd BB, Slavin JL. Health effects and sources of prebiotic dietary fiber. Curr Develop Nutr. 2018;2:nzy005. https://doi.org/10.1093/cdn/nzy005.
99. Ford AC, Harris LA, Lacy BE, Quigley EMM, Moayyedi P. Systematic review with meta-analysis: the efficacy of prebiotics, probiotics, synbiotics and antibiotics in irritable bowel syndrome. Aliment Pharmacol Ther. 2018;48(10):1044–60. https://doi.org/10.1111/apt.15001.
100. El-Salhy M, Ystad SO, Mazzawi T, Gundersen D. Dietary fiber in irritable bowel syndrome (review). Int J Mol Med. 2017;40(3):607–13. https://doi.org/10.3892/ijmm.2017.3072.
101. Suez J, Zmora N, Segal E, Elinav E. The pros, cons, and many unknowns of probiotics. Nat Med. 2019;25(5):716–29. https://doi.org/10.1038/s41591-019-0439-x.
102. Dale HF, Rasmussen SH, Asiller OO, Lied GA. Probiotics in irritable bowel syndrome: an up-to-date systematic review. Nutrients. 2019;11(9):2048. https://doi.org/10.3390/nu11092048.
103. Abdelghafar YA, Qadir YHA, Motawea KR, Nasr SA, Omran HAM, Belal MM, Elhashash MM, Azim AAA, Shah J. Efficacy and safety of fecal microbiota transplant in irritable bowel syndrome: an update based on meta-analysis of randomized control trials. Health Sci Rep. 2022;5(5):e814. https://doi.org/10.1002/hsr2.814.
104. Ianiro G, Eusebi LH, Black CJ, Gasbarrini A, Cammarota G, Ford AC. Systematic review with meta-analysis: efficacy of Faecal microbiota transplantation for the treatment of irritable bowel syndrome. Aliment Pharmacol Ther. 2019;50(3):240–8. https://doi.org/10.1111/apt.15330.
105. Xu D, Chen VL, Steiner CA, Berinstein JA, Eswaran S, Waljee AK, Higgins PDR, Owyang C. Efficacy of fecal microbiota transplantation in irritable bowel syndrome: a systematic review and meta-analysis. Am J Gastroenterol. 2019;114(7):1043–50. https://doi.org/10.14309/ajg.0000000000000198.
106. Rokkas T, Hold GL. A systematic review, pairwise meta-analysis and network meta-analysis of randomized controlled trials exploring the role of fecal microbiota transplantation in irritable bowel syndrome. Eur J Gastroenterol Hepatol. 2023;35(4):471–9. https://doi.org/10.1097/MEG.0000000000002519.
107. El-Salhy M, Hatlebakk JG, Gilja OH, Kristoffersen AB, Hausken T. Efficacy of faecal microbiota transplantation for patients with irritable bowel syndrome in a randomised, double-blind, placebo-controlled study. Gut. 2019a;69:859. https://doi.org/10.1136/gutjnl-2019-319630.
108. Mazzawi T, Hausken T, Hov JR, Valeur J, Sangnes DA, El-Salhy M, Gilja OH, Hatlebakk JG, Lied GA. Clinical response to fecal microbiota transplantation in patients with diarrhea-predominant irritable bowel syndrome is associated with normalization of fecal microbiota composition and short-chain fatty acid levels. Scand J Gastroenterol. 2019;54(6):690–9. https://doi.org/10.1080/00365521.2019.1624815.
109. El-Salhy M, Hausken T, Hatlebakk JG. Increasing the dose and/or repeating faecal microbiota transplantation (FMT) increases the response in patients with irritable bowel syndrome (IBS). Nutrients. 2019b;11(6):1415. https://doi.org/10.3390/nu11061415.
110. Singh P, Lee H-N, Rangan V, Ballou S, Lembo J, Katon J, McMahon C, et al. Similarities in clinical and psychosocial characteristics of functional diarrhea and irritable bowel syndrome with diarrhea. Clin Gastroenterol Hepatol. 2019;18:399–405.e1. https://doi.org/10.1016/j.cgh.2019.08.020.
111. Zhao Y, Yan-Bo Y. Intestinal microbiota and chronic constipation. Springerplus. 2016;5(1):1130. https://doi.org/10.1186/s40064-016-2821-1.
112. Triantafyllou K, Chang C, Pimentel M. Methanogens, methane and gastrointestinal motility. J Neurogastroenterol Motil. 2014;20(1):31–40. https://doi.org/10.5056/jnm.2014.20.1.31.
113. Low K, Hwang L, Hua J, Zhu A, Morales W, Pimentel M. A combination of Rifaximin and neomycin is most effective in treating irritable bowel syndrome patients with methane on lactulose breath test. J Clin Gastroenterol. 2010;44(8):547–50. https://doi.org/10.1097/MCG.0b013e3181c64c90.

How the Microbiome Affects the Risk for Colon Cancer

Julia L. Drewes and Aadra P. Bhatt

Highlights Enrichments in specific gut microbial populations have been implicated in the development of colorectal cancer (CRC), including enterotoxigenic *Bacteroides fragilis,* DNA-damaging colibactin-producing strains of *Escherichia coli (Ec), Fusobacterium nucleatum (Fn)*, and mucus-invasive biofilms.

- *Fn* may confer chemoresistance by triggering TLR4 and MYD88 innate immune signaling, leading to autophagy. *Fn* can also bind to and activate the inhibitory receptor CEACAM1, which inhibits T and NK cell function, protecting CRC cells from immune surveillance.
- Enterotoxigenic *B. fragilis* (ETBF) induces colitis and colon tumors via STAT3 activation, forms biofilms with *Ec*, and drives barrier dysfunction by cleaving the zonula adherens protein E-cadherin and stimulating cell proliferation.
- Polyketide synthase *(pks)* + *Ec* strains promote inflammation, and express the oncotoxin colibaction. *Pks* + *Ec*-infected cells senesce rapidly, promoting proliferation of uninfected cells and increasing tumor growth.
- Consumption of a high-fat diet (HFD) may increase CRC risk by selecting for bile-resistant pro-inflammatory and pro-carcinogenic Proteobacteria and Clostridia microbes and the hydrogen sulfide producer *Bilophila wadsworthia,* as well as Lachnospiraceae (e.g. *Clostridium scindens*) and Peptostreptococcaceae, which produce secondary bile acids that induce oxidative stress and DNA damage.
- Inflammatory bowel disease is a strong risk factor for developing CRC, due at least in part to reduced mucus production and loss of compartmentalization between gut microbes and the epithelium. The LY6/PLAUR domain containing 8 (LYPD8) protein, which binds to flagellated bacteria, is downregulated in sporadic CRC, leading to increased abundance of pathogenic *Proteus, Helicobacter*, and *Escherichia* spp. Lipocalin-2 deficiency increases Proteobacteria and facultative pathogen *Alistipes*, resulting in IL-6 driven inflammation and tumorigenesis.

Microbiota residing within the gastrointestinal tract (among other anatomical sites) are an environmental factor to which we have continual exposure throughout our lifespan. Previously, it was difficult to study intestinal bacteria and their diversity as most gut bacteria are fastidious anaerobic organisms. However, the advent of high-throughput, next generation sequencing technologies combined with computational analysis has allowed for a deeper understanding of intestinal microbiota.

Microbiota have been implicated in colorectal cancer (CRC) development as early as the 1960s, when researchers observed differential CRC outcomes in germ-free vs. conventionally raised rodents. Administration of the carcinogenic glucoside cycasin to conventional, but not germ-free rodents led to the development of colon tumors [1]. This phenomenon was also observed with another carcinogen 1,2-dimethylhydrazine, which induced colon tumors in 93% of conventional rats compared to only 20% of germ-free rats [2]. In another study, mice were first exposed to the colon-tropic carcinogen azoxymethane (AOM), followed by antibiotic treatment and oral gavage of stools derived from either healthy controls or CRC patients. Mice colonized with CRC microbiota developed more polyps in the distal colon, with a

Julia L. Drewes and Aadra P. Bhatt contributed equally with all other contributors.

J. L. Drewes
Division of Infectious Diseases, Department of Medicine, Johns Hopkins University School of Medicine, Baltimore, MD, USA
e-mail: jdrewes2@jhmi.edu

A. P. Bhatt (✉)
Division of Gastroenterology and Hepatology, Department of Medicine, Center for Gastrointestinal Biology and Disease, Chapel Hill, NC, USA

Lineberger Comprehensive Cancer Center, University of North Carolina at Chapel Hill, Chapel Hill, NC, USA
e-mail: Aadra_Bhatt@med.unc.edu

M. Pimentel et al. (eds.), *Clinical Understanding of the Human Gut Microbiome*, https://doi.org/10.1007/978-3-031-46712-7_7

higher grade of dysplasia. Furthermore, the CRC-stool colonized mice had enhanced epithelial proliferation, higher levels of inflammation, and increased expression of genes involved in angiogenesis, proliferation, stemness, apoptosis, angiogenesis, invasion, and metastasis [3].

Although laboratory rodent models are critical for unraveling mechanisms of various biological processes, they have inherent limitations to modeling human diseases with complex, multifactorial etiologies, and predicting responses. A recent study found genetic similarities between laboratory and free-range, wild-caught mice, but divergent microbiota resulting from dissimilar evolutionary pressures. Transferring the highly diverse wild mouse microbiota into their laboratory counterparts conferred increased resistance to inflammation and AOM-dependent tumorigenesis [4], underscoring the importance of microbiota in affecting host immune responses and neoplasia [5].

CRC in humans may be hereditary or sporadic, with the vast majority of cases (>70%) being the latter. Hereditary cases have germ-line mutations in genes such as mismatch repair (e.g., *MLH1* in Lynch syndrome) and tumor suppressors (*APC* in familial adenomatous polyposis, *SMAD4* in juvenile polyposis), amongst others. However, limited data exists regarding the microbial composition of inherited forms of CRC. The microbiota of Lynch syndrome patients is characterized by elevated levels of pathobionts such as *Escherichia* and *Klebsiella*. *Fusobacterium* is found in very low levels, and with certain co-occurring bacteria, in primary tumors and distant metastases. Reduced abundance of certain Clostridiaceae were observed in baseline adenomas. *Desulfovibrio* were found to be increased in stool and mucosal biopsies. Metagenomic and metatranscriptomic analyses of Lynch syndrome patient stools uncovered a change in the oxidative metabolic microenvironment, and a shift in flagellin production, which could be potential factors in CRC pathogenesis [6]. In contrast, FAP patients harbor mucus-invasive biofilms that are enriched in enterotoxigenic *Bacteroides fragilis* and the DNA-damaging colibactin-producing strains of *Escherichia coli* [7], which synergistically promote tumorigenesis in mouse models [7]. Ongoing efforts to elucidate the microbial contributions to familial CRC will be instrumental in efforts to improve outcomes in these patients, as the wide age ranges in which these diseases manifest as overt cancer indicate that environmental factors can still modify these germline CRC associations.

In contrast to the limited data on inherited forms of CRC, tremendous efforts over the past two decades have thoroughly examined the microbial composition of the more common sporadic CRC with increasingly sensitive measures. Multiple metagenomic studies have investigated the role of gut microbiota in sporadic CRC compared to healthy controls (reviewed in [8–10]). The common finding of these studies is a differential ecological landscape of microbiota between CRC patients and healthy individuals, which is termed *dysbiosis*. Dysbiosis in CRC has a characteristic decrease in microbial diversity and lowered community stability, with corresponding alterations in functional output, and a significantly different microenvironment. CRC microbiota are predominated by potentially pro-carcinogenic taxa such as enterotoxigenic *B. fragilis,* colibactin-producing clades of *E. coli,* and *Fusobacterium nucleatum* [11, 12]. Oral microbes as a whole, of which *F. nucleatum* is often touted as the flagship organism, are consistently associated with CRC in metagenomic and 16S amplicon sequencing studies [13]. Various association studies also implicate *Enterococcus fecalis, Streptococcus gallolyticus, Porphyromonas,* and mucus-invasive biofilms in CRC. In addition, CRC microbiota have a depletion, or marked reduction, of beneficial, protective taxa such as *Roseburia* and *Faecalibacterium prausnitzii* (reviewed in [8, 9, 14]).

Below, we highlight four model organisms as well as mucus-invasive biofilms with strong preclinical molecular and mechanistic evidence linking them to one of the three steps of carcinogenesis: initiation, promotion, and progression. This is illustrated in Fig. 1. Further along, we delineate multiple, robust lines of evidence for how the microbiome impacts two independent CRC risk factors, one of which is modifiable (diet), and another that is unmodifiable (inflammatory bowel diseases).

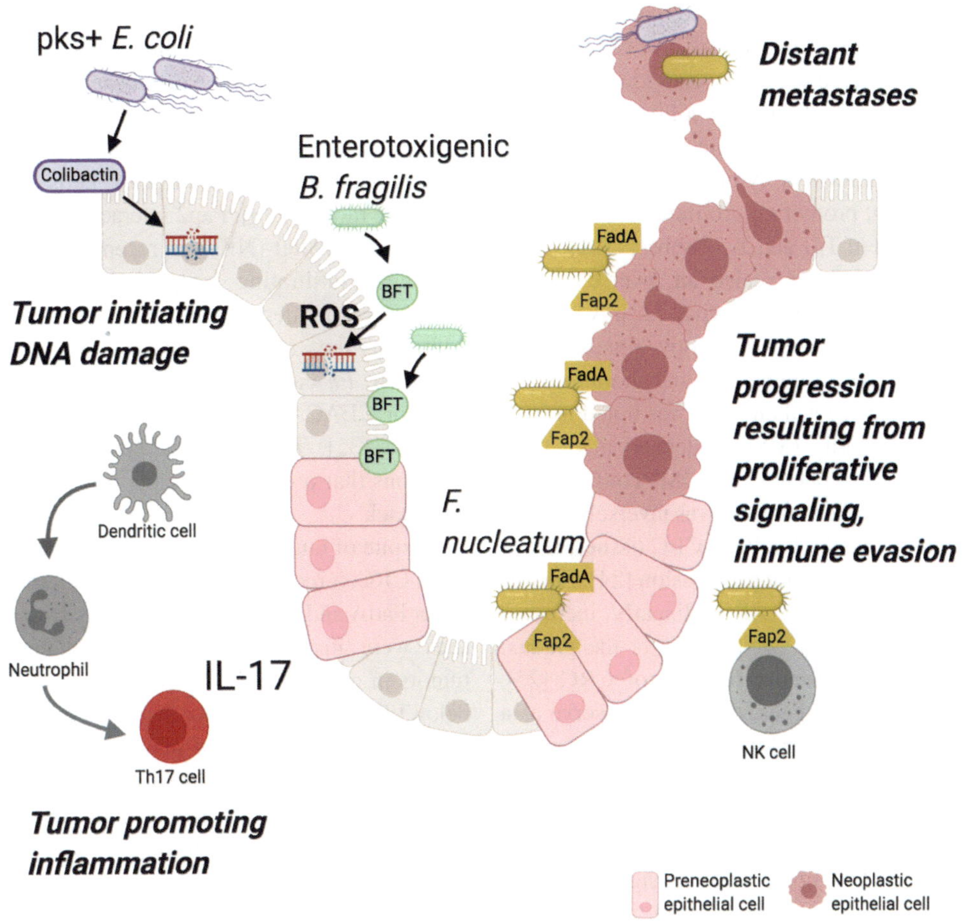

Fig. 1 Microbiota influence all phases of carcinogenesis: Tumor initiating DNA damage occurs from colibactin-expressing *E. coli*, and reactive oxygen species (ROS) generated via *B. fragilis* toxin (BFT) encoded by Enterotoxigenic *B. fragilis*. The resulting genotoxicity induces DNA damage signaling and disrupted epithelial integrity, causing epithelial cells, neutrophils, and dendritic cells to secrete cytokines that activate T cells, resulting in Th17-driven inflammation. Preneoplastic cells endure a further hit via *F. nucleatum* which uses Fap2 to bind Gal-GalNAc on cancerous cells to activate proliferative signaling, which is further augmented by FadA binding to E-cadherin. Fap2 also promotes immune evasion by binding the negative regulator TIGIT which limits NK cell surveillance of neoplastic tissues. CEACAM binding by an unknown fusobacterial protein also promotes immune evasion. Collectively, this results in tumor progression resulting from proliferative signaling and immune evasion. *F. nucleatum* and colibactin mutagenic signature are also detected outside the intestines, indicating they may influence distant metastases

1 Specific Bacteria Implicated in CRC

Proposed models for specific bacterial roles in CRC include the driver/passenger model and the hit-and-run model. In the driver/passenger model, "alpha bugs" drive tumorigenesis, but other microbes that thrive in the tumor microenvironment can enhance, inhibit, or not impact tumor progression at all [15]. The hit-and-run model proposes that even a brief exposure to some microbes may lead to pro-carcinogenic changes; due to the long time from possible exposure to onset of sporadic CRC (>10 years), initial driver bacteria may no longer be enriched in CRC tissues [16]. Importantly, none of the putative pro-carcinogenic bacteria below have been definitely linked to CRC causality using prospective studies. Nevertheless, pre-clinical animal models and association studies in CRC patients point to a strong link between these bacteria and CRC risk.

1.1 Fusobacterium nucleatum (Fn)

A multitude of clinical and preclinical evidence supports a role for *Fn* in colorectal cancer. *Fn* is an anaerobic commensal associated with periodontitis and appendicitis, and is rarely found within the intestine. Subsequent studies have linked periodontal disease with increased risk for CRC [17]. Using RNASeq and host sequence subtraction, Castellarin *et al* found *Fn* sequences overrepresented in CRC tumors compared to controls, with positive association with lymph node metastasis [11]. Kostic *et al* also found *Fusobacterium* spp. enriched in colon adenomas compared to healthy tissue, and *Fusobacterium* spp. enrichment in stools of adenoma and carcinoma patients, compared to healthy individuals [18–20]. *Fn* can exacerbate experimental murine colitis by impairing the epithelial barrier with accompanying increased permeability, enhanced proinflammatory cytokine secretion, activation of STAT3 signaling, promoting proliferation of $CD4^+$ T cells, and differentiation of T_H1 and T_H17 subsets [21].

Analysis of CRC biopsies revealed an inverse relation between *Fn* abundance and overall survival [22]. A separate study also correlated increased *Fn* copy number with advanced tumor stages, *KRAS* mutations, and shorter overall survival [23]. Interestingly, *Fn* abundance was noted in CRC with microsatellite instability (MSI), independent of CRC with CpG island methylator phenotype (CIMP) or CRC with *KRAS* mutations [24]. Furthermore, *Fn* was inversely associated with the number of $CD3^+$ T cells in CRC tissues, implicating immune regulation in tumor progression [24]. Finally, the proportion of *Fn* enriched CRCs linearly increased between the rectum to cecum, which reflects the gradual geographical change in molecular features of CRC [25]. Collectively, these findings suggest that *Fn* may serve as a potential non-invasive biomarker for disease stage and severity, as well as a prognostic biomarker for CRC.

Subsequent studies have implicated *Fn* in conferring chemoresistance and recurring disease in CRC tumors. *Fn* triggered TLR4 and MYD88 innate immune signaling, leading to miR-18a and miR-4802 dependent activation of autophagy, a mechanism known to bestow chemoresistance. These findings suggest that CRC patients with high *Fn* burden may not respond to traditional chemotherapy regiments, unless *Fn* is eradicated using antibiotics [26]. In a study of paired biopsies, *Fn* was detected in both the primary colorectal tumor and in distant liver metastases. In addition, other Gram-negative anaerobes co-occurred with *Fn* in both primary and metastatic tumors. Treating CRC tumor-bearing mice with metronidazole decreased *Fn* load, and diminished tumor growth with reduced cancer cell proliferation [27].

Mechanistically, Fn does not encode any known toxins. However, *Fn* encodes the FadA adhesin which facilitates host-cell binding, and promotes invasion into epithelial cells [28]. FadA stimulates proliferation of human colon cancer cell lines [29, 30]. FadA binds to E-cadherin, a tumor suppressor belonging to a large family of calcium dependent cell adhesion glycoproteins present on epithelial cells including CRC cells. FadA binding inhibits E-cadherin's tumor suppressor activity, leading to activation of β-catenin regulated transcription, and inflammatory and proliferative gene expression. Colorectal tumors and precancerous adenomas displayed increased FadA expression levels compared to healthy tissues [29]. *Fn* can also promote epithelial-to-mesenchymal transition (EMT) in oral epithelial cells [31], and was demonstrated to inflict DNA damage via FadA [32].

Human colon adenocarcinomas overexpress Gal-GalNAc, a polysaccharide that is bound by a fusobacterial protein Fap2. *Fn* binding to colon adenocarcinomas correlated with host Gal-GalNac expression, and was inhibited by treating CRC cells with *O*-glycanase. *Fn* mutants lacking functional Fap2 were unable to attach to CRC cells or establish tumors in mouse models [33]. Fap2 interacts with TIGIT, an inhibitory Natural Killer (NK) cell receptor that is also expressed by tumor infiltrating lymphocytes (TILs) found within human colon adenocarcinomas, and other T cells [34]. Although the fusobacterial binding candidate remains unknown, *Fn* can also bind to and activate the inhibitory receptor CEACAM1, also known to inhibit T and NK cell function [35]. Thus, *Fn* can perform immunomodulatory functions that protect CRC cells from immune surveillance and attack.

In spite of multiple lines of evidence squarely implicating Fn in CRC, a recent study demonstrated that *Fn* strains with intact FadA and Fap2 cultured directly from human CRC tissues were insufficient to induce inflammation or colon tumors in germ-free $Apc^{Min/+}$ mice, unlike *pks + E. coli* (see below) [36]. This suggests that only a subset of *Fn* strains may be carcinogenic, and may be so only in context with other members of the microbiota to cooperatively promote (or prevent) tumorigenesis *in vivo*. Further studies are necessary to address whether *Fn* truly drives CRC, or is merely a passenger.

1.2 Bacteroides fragilis (Bf)

Bf is an important member of the gut microbiota, and certain beneficial non-toxigenic *Bf* strains promote development of regulatory T cells [37]. Non-toxigenic *Bf* secretes polysaccharide A, which reduces proinflammatory IL-17 cytokine production, and protects mice from experimental colitis resulting from *Helicobacter hepaticus* infection [38]. In contrast, strong preclinical evidence links CRC with enterotoxigenic *B. fragilis* (ETBF) that encodes and secretes a toxin called *Bf* toxin (BFT), which induces colitis and colon tumors in multiple intestinal neoplasia (Min) mice. ETBF induced STAT3 activation, and colitis was largely driven by Th17 inflammatory cells. Blockade of IL-17 or IL-23 receptor inhibited colitis and tumorigenesis [39]. ETBF in humans is largely associated with childhood diarrheal disease, but the *bft* gene is also elevated in CRC mucosal samples, compared to normal tissues [39]. *Bft* is highly enriched in the colonic mucosa of FAP patients, which contain patchy bacterial biofilms comprised of *E. coli* and *Bf*. Tumor prone mice colonized with colibactin-producing *E. coli* and ETBF had increased colonic IL-17 and DNA damage, resulting in accelerated tumor onset and increased mortality [7]. Clearance of ETBF with the antibiotic cefoxotin reduced mucosal IL-17 expression and prevented colon adenoma formation.

BFT can also drive barrier dysfunction as it rapidly cleaves the zonula adherens protein E-cadherin [40], releasing β-catenin and driving *c-myc* expression which stimulates cell proliferation [41]. In addition, BFT transcriptionally upregulates host spermine oxidase (SMO), a polyamine catabolic enzyme. SMO is an epithelial source for inflammation-induced reactive oxygen species (ROS), which can also damage host DNA [42]. Thus, *Bf* is strongly implicated in the initiation phase of CRC.

1.3 Escherichia coli (Ec)

Although *Ec* is a member of the commensal microbiota, a western diet comprised of high amounts of fat promotes the growth of pathogenic adherent invasive *Ec* (AIEC) [43, 44]. Compared to healthy tissue, CRC is characterized by increased amounts of mucosal-adherent *Ec*, with colonization depth strongly correlating with mucosal proliferation index [45]. Transferring CRC-associated *Ec* into tumor-prone *Apc*$^{Min/+}$ mice increased colon polyp numbers, linking pathogenic *Ec* with CRC.

Preclinical evidence implicates *Ec* subsets that promote inflammation, and express an oncotoxin called colibactin, a secondary metabolite produced by the polyketide synthase (*pks*) gene cluster. *Pks* + *Ec* strains are prevalent amongst CRC patients [46], and induce CRC in mouse models [36, 46–48]. Colibactin inflicts a specific DNA damage signature found to be enriched in human CRC genomes, in regions with previously known mutational hotspots [49, 50]. Colibactin-induced DNA double strand breaks (DSBs) result in γH2AX foci, G2/M cell cycle arrest, and resulting activation of ATM/CHK/CDC25/CDK1 DNA damage signaling [46, 51–53]. Cells infected with *pks* + *Ec* rapidly senesce through post-translational modification of the p53 tumor suppressor, resulting in growth factor production that promotes proliferation of uninfected cells [54]; increased tumor growth is a collateral consequence of this senescence-associated secretory phenotype [54]. Pharmacologic inhibitors of the colibactin biosynthetic pathway inhibited its production, with corresponding decrease in DNA damage and CRC cell proliferation *in vitro* [48]. Treating *pks* + *Ec* infected mice with this inhibitor diminished tumorigenesis, highlighting the importance of colibactin in driving cancer initiation *in vivo*.

Host inflammation is inextricably linked to *Ec*'s cancer-inducing properties [46]. Moreover, intestinal inflammation drives *Ec* adaption to these hostile conditions *via* activation of distinct gene expression profiles, e.g. upregulating *pks* to increase colibactin expression [47]. Tumors containing *pks* + *Ec* promoted massive recruitment of inflammatory cells, leading to increased inflammation and production of reactive oxygen species (ROS) [55].

The colonic mucosa of *pks* + *Ec* colonized CRC patients have higher expression of autophagy-associated genes, compared to bacteria-free mucosa. Intestinal epithelial cells infected with *pks* + *Ec* demonstrated canonical DNA damage response and repair, accompanied by autophagy, whereas *Ec* lacking functional colibactin did not. Molecular studies indicated that autophagy was necessary for degrading intracellular bacteria, promoting repair of colibactin-induced DNA damage, and limiting bacteria-induced colonic inflammation. Dysfunctional autophagy prevented DNA damage repair, induced cellular senescence, leading to increased cell proliferation and subsequent colorectal carcinogenesis in tumor prone *Apc*$^{Min/+}$ mice infected with *pks* + *Ec*. Thus, host autophagy is critical for repairing colibactin-induced DNA damage that is induced by colibactin [56]. Modulating autophagy in CRC associated with *pks* + *Ec* may have therapeutic benefit. Intrarectal thiopurine analogs have recently been reported to alleviate colitis and reduce tumor burden in the AOM model, in an autophagy-dependent manner [57].

The cytolethal distending toxin (CDT) is another genotoxin encoded by *Ec* and *Campylobacter* spp. also known to induce DNA DSBs. In vitro studies with human cell lines suggest that CDT may be important for promoting cancer in premalignant colorectal cancer cell lines with underlying mutations in major genes involved in CRC [58]. Thus, *Ec* plays a critical role in the initiation and promotion of CRC.

1.4 Enterococcus fecalis (Ef)

Ef also appears to play a role in the initiation of CRC. *Ef* produces extracellular superoxide that induces chromosomal instability via macrophage-induced bystander effects [59]. A diffusible metabolite 4-hydroxy-2-nonenal produced by *Ef* infected macrophages generated γH2AX foci [60]. This was accompanied by G_2M cell cycle arrest, damaged mitotic spindle, and failure to complete cytokinesis. Thus, this *Ef* metabolite links macrophage-induced bystander effects to CRC.

1.5 Mucus-Invasive Biofilms

Microbial biofilms are ubiquitous in nature, representing the dominant growth state of nearly all bacteria. In the colon, "healthy" biofilms are likely present in most individuals, but are restricted to the lumen of the gut (stool, outer mucus). In contrast, in disease states like IBD and CRC, biofilms invade the normally sterile inner mucus layer and interact extensively with the colonic epithelium [61, 62]. These interactions have been shown to be associated with a number of pro-carcinogenic pathways, including enhanced Ki67, pSTAT3, and alterations in E-cadherin [62]. Mucus-invasive biofilms are also associated with enrichment of the polyamine N [1],N(12)-diacetylspermine, which may play a role in promotion of cellular proliferation in the host [63]. Intriguingly, CRC-associated biofilms are particularly enriched in tumors proximal to the hepatic flexure, where nearly all tumors are biofilm positive, in contrast to distal tumors which range from ~15% - 40% biofilm positive in US and Malaysian CRC cohorts, respectively [13, 62]. However, to date these biofilms have not been associated with other common features of right-sided CRC, such as MSI status, *KRAS* mutations, etc. Additionally, biofilms in healthy screening colonoscopy subjects are present in equal proportions on the left and right sides of the colon, suggesting that at baseline the proximal side is not inherently predisposed to mucus invasion [62].

Importantly, a mixture of mucosal slurries from five biofilm-containing CRC patients was recently shown to be directly tumorigenic in multiple mouse models of colon tumorigenesis [64]. Slurries reassembled into colonic biofilms within 1 week, and concurrently induced potent IL-17 responses. Biofilm-containing slurries from healthy colonoscopy patients and flanking normal tissues from CRC patients also induced colon tumors, whereas slurries from biofilm-negative patient samples did not [64].

Several key pro-carcinogenic microbes may in fact be critical to CRC biofilms. *Fn* is present in a majority of the biofilms by FISH staining, and in approximately 25% of cases large blooms or microcolonies of *Fn* can be observed [7, 13]. In FAP patients, biofilms were dominated ETBF and *pks + Ec*. These two microbes synergistically increased colon tumorigenesis in the $Apc^{Min/+}$ mouse model [7]. Environmental microbial biofilms have enhanced sorption sites provided by the extracellular polymeric matrix, which can be both sources and sinks of carcinogenic toxins, presenting another potential source of pro-carcinogenic signals; this has not been observed to date in human colonic biofilms [65]. It remains elusive whether invasive biofilms precede colon tumorigenesis in humans, or merely represent a marker of aberrant mucosal responses after the adenoma/carcinoma sequence has been initiated.

Modifiable Risk Factor for CRC: Diet and the Microbiome

While colonization with specific, pro-carcinogenic bacteria mentioned in the preceding sections provides potential direct links between the microbiota and risk of CRC, microbiota may also modulate CRC risk via its vast repertoire of metabolic activities that respond to host dietary changes. In the subsequent sections, we will review the evidence for microbial involvement in dietary risk factors for CRC, which is illustrated in Fig. 2.

Prudent Diet Vs. Western Diet

The modern human diet is often described as either a Western diet or a prudent/Mediterranean diet. A Western diet is typically low in fiber and high in red and processed meat, added fats and sugar, and refined grains. Prudent/Mediterranean diets are replete with vegetables, fish, poultry, and whole grains. Large prospective cohorts demonstrated strong beneficial effects of prudent/Mediterranean or vegetarian diets compared to the Western and non-vegetarian diets in terms of CRC risk [66, 67]. Amongst vegetarians, pesco-vegetarians (vegetarians who consumed fish) had the lowest CRC rates (40% lower risk of CRC than non-vegetarians), suggesting a potential additive protective effect of avoiding red and processed meats while consuming fish rich in Ω-3-fatty acids [66, 67].

Importantly, both short- and long-term dietary changes influence gut microbiota composition; therefore, the microbiome may play a key intermediary between diet and CRC risk [68–71]..While some microbial changes result from transmission of bacteria from food (e.g. lactic acid-producing bacteria commonly found in fermented foods like cheese and sauerkraut) [69], the vast majority of diet-responsive microbiome changes are thought to arise from changing preferences for energy-generating substrates for endogenous host intestinal bacteria. Gut microbes are highly specialized at breaking down host dietary components, while synthesizing critical absent elements. Not surprisingly, varied diets differentially alter microbiota. More specifically, a carnivore's microbiome encodes more genes for digesting proteins into amino acids, and is enriched for Bacteroidetes and Actinobacteria phyla, while an herbivore's microbiome tends to be enriched with Firmicutes and Proteobacteria phyla that are highly specialized for amino acid biosynthetic pathways [71, 72]. Finally, not all dietary effects on the microbiome are reversible, as consumption of a chronic low-fiber diet in mice gradually led to extinction of species that did not return after resuming a high-fiber diet [73]. Similarly, obesity induction in mice resulted in long-term microbial compositional changes that did not precisely revert to normal following resumption of a regular diet [74].

Thus, both short- and long-term dietary patterns may strongly influence the gut microbial composition, although these changes may be more dramatic in some individuals than others. The next sections will break down overall dietary patterns into specific dietary components associated with changes in the microbiome and CRC risk. These factors are also summarized in Fig. 2.

High-Fat Diet (HFD) and Bile Acids (BA)

HFD may alter risk of CRC primarily via primary and secondary BA-induced changes. The primary BAs cholic acid and chenodeoxycholic acid are detergent-like molecules synthesized in the liver from cholesterol following consump-

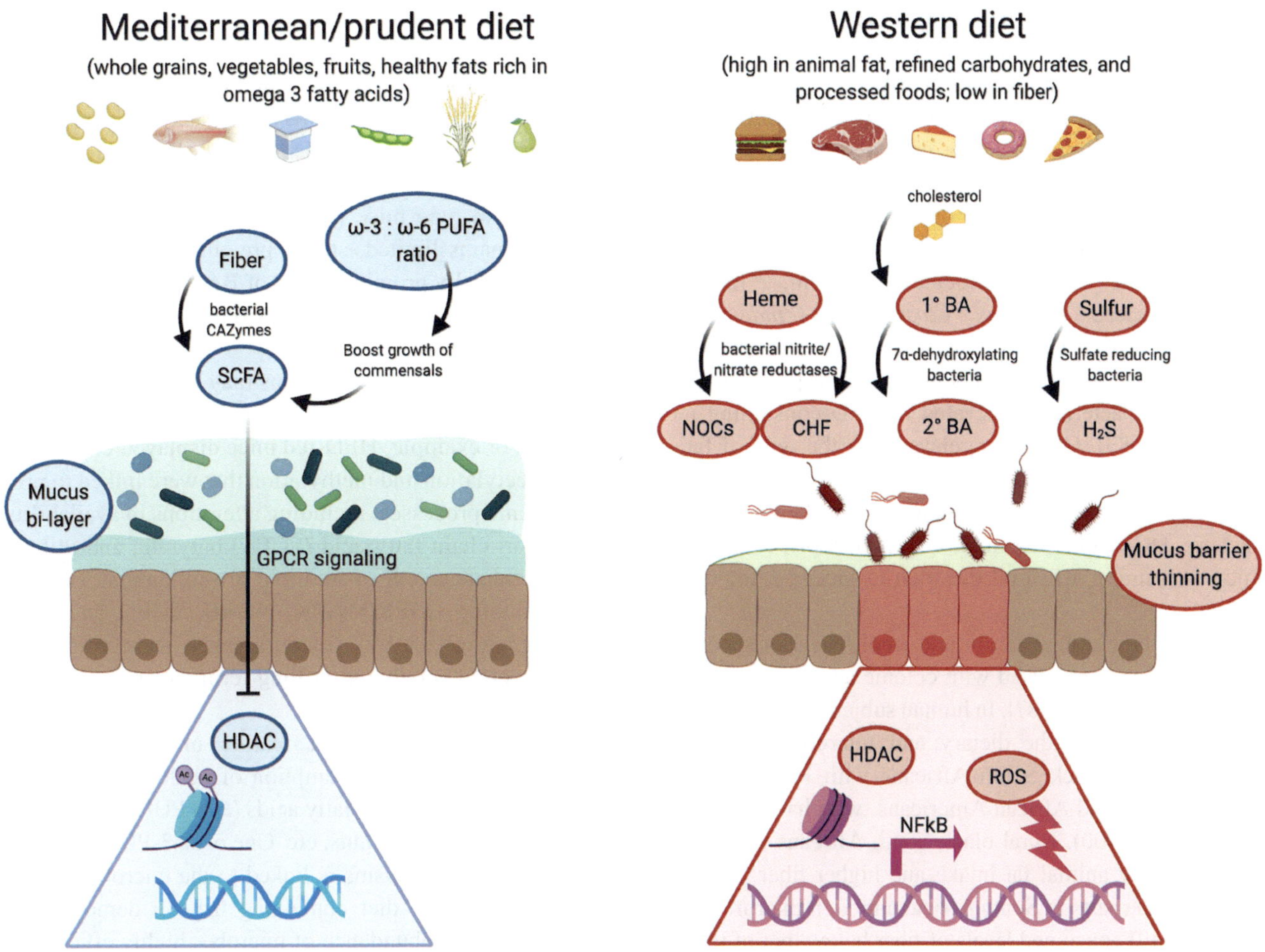

Fig. 2 Microbial mediators of diet and CRC risk. The microbiome is required for many of the beneficial and deleterious effects of healthy and unhealthy diets. **(LEFT)** Mechanisms of homeostasis/protection due to the Mediterranean/prudent diets. 1) **Soluble fiber** is metabolized by bacterial carbohydrate-active enzymes (CAZymes) into SCFAs and other metabolites. Insoluble fiber increases luminal bulk and accelerates transit times to reduce epithelial exposure to carcinogens. 2) **Short chain fatty acids** (SCFAs) provide energy to colonocytes, promote an anti-inflammatory milieu through GPCR signaling and HDAC inhibition, and promote barrier health via stimulation of MUC2. 3) The colon is protected by a dense **mucus bi-layer**, the inner layer of which is typically not infiltrated by bacteria. While germ-free animals still produce mucus, gut microbiota derived SCFAs from dietary mucus, inducing maturation of the mucus layer, and preventing susceptibility to pathogenic and pro-carcinogenic organisms. 4) **Omega-3 polyunsaturated fatty acids** (ω-3 PUFA), highly abundant in healthy foods e.g. fish increase abundance of SCFA-producing bacteria. In contrast, ω-6 PUFA are a hallmark of a Western diet and may contribute to pro-inflammatory mechanisms. **(RIGHT) D**eleterious mechanisms attributable to a Western diet. 1) Excess **Heme** derived from red meat accumulates in the colon, where it can induce ROS directly and/or form **NOC**s and cytotoxic heme factor (**CHF**) which can also damage epithelial cells. 2) **N-nitroso compounds** (NOCs) are produced by bacterial nitrite and nitrate reductase enzymes in response to nitrates and nitrites, commonly found in red meat and preservatives. NOCs can induce DNA adducts and subsequent DNA damage. 3) **Primary bile acids** (1° BA) are synthesized in the liver from cholesterol following consumption of high-fat diets. 1° BA have bactericidal activity that alters the microbiota, killing commensals and promoting expansion of bile-resistant organisms. Excess 1° BA reach the colon and are converted into 2° BA e.g. deoxycholate and lithocholate by 7a-dehydroxylating bacteria. 2° BA can trigger oxidative stress (**ROS**), resulting in DNA damage. 4) Animal proteins are rich in sulfur-containing amino acids e.g. cysteine and methionine, broken down by sulfate-reducing bacteria (SRB) into **hydrogen sulfide** (H_2S) that can induce DNA damage in colonocytes. 5) **Mucus deficits:** In the context of a low-fiber diet, the microbiota switches from consuming complex dietary fiber carbohydrates to degrading mucus layer glycans, resulting in thinning of the protective mucus bi-layer, rendering the colonic epithelium susceptible to pathogens, colitis, and resulting pro-carcinogenic inflammation. 6) Reduced butyrate production by the gut microbiota results in pro-inflammatory epigenetic changes such as increased activity of **histone deacetylases** (HDACs) in colonic epithelial cells. This activity can turn on pro-inflammatory, potentially tumorigenic pathways including NF-κB

tion of HFDs, and are essential for emulsification and solubilization of fats, particularly in the small intestine [75]. However, this ability to disrupt phospholipids also renders them potent antimicrobials that alter microbial composition [75]. Thus, HFD consumption can lead to selection of bile-resistant organisms, which include pro-inflammatory and pro-carcinogenic organisms belonging to the Proteobacteria and Clostridia groups, and/or the hydrogen sulfide-producing

Bilophila wadsworthia [76]. Perhaps even more importantly, approximately 5% of primary BAs reach the colon, where they are converted into secondary BAs including deoxycholate (DCA) and lithocholate (LCA) by select bacteria expressing 7α-dehydroxylating enzymes. These secondary BA induce oxidative stress and DNA damage in vitro [77–81], which consequently may promote tumorigenesis. DCA and LCA are predominantly synthesized by Gram-positive members of the Firmicutes phylum, including Lachnospiraceae such as *Clostridium scindens*, and Peptostreptococcaceae members. Despite comprising an estimated <1% of the normal gut microbiome [82], 7α -dehydroxylating bacteria are elevated in HFD consuming individuals [69, 83]. Of further relevance is the type of fat, as mice consuming a corn oil diet enriched for omega-6 fatty acids exhibited increased abundance of 7α -dehydroxylating bacteria, DCA, and LCA in the ceca and stools, compared to mice consuming a fish oil diet rich in omega-3 fatty acids [67, 84].

Clinically, elevated serum and stool levels of secondary BAs have been associated with colonic adenomas and CRC in patient studies [85–87]. In human subjects, extensive studies have analyzed the dietary and microbial differences between rural black South Africans with low CRC rates (<5:100,000), and African Americans, who have high CRC rates (65:100,000). Rural black South Africans were found to have a low animal fat intake and higher fiber than their white counterparts in South Africa, although the rural South Africans still consumed levels of fiber below that of the recommended daily allowance, suggesting that their relative CRC protection was due to the absence of animal fat [88]. An experimental 2-week dietary switch between rural South Africans and African Americans revealed that consumption of an animal-based diet (high fat, low fiber) for 2 weeks led to an increase in DCA and 7 α-dehydroxylating bacteria in the rural Africans, while consumption of a low-fat/high fiber diet for 2 weeks in African Americans showed the reverse: lowered fecal DCA and fewer 7 α-dehydroxylating bacteria [83]. Furthermore, dietary switches were associated with reciprocal shifts in pro-carcinogenic markers including upregulated proliferation, $CD3^+$ T cells, and $CD68^+$ macrophages in the colonic mucosa of the rural Africans following 2 weeks on HFD [83]. However, a long-term dietary intervention clinical trial (~50,000 post-menopausal women) wherein women were assigned to either consume their typical diet (~35% of calories from fat) or a low-fat diet (~25% of calories from fat) surprisingly demonstrated no protective effect of the low-fat diet on CRC risk, with a median of nearly 20 years of follow-up [89, 90].

Additional mechanistic data on the role of BA and the CRC microbiome have been demonstrated in susceptible mouse models. Oral or rectal administration of DCA and knockout of BA re-uptake receptor in the distal ileum (*Slc10a2*$^{-/-}$ mice) resulted in higher colonic DCA, enhanced colon tumorigenesis in susceptible mouse models (*Apc*$^{Min/+}$ or AOM/DSS models) [91–94]. Additionally, long-term consumption of a Western diet was associated with increased weight gain, increased crypt lengths, aberrant BA metabolism, and an increase in colonic tumors in long-term experiments in wild type mice [95]. Thus, while causal data in patients remains limited, strong pre-clinical evidence suggests a potentially prominent role of BA (especially secondary BA) and the microbiota in pro-tumorigenic processes associated with high fat diets.

Other microbe:diet interactions may contribute to HFD-induced obesity and CRC risk, including epigenetic changes. For example, HFD-fed mice displayed changes in histone acetylation and methylation that were linked to pro-carcinogenic processes, including alterations in availability of the short chain fatty acid (SCFA) butyrate, and utilization and enhancement of multiple pro-inflammatory pathways including EGFR/MAPK, NF-κB, TGFβ, and JAK/STAT signaling [96, 97]. Encouragingly, these epigenetic changes were reversible after long-term dietary changes in the mice [96].

Conversely, an important feature of the anti-cancer Mediterranean diet is consumption of healthy fats such as omega-3-polyunsaturated fatty acids (Ω-3 PUFA), which are found in fish, avocados, nuts, etc. One of Ω-3-PUFA's potential beneficial mechanisms is linked to the microbiome and SCFAs: mice fed a diet containing fish oil demonstrated increased relative abundance of microbes highly efficient in butyrate production, including *Bifidobacterium* and *Lactobacillus* [98, 99]. Similarly, in a dietary intervention study in patients, supplementation with Ω-3-PUFAs led to increases in butyrate producers *Bifidobacterium, Lachnospira, Roseburia,* and *Lactobacillus* [100].

Red and Processed Meat

The International Agency for Research on Cancer has labeled processed meat as a carcinogen, and red meat as a probable carcinogen [101]. The potential mechanisms for their tumorigenic effects include formation of N-nitroso compounds (NOCs) from nitrates and nitrites found in food preservatives, heme iron, production of mutagenic heterocyclic aromatic amines (HAA), and fermentation products of dietary protein.

NOCs underlie some of the strongest data supporting a role for red/processed meat in CRC. Bacterial nitrite and nitrate reductase enzymes are responsible for freeing nitrogenous substances from red meat for N-nitrosation, resulting in NOCs. NOCs frequently induce G → A transitions in *KRAS*, and thus are strongly associated with *KRAS* -mutant CRCs. NOCs consequently form DNA adducts such as O(6)-carboxymethyl guanine (O(6)CMG) that result in DNA damage and mutations [102]. In controlled dietary studies in humans, two-week consumption of a high red meat diet dramatically increased fecal NOCs compared to low red meat diet or vegetarian controls [103]. Importantly, fecal NOC levels generated by high red meat diet (600 g/day) were equivalent to the fecal NOCs generated by cigarette smoking [104]. High red meat diet with high fiber demonstrated intermediate levels of NOCs compared to red meat alone or vegetarian diet controls, suggesting that fiber can modulate the deleterious effects of high red meat diets [102].

Hemoglobin and myoglobin in red meat are the major dietary sources of heme. Excess heme escapes absorption in the small intestine and accumulates in the colon, inducing colonic ROS that triggers formation of cytotoxic heme factor, which damages epithelial cells [105]. Heme also contributes to NOC production, thus driving red meat's tumorigenicity [103]. Finally, heme is also directly cytotoxic and can induce colonic proliferation in a microbially dependent manner [106].

Bacterial fermentation of animal meat proteins generates H_2S, polyamines, secondary BAs, ammonia, and phenolic compounds [107]. Sulfur-containing amino acids (cysteine, methionine) are broken down by sulfate-reducing bacteria into H_2S, which has myriad effects including inhibiting epithelial cell respiration and inducing DNA damage [108]. Multiple amino acids are converted by either host or bacterial enzymes into polyamines, which can alter cell proliferation and contribute to CRC biofilms [63]. Finally, protein can also be metabolized into SCFA, similarly to dietary fiber [108], marking a beneficial role of protein fermentation in the colon. Thus, while microbial digestion of animal protein has some deleterious effects, these may be counteracted by SCFA production.

Finally, high-temperature cooking, such as grilling or pan-frying generates potentially genotoxic HAAs, whose exposure to the intestinal epithelium can be enhanced by bacterial ß-glucuronidase enzymes. For example, *E. coli* ß-glucuronidase dramatically increased the DNA damaging activity of the HAA 2-amino-3-methylimidazo[4,5-f]quinoline (IQ) [109]. Similarly, *F. prausnitzii* ß-glucuronidase promoted the hydrolysis of the inactive HAA PhiP-glucuronide into the DNA-damaging PhiP metabolite [110].

Dietary Fiber

Dietary fiber is a prominent component of the healthy Mediterranean/prudent diet and is intimately intertwined with the gut microbiota. While insoluble fiber (e.g. cellulose from plant cell walls) bulks stool and reduces colonic transit time thereby limiting the exposure of gastrointestinal mucosa to potential carcinogens, soluble fiber and resistant starches are fermented in the colon by a diverse array of carbohydrate-active enzymes (CAZymes) expressed by the gut microbiota. The dominant output of these microbial CAZymes are SCFAs e.g. acetate, propionate, and butyrate, whose luminal concentrations reach approximately 50–200 mM [111]. SCFAs have several critical roles in gut homeostasis, including being an energy source for the colonic epithelium, regulating gene expression, inducing MUC2 mucin production, and initiating myriad signaling pathways (reviewed in [9, 111–114]).

Acetate is produced by most gut anaerobes, while propionate and butyrate have more specialized microbial sources. The major butyrate producers are Firmicutes phylum members, including *F. prausnitzii, Clostridium reptum, Eubacterium rectale*, and *Roseburia* spp. [115]. In contrast, major propionate producers belong to phylum Bacteroidetes and the Negativicutes class of Firmicutes, amongst others [115]. SCFA functions vary dramatically: propionate plays an important role in satiety through signaling pathways. Butyrate is a ligand for certain GPRs implicated in tumor suppression [116], and is absorbed by colonocytes, where it localizes to both the mitochondria and nucleus. Mitochondrial butyrate provides >70% of the energy demands of the cell [117], and its oxidation triggers a hypoxia-inducible factor 1α-based mechanism to preserve the epithelial barrier [118]. Nuclear butyrate acts as a histone deacetylase inhibitor (HDACi) [119, 120], thereby upregulating expression of tight junction genes to further maintain barrier integrity. Thus, butyrate is considered anti-inflammatory and important for gut barrier integrity.

Consumption of a low fiber diet alters gut microbiota and consequent SCFA production in several important ways, including an overall reduction in microbial diversity, expansion of microbes unreliant on dietary fiber, and alterations in barrier integrity/mucus layer, all of which may precipitate pro-carcinogenic events in the colon. A switch in fermentation of dietary fiber to mucus glycans on a low fiber diet led to thinning of the mucus layer itself, which in turn rendered mice susceptible to pathogenic infections with *Citrobacter rodentium* [121]. Mucus degradation can occur even in mice on regular (high fiber) diets: in the *Apc*$^{Min/+}$ mouse model of CRC, degradation of host mucus by ETBF led to enhanced mucosal colonization of *pks + E. coli* and enhanced colon tumorigenesis compared to mono-colonized and sham mice [7].

Conversely, adding additional fiber to the diet can be protective. Tumor-prone mice fed high fiber diet had fewer colon polyps and elevated SCFAs compared to regular chow fed mice [122]; colonization with the butyrate producer *Butyrivibrio fibrisolvens* further augmented this protection [123]. Furthermore, while normal colonocytes rapidly utilize butyrate as an energy source, tumors are known for switching from mitochondrial butyrate oxidation to glycolysis, a phenomenon known as the Warburg effect. Colon tumors from the fiber-fed, *B. fibrisolvens* colonized mice displayed nuclear butyrate accumulation leading to potent HDACi activity resulting in increased apoptosis and reduced cell proliferation [123].

Despite the intriguing animal data on fiber and human studies on the protective effects overall of a prudent diet vs. a Western diet against CRC risk [124], large, prospective studies that have examined the effects of fiber by itself have reported inconsistent results. A study conducted over 3–5 years found that fruit- and cereal fibers drive protection against distal colorectal adenomas and distal CRC [125]. Another study found no protective effect of total dietary fiber, although there was a protective trend towards women who consumed the lowest amounts of folate, and high amounts of fruit fiber [126]. In a recent follow-up study, when tumors were stratified by *F. nucleatum* status, fiber was found to be protective, but only in *F. nucleatum*-containing CRCs [66]. A separate study revealed that a high fiber diet protected against Crohn's disease but not ulcerative colitis [127], and once again fruit fibers were the most protective. In a cross-sectional cohort, patients with advanced colorectal adenomas at colonoscopy had more acetate, propionate, and butyrate in their stool than healthy controls, which paralleled lower relative abundances of the SCFA producers *Clostridium, Roseburia*, and *Eubacterium* spp. in the adenomas as well [128]. Finally, several interventional studies have tested whether fiber supplements can reduce risk of polyps or CRC: a protective effect was found in a population of FAP patients over a 4 year course [129]. In a separate study of patients with a history of adenomas, intervention with a low fat/high fiber diet significantly reduced the incidence of new large adenomas 24–48 months later [130], although fiber alone or a low-fat diet alone did not have statistically significant impacts, indicating the importance of holistic dietary changes.

Vitamins, Minerals, and Antioxidants

Additional vitamins and minerals including Vitamins C, D, E, calcium, and selenium demonstrate protection against CRC, but mechanistic data remain limited [131]. Three separate cohorts suggest yogurt consumption protects from developing colon polyps, also via unelucidated mechanisms [132, 133]. A posited mechanism is that calcium counteracts deleterious effects of heme contained in red meat [134], similar to the role that dietary fiber can play against high-fat diets. Gut microbiota can also convert other dietary components into metabolites with potential tumor-suppressor functions [9, 113]. Broccoli and other cruciferous vegetables contain glucosinolates which are converted into sulforaphane and other isothiocyanates which have epigenetic and anti-inflammatory effects. Soy-based foods contain daidzein which is converted into equol, an antioxidant. Finally, pomegranates and certain berries contain ellagitannins which are converted into urolithins, that impact cellular physiology and improve mitochondrial health [135].

1.6 Non-modifiable Risk Factor for CRC: Inflammatory Bowel Diseases and the Microbiome

Inflammatory bowel diseases (IBD), consisting of Crohn's disease (CD) and ulcerative colitis (UC) are chronic, progressive, inflammatory diseases of the gastrointestinal tract. A recent study revealed that 6.8 million cases of IBD occur globally, with significant geographic variation in disease levels and trends [136]. IBD prevalence rises in tandem with increased sociodemographic index, with the USA and UK having the first and second highest age-standardized prevalence in the world. IBD is on the rise in newly industrialized countries in South America, Asia, eastern Europe, and Africa. IBD is one of the known risk factors for developing CRC [137, 138], and the risk is driven by both patient- and disease-related factors. Patient-related factors include male sex, age at which IBD onset occurred, a family history of CRC, and primary sclerosing cholangitis. Disease-related factors include the extent and duration of colitis, severity of inflammation and aggregate inflammatory burden, and structural alterations that result from chronic inflammation e.g. strictures. Colitis-associated cancer (CAC) arises due to recurrent intestinal epithelial inflammation and repair, with accompanying prolonged exposure to inflammatory cytokines and reactive oxygen species [139, 140].The increased risk of CRC is in part due to every stage of the disease characterized by chronic inflammation, which influences nearly every stage of tumorigenesis including transformation, proliferation, angiogenesis, invasion, and metastasis. Indeed, inflammation is a well-described hallmark of cancer [141].

Here, we review preclinical evidence utilizing sophisticated mouse models to unravel the reciprocal mechanisms between colitis/IBD and the microbiome that impact intestinal barrier, inflammation, and immune regulation, which relate to the development or progression of CRC.

2 Intestinal Barrier Defects in IBD

2.1 Mucins

As with CRC, reduced mucus production and loss of compartmentalization between microbiota and the epithelium occurs with IBD. Bacterial spatial distribution within the mucosa of IBD patients differed from that of controls [142], and their mucosa also contains fewer goblet cells and a thinner mucus layer [143]. Mucins, differentially expressed throughout the gut [144], play varied roles in IBD. MUC2-deficient mice are highly susceptible to DSS-mediated colitis and spontaneous CRC [145]. Characterized by enhanced proliferation and protection from apoptosis, MUC2-deficient mice develop rectal tumors and small intestinal adenomas that progress to invasive adenocarcinoma [146].

The *il10*-deficient mouse (*il10*KO) is a widely used model of experimental colitis [147, 148]. Compared to germ-free (GF) wild-type (WT) mice, GF-*il10*KO mice had significantly lower MUC2 synthesis. Colonization with enteric bacteria induced MUC2 biosynthesis in *il10*KO but not WT mice [149]. Colitis-prone Winnie mice also bear a specific mutation in MUC2, leading to altered mucus production and dysbiosis characterized by increased abundance of Bacteroidetes and Verrucomicrobia [150].

Thus, a compromised mucus barrier with aberrant MUC2 levels renders intestinal mucosa susceptible to colitis resulting from non-pathogenic commensals. In contrast, MUC4 deficiency induces compensatory MUC2 and MUC3 expression, reduced inflammatory infiltrates, and reduced gene expression of inflammatory cytokines and antimicrobial peptides, resulting in significantly reduced tumor burden in the AOM/DSS model compared to WT mice [151]. Collectively, these studies highlight the importance of the mucus barrier in compartmentalization of microbiota and the epithelium, and protection from colitis.

2.2 α6β4 Integrin

Apical mucus and the basolateral basement membrane form a barrier that divides the host from commensal and pathogenic bacteria. Polarized intestinal epithelial cells are anchored to the basement membrane via α6β4 Integrin, the lack of which results in misshapen epithelial cells and epithelial detachment leading to increased intestinal barrier permeability and subsequent colitis, in addition to a defective mucus barrier that allowed bacterial attachment to the colonic epithelium [152]. Ultimately, these mice developed spontaneous adenocarcinomas whose severity was driven by the extent of activation of gene networks involved in extracellular matrix composition and remodeling, and inflammation, molecular signatures well-described in colorectal carcinogenesis.

2.3 LY6/PLAUR Domain Containing 8 (LYPD8)

LYPD8 is among the top 10 upregulated transcripts in the transcriptome of long-course UC patients that develop CAC [153]. LYPD8 is a highly glycosylated glycosylphosphatidylinositol-anchored protein expressed on differentiated epithelial cells, and is also secreted into the intestinal lumen, wherein it binds to flagellated bacteria and impairs bacterial motility. LYPD6 deficiency results in increased abundance of pathogenic bacteria such as *Proteus* spp., *Helicobacter* spp., and *Escherichia* spp., which exert colitogenic effects deep within the inner mucus layer of intestinal crypts; these mice are also highly susceptible to DSS colitis [154]. In contrast, LYPD8 is downregulated in sporadic CRC [155]. Moreover, LYPD8 exerted tumor inhibitory activity via reduced phosphorylation of STAT3 and the NF-κB P65 subunit, leading to reduced secretion of proinflammatory IL-6 and TNFα.

2.4 Lipocalin-2 (Lcn-2)

Lcn-2 is a secreted siderophore-binding protein upregulated in murine experimental colitis, and neutrophil gelatinase-associated lipocalin (NGAL) is its human ortholog. Lcn-2/NGAL is dramatically induced upon inflammation in a MyD88-dependent manner, and its deficiency enhanced colitogenic T cells and exacerbated symptoms in the murine DSS colitis model in a microbiota-dependent manner; antibiotic treatment alleviated this phenotype [156]. Lcn-2 deficiency promotes dysbiosis with an expansion of Proteobacteria [156], which is also observed in IBD [157] and CRC [9, 46, 47]. Concurrently, Lcn-2 is among the top 10 downregulated coding transcripts in the mucosal transcriptome of patients with long-course UC, who frequently develop colitis-associated cancer (CAC) [153]. Furthermore Lcn-2 has a protective role in early-onset colitis and right-sided colon tumors, as double deficient *il10*KO, Lcn-2KO mice have increased IL-6 driven inflammation and tumorigenesis, and an accompanying bloom of the facultative pathogen *Alistipes* spp. This colitogenic, inflammatory microbiota induced colitis and right-sided tumors when transferred into il10KO susceptible mice [158].

2.5 Notch-1

Intestinal crypt cell proliferation and the fate of secretory lineages is controlled by Notch signaling which activates downstream *Hes* family genes. Many signaling pathways impinge upon Notch-1, whose expression and downstream signaling are aberrant in CRC [159]. Moreover, the fucosylated extracellular ligand-binding domains of Notch are aberrantly glycosylated in CRC, another signature of malignant transformation [160]. Mice with incorrectly fucosylated Notch develop colitis that progresses to adenocarcinoma, resulting from loss of Notch signaling [161].

2.6 Keratin-8 (K8)

K8 is one of the main keratin filaments, which guard against cellular stress within the intestinal epithelium. K8-deficient mice display epithelial hyperproliferation and chronic colonic inflammation resembling UC. K8-deficiency leads to increased distal colon tumors resulting from AOM, or when combined with the *Apc*$^{Min/+}$ model. The increased tumor number was due to aberrant IL-22 signaling, and epithelial inflammasome activation leading to increased cleaved caspase-1 and IL-18 [162].

K8 heterozygous KO mice develop cancer following AOM/DSS treatment, and have characteristic dysbiosis with increased Firmicutes and Proteobacteria, with concurrent decrease in Bacteroidetes and Verrucomicrobia. Tumorigenesis was inhibited by antibiotic treatment, underscoring importance of microbiota in this model. Thus, Keratin 8 is critical for modulating colonic permeability and regulating gut microbiota composition [163].

2.7 Calcineurin/Nuclear Factor of Activated T-Cells (NFAT) Signaling

Calcineurin, which is upregulated in CRC, is a phosphatase important for activation of NF-AT transcription factors; both are constitutively expressed in the intestinal epithelium. Detection of microbiota by innate immune sensors activates calcineurin, resulting in calcineurin-dependent intestinal tumor development in *Apc*$^{Min/+}$ mice [164]. The pronounced activation of calcineurin-NF-AT signaling was due to impaired barrier function in tumor-bearing mice which had a dysbiotic microbiome characteristic of colorectal tumors: increased abundance of *Escherichia*, *Shigella*, and *Citrobacter* genera, and overall increases at the family and phylum levels of Enterobacteriaceae and Proteobacteria respectively. Dysbiosis alters epithelial responses to bacteria. Thus NF-AT bound in a microbiota- and calcineurin-dependent manner and activated the promoters of *Cd44*, *Lgr5*, *Olfm4*, and *Dclk4*, which are critical for intestinal stem cells [165], thereby supporting tumor stem cell function.

3 Immune Signaling and Modulation in IBD

3.1 Innate Immune System Sensors

Pattern recognition receptors (PRRs) comprised of Toll Like Receptors (TLRs), Nod Like Receptors (NLRs), and Retinoic Acid Inducible Gene Like Receptor (RLRs) detect pathogen associated molecular patterns (PAMPs), which are conserved, non-host molecular motifs such as cell wall components (peptidoglycan, lipoteichoic acid, lipopolysaccharide, flagellin) and nucleic acids (double stranded RNA) shared by classes of microbes. TLR4 recognizes lipopolysaccharide from Gram-negative bacteria, and is upregulated in epithelium of IBD, and both CAC and sporadic CRC [140]. TLR4-deficient mice were significantly protected in an AOM/DSS model, as TLR4-replete mice had activated EGFR signaling and cyclooxygenase-2 induction, resulting in elevated prostaglandin E2 production, which drove CAC [166]. NLRP12 deficiency also promotes growth of colitogenic bacteria (Erysipelotrichaceae) and loss of beneficial, protective commensals like Lachnospiraceae. Supplementation with the latter, or blocking inflammatory cytokines, reversed dysbiosis and colitis [167]. NLRP12 deficiency drives colitis and subsequent CAC via robust, unregulated NF-kB activation and ERK signaling, resulting in hyperproliferation and inflammation [168].

Crossing *Apc*$^{Min/+}$ mice with mice lacking the TLR adaptor MyD88, thus incapable of TLR signaling, resulted in significantly fewer tumors, resulting from reduced expression of genes involved in inflammation and epithelial repair [169]. Subsequent studies revealed that myeloid cells upregulate STAT3 [170] and IL-23 [171] following TLR signaling in response to microbes, which drives tumor growth. Overexpression of TLR4 in the intestinal epithelium increases proliferation, expands crypts, and aged mice develop duodenal dysplasia. AOM administered to TLR4 overexpressing mice increased colon tumors and enhanced beta-catenin, which was found to be PI3K dependent. Thus, TLR4 activation triggers neoplasia via Wnt/β-catenin activation, linking innate immunity to CRC [172].

3.2 Innate Immune Signaling Modulators

Combined TLR and Wnt activation results in intratumoral expression of IRAK-M, a negative regulator of TLR signaling that also modulates microbial diversity, and stabilizes STAT3 to promote tumor growth [173]. Activating Transcription Factor 6 (ATF6) is a regulator of endoplasmic stress whose upregulation correlates with abbreviated survival in CRC patients. Intestinal epithelial specific activation of ATF6 led to dysbiosis and bacterial penetration into the inner mucus layer, which activated TLR signaling and downstream JAK-STAT pathway activation, resulting in increased tumors [174].

Multiple PRR pathways including TLRs, RLRs, and the cGAS/STING pathway impinge upon Interferon Regulatory Factor 3 (IRF3), a signaling mediator and a transcription factor. A recent study discovered that cytoplasmic IRF inhibits proliferative signaling and tumor formation by sequestering beta-catenin within the cytoplasm, which is relieved by microbial IRF3 activation. Thus, IRF3 integrates microbiota and CRC via negative regulation of the beta-catenin pathway [175].

3.3 Interleukin-22 (IL-22)

Immune cells at the mucosal surface produce Interleukin-22 (IL-22), which promotes intestinal stem cell mediated epithelial regeneration, and maintains barrier integrity. Inflammatory conditions of CD and UC upregulate colonic IL-22 expression [176]. Whereas it protects intestinal stem cells from genotoxic stress, IL-22's extended and deregulated activity leads to intestinal tumorigenesis [177]. IL-22 is highly expressed in human colorectal cancer [178] and is correlated with chemoresistance [179]. IL-22 stimulates the production of antibacterial peptides e.g. regenerating gene (Reg)-IIIγ and RegIIIβ, which are important for selectively controlling Gram-positive bacteria [180, 181], underscoring the mutualism that links resident microbiota with mucosal immunity.

The soluble receptor of IL-22, IL-22BP is critical for regulating colon epithelial cell proliferation and subverting tumorigenesis. Excessive IL-22 expression without careful control by IL-22BP can promote tumor development. In fact, IL-22BP is significantly downregulated in colorectal tumors with a concurrent increase in IL-22 [182]. IL-22BP was found to be regulated via NF-κB signaling by the TNF superfamily member lymphotoxin which prevents the overgrowth of certain intestinal microbiota, including segmented filamentous bacteria [183].

4 Gut Microbiota and Cancer Treatment

Microbiota encode a plethora of enzymes with immense metabolic potential, and the capability to carry out innumerable, highly sophisticated biotransformations. Bacterial enzymes can alter the metabolism of diverse pharmaceuticals including chemotherapeutic drugs, which can result in altered efficacy, or varying side effects, sometimes severe enough to require suspending treatment. Furthermore, intestinal bacterial composition and translocation to lymphoid organs can also impact the efficacy of immunotherapies such as checkpoint inhibitors. We encourage readers to refer to recent reviews describing how the microbiome impacts the efficacy of a number of widely-used drugs, including treatments for cancer [9, 184–192].

5 Summary

Microbiota have both tumor suppressive and oncogenic effects, carried out by several molecular mechanisms summarized in Fig. 3. These mechanisms, in most cases, are diametrically opposed. Microbial proteins may exert genotoxic effects, which can be remedied by microbiota-dependent repair occurring through host processes such as autophagy. Oxidative stress induced by reactive oxygen species can be reversed by antioxidant molecules synthesized by the microbiota from the host diet. Immune activation due to bacterial translocation, or certain microbial metabolites such as secondary bile acids can promote inflammation, which is beneficial in certain cases, e.g. with cancer immunotherapy. Inflammation can be dampened by immunomodulatory microbial metabolites such as short chain fatty acids. Barrier integrity can either be improved or attenuated by bacteria-driven mucus synthesis or digestion, or other signaling pathways that additionally promote cellular proliferation. Beneficial and highly diverse microbiota may deteriorate into dysbiosis, with lack of spatial organization and development of biofilms containing pro-carcinogenic bacteria. Each of these molecular mechanisms likely occurs in combination, rather than in isolation. Combinatorial mechanisms following an initiating event may accelerate transformation, and result in tumorigenesis.

Acknowledgements Figures were created with BioRender.com. JLD is supported by R00CA230192. APB is supported by the Crohn's and Colitis Foundation, the Global Probiotics Council, and the University Cancer Research Fund.

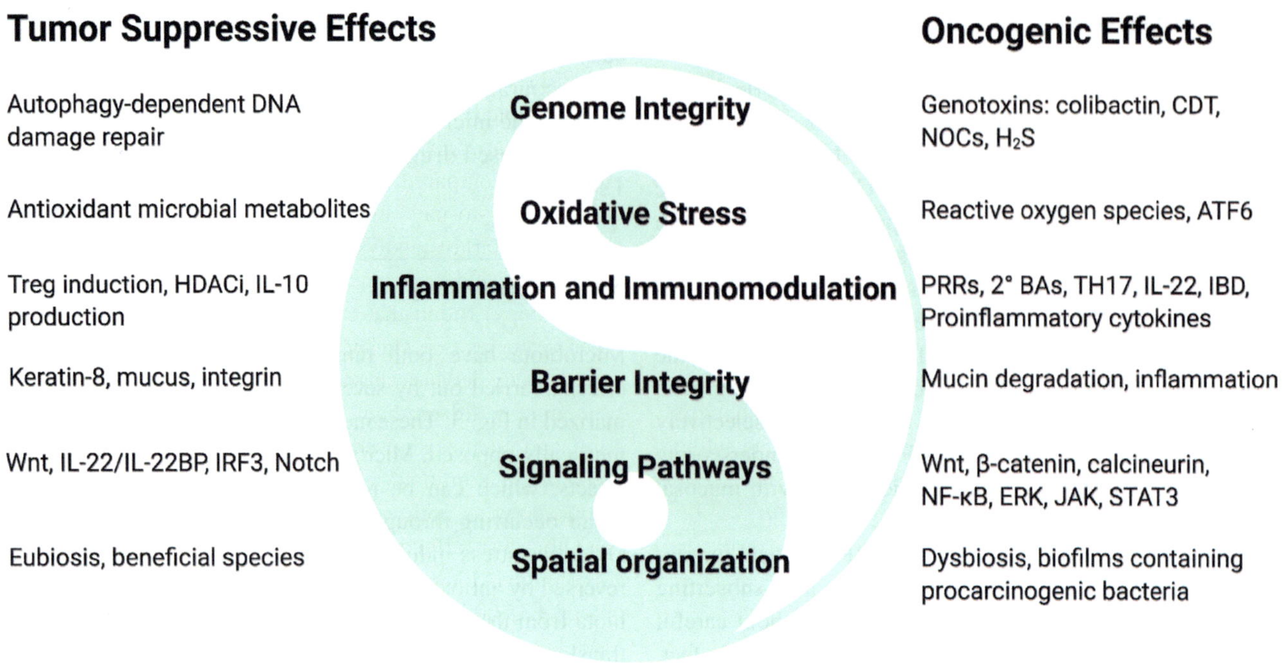

Fig. 3 Summary of microbial mechanisms with tumor suppressive or oncogenic effects. Various molecular mechanisms contribute to ways in which the microbiome influences CRC. These mechanisms are driven by microbial gene products and metabolites with genotoxic, immunomodulatory, and signaling capabilities, in addition to microbial composition, diversity, and spatial organization. Each mechanism is described in detail in the main text. *Treg* regulatory T cell, *HDACi* histone deacetylase inhibitor, *IL-10, −22* interleukin-10, −22, *IL-22BP* interleukin-22 binding protein, *IRF3* interferon regulatory transcription factor 3, *CDT* cytolethal distending toxin, *NOC* N-nitroso compounds, *PRR* pattern recognition receptor, *BA* bile acids, *IBD* inflammatory bowel diseases, *NF- κB* Nuclear Factor κB, *ERK* extracellular signal-regulated kinase, *JAK* janus kinase, *STAT3* signal transducer and activator of transcription 3

References[1]

1. Laqueur GL, McDaniel EG, Matsumoto H. Tumor induction in germfree rats with methylazoxymethanol (MAM) and synthetic MAM acetate. J Natl Cancer Inst. 1967;39:355–71.
2. Reddy BS, Weisburger JH, Narisawa T, et al. Colon carcinogenesis in germ-free rats with 1,2-dimethylhydrazine and N-methyl-n'-nitro-N-nitrosoguanidine. Cancer Res. 1974;34:2368–72.
3. Wong SH, Zhao L, Zhang X, et al. Gavage of fecal samples from patients with colorectal cancer promotes intestinal carcinogenesis in germ-free and conventional mice. Gastroenterology. 2017;153:1621–1633.e6.
4. Rosshart SP, Vassallo BG, Angeletti D, et al. Wild mouse gut microbiota promotes host fitness and improves disease resistance. Cell. 2017;171:1015–1028.e13.
5. Rosshart SP, Herz J, Vassallo BG, et al. Laboratory mice born to wild mice have natural microbiota and model human immune responses. Science. 2019;365:eaaw4361.
6. Yan Y, Drew DA, Markowitz A, et al. Structure of the mucosal and stool microbiome in lynch syndrome. Cell Host Microbe. 2020;27:585–600.e4.
7. Dejea CM, Fathi P, Craig JM, et al. Patients with familial adenomatous polyposis harbor colonic biofilms containing tumorigenic bacteria. Science. 2018;359:592–7.
8. Wong SH, Yu J. Gut microbiota in colorectal cancer: mechanisms of action and clinical applications. Nat Rev Gastroenterol Hepatol. 2019;16:690–704.
9. Bhatt AP, Redinbo MR, Bultman SJ. The role of the microbiome in cancer development and therapy. CA Cancer J Clin. 2017;67:326–44.
10. Loke YL, Chew MT, Ngeow YF, et al. Colon carcinogenesis: the interplay between diet and gut microbiota. Front Cell Infect Microbiol. 2020;10:603086.
11. Castellarin M, Warren RL, Freeman JD, et al. Fusobacterium nucleatum infection is prevalent in human colorectal carcinoma. Genome Res. 2012;22:299–306.
12. Feng Q, Liang S, Jia H, et al. Gut microbiome development along the colorectal adenoma-carcinoma sequence. Nat Commun. 2015;6:6528.
13. Drewes JL, White JR, Dejea CM, et al. High-resolution bacterial 16S rRNA gene profile meta-analysis and biofilm status reveal common colorectal cancer consortia. NPJ Biofilms Microbiomes. 2017;3:34.
14. Lopez LR, Bleich RM, Arthur JC. Microbiota effects on carcinogenesis: initiation, promotion and progression. Annu Rev Med. 2020;72:243–61.
15. Tjalsma H, Boleij A, Marchesi JR, et al. A bacterial driver-passenger model for colorectal cancer: beyond the usual suspects. Nat Rev Microbiol. 2012;10:575–82.
16. Sears CL, Garrett WS. Microbes, microbiota, and colon cancer. Cell Host Microbe. 2014;15:317–28.
17. Momen-Heravi F, Babic A, Tworoger SS, et al. Periodontal disease, tooth loss and colorectal cancer risk: results from the Nurses' health study. Int J Cancer. 2017;140:646–52.
18. Kostic AD, Chun E, Robertson L, et al. Fusobacterium nucleatum potentiates intestinal tumorigenesis and modulates the tumor-immune microenvironment. Cell Host Microbe. 2013;14:207–15.
19. Ohkusa T, Okayasu I, Ogihara T, et al. Induction of experimental ulcerative colitis by fusobacterium varium isolated from colonic mucosa of patients with ulcerative colitis. Gut. 2003;52:79–83.

[1]We sincerely apologize to colleagues whose work we were unable to include due to space constraints.

20. Ohkusa T, Yoshida T, Sato N, et al. Commensal bacteria can enter colonic epithelial cells and induce proinflammatory cytokine secretion: a possible pathogenic mechanism of ulcerative colitis. J Med Microbiol. 2009;58:535–45.
21. Liu H, Hong XL, Sun TT, et al. Fusobacterium nucleatum exacerbates colitis by damaging epithelial barriers and inducing aberrant inflammation. J Dig Dis. 2020;21:385–98.
22. Flanagan L, Schmid J, Ebert M, et al. Fusobacterium nucleatum associates with stages of colorectal neoplasia development, colorectal cancer and disease outcome. Eur J Clin Microbiol Infect Dis. 2014;33:1381–90.
23. Yamaoka Y, Suehiro Y, Hashimoto S, et al. Fusobacterium nucleatum as a prognostic marker of colorectal cancer in a Japanese population. J Gastroenterol. 2018;53:517–24.
24. Mima K, Sukawa Y, Nishihara R, et al. Fusobacterium nucleatum and T cells in colorectal carcinoma. JAMA Oncol. 2015;1:653–61.
25. Mima K, Cao Y, Chan AT, et al. Fusobacterium nucleatum in colorectal carcinoma tissue according to tumor location. Clin Transl Gastroenterol. 2016;7:e200.
26. Yu T, Guo F, Yu Y, et al. Fusobacterium nucleatum promotes chemoresistance to colorectal cancer by modulating autophagy. Cell. 2017;170:548–563.e16.
27. Bullman S, Pedamallu CS, Sicinska E, et al. Analysis of fusobacterium persistence and antibiotic response in colorectal cancer. Science. 2017;358:1443–8.
28. Xu M, Yamada M, Li M, et al. FadA from fusobacterium nucleatum utilizes both secreted and nonsecreted forms for functional oligomerization for attachment and invasion of host cells. J Biol Chem. 2007;282:25000–9.
29. Rubinstein MR, Baik JE, Lagana SM, et al. Fusobacterium nucleatum promotes colorectal cancer by inducing Wnt/beta-catenin modulator Annexin A1. EMBO Rep. 2019;20:e47638.
30. Dadashi M, Hajikhani B, Faghihloo E, et al. Proliferative effect of FadA recombinant protein from fusobacterium nucleatum on SW480 colorectal cancer cell line. Infect Disord Drug Targets. 2020;21:623–8.
31. Zhang S, Li C, Liu J, et al. Fusobacterium nucleatum promotes epithelial-mesenchymal transiton through regulation of the lncRNA MIR4435-2HG/miR-296-5p/Akt2/SNAI1 signaling pathway. FEBS J. 2020;287:4032–47.
32. Guo P, Tian Z, Kong X, et al. FadA promotes DNA damage and progression of fusobacterium nucleatum-induced colorectal cancer through up-regulation of chk2. J Exp Clin Cancer Res. 2020;39:202.
33. Abed J, Emgard JE, Zamir G, et al. Fap2 mediates fusobacterium nucleatum colorectal adenocarcinoma enrichment by binding to tumor-expressed gal-GalNAc. Cell Host Microbe. 2016;20:215–25.
34. Gur C, Ibrahim Y, Isaacson B, et al. Binding of the Fap2 protein of fusobacterium nucleatum to human inhibitory receptor TIGIT protects tumors from immune cell attack. Immunity. 2015;42:344–55.
35. Gur C, Maalouf N, Shhadeh A, et al. Fusobacterium nucleatum supresses anti-tumor immunity by activating CEACAM1. Onco Targets Ther. 2019;8:e1581531.
36. Tomkovich S, Yang Y, Winglee K, et al. Locoregional effects of microbiota in a preclinical model of colon carcinogenesis. Cancer Res. 2017;77:2620–32.
37. Round JL, Mazmanian SK. Inducible Foxp3+ regulatory T-cell development by a commensal bacterium of the intestinal microbiota. Proc Natl Acad Sci U S A. 2010;107:12204–9.
38. Mazmanian SK, Round JL, Kasper DL. A microbial symbiosis factor prevents intestinal inflammatory disease. Nature. 2008;453:620–5.
39. Boleij A, Hechenbleikner EM, Goodwin AC, et al. The Bacteroides fragilis toxin gene is prevalent in the colon mucosa of colorectal cancer patients. Clin Infect Dis. 2015;60:208–15.
40. Wu S, Lim KC, Huang J, et al. Bacteroides fragilis enterotoxin cleaves the zonula adherens protein, E-cadherin. Proc Natl Acad Sci U S A. 1998;95:14979–84.
41. Wu S, Morin PJ, Maouyo D, et al. Bacteroides fragilis enterotoxin induces c-Myc expression and cellular proliferation. Gastroenterology. 2003;124:392–400.
42. Goodwin AC, Destefano Shields CE, Wu S, et al. Polyamine catabolism contributes to enterotoxigenic Bacteroides fragilis-induced colon tumorigenesis. Proc Natl Acad Sci U S A. 2011;108:15354–9.
43. Martinez-Medina M, Denizot J, Dreux N, et al. Western diet induces dysbiosis with increased E coli in CEABAC10 mice, alters host barrier function favouring AIEC colonisation. Gut. 2014;63:116–24.
44. Agus A, Denizot J, Thevenot J, et al. Western diet induces a shift in microbiota composition enhancing susceptibility to adherent-Invasive E. Coli infection and intestinal inflammation. Sci Rep. 2016;6:19032.
45. Bonnet M, Buc E, Sauvanet P, et al. Colonization of the human gut by E. Coli and colorectal cancer risk. Clin Cancer Res. 2014;20:859–67.
46. Arthur JC, Perez-Chanona E, Muhlbauer M, et al. Intestinal inflammation targets cancer-inducing activity of the microbiota. Science. 2012;338:120–3.
47. Arthur JC, Gharaibeh RZ, Muhlbauer M, et al. Microbial genomic analysis reveals the essential role of inflammation in bacteria-induced colorectal cancer. Nat Commun. 2014;5:4724.
48. Cougnoux A, Delmas J, Gibold L, et al. Small-molecule inhibitors prevent the genotoxic and protumoural effects induced by colibactin-producing bacteria. Gut. 2016;65:278–85.
49. Dziubanska-Kusibab PJ, Berger H, Battistini F, et al. Colibactin DNA-damage signature indicates mutational impact in colorectal cancer. Nat Med. 2020;26:1063–9.
50. Pleguezuelos-Manzano C, Puschhof J, Rosendahl Huber A, et al. Mutational signature in colorectal cancer caused by genotoxic pks(+) E. Coli. Nature. 2020;580:269–73.
51. Lopez LR, Bleich RM, Arthur JC. Microbiota effects on carcinogenesis: initiation, promotion, and progression. Annu Rev Med. 2021;72:243–61.
52. Nougayrede JP, Homburg S, Taieb F, et al. Escherichia coli induces DNA double-strand breaks in eukaryotic cells. Science. 2006;313:848–51.
53. Cuevas-Ramos G, Petit CR, Marcq I, et al. Escherichia coli induces DNA damage in vivo and triggers genomic instability in mammalian cells. Proc Natl Acad Sci U S A. 2010;107:11537–42.
54. Dalmasso G, Cougnoux A, Delmas J, et al. The bacterial genotoxin colibactin promotes colon tumor growth by modifying the tumor microenvironment. Gut Microbes. 2014;5:675–80.
55. Veziant J, Gagniere J, Jouberton E, et al. Association of colorectal cancer with pathogenic Escherichia coli: focus on mechanisms using optical imaging. World J Clin Oncol. 2016;7:293–301.
56. Lucas C, Salesse L, Hoang MHT, et al. Autophagy of intestinal epithelial cells inhibits colorectal carcinogenesis induced by Colibactin-producing Escherichia coli in Apc(min/+) mice. Gastroenterology. 2020;158:1373–88.
57. Sheng YH, Giri R, Davies J, et al. A nucleotide analog prevents colitis-associated cancer via Beta-catenin independently of inflammation and autophagy. Cell Mol Gastroenterol Hepatol. 2021;11:33–53.
58. Graillot V, Dormoy I, Dupuy J, et al. Genotoxicity of Cytolethal distending toxin (CDT) on isogenic human colorectal cell lines: potential promoting effects for colorectal carcinogenesis. Front Cell Infect Microbiol. 2016;6:34.
59. Wang X, Huycke MM. Extracellular superoxide production by Enterococcus faecalis promotes chromosomal instability in mammalian cells. Gastroenterology. 2007;132:551–61.

60. Wang X, Yang Y, Moore DR, et al. 4-hydroxy-2-nonenal mediates genotoxicity and bystander effects caused by Enterococcus faecalis-infected macrophages. Gastroenterology. 2012;142:543–551.e7.
61. Swidsinski A, Loening-Baucke V, Lochs H, et al. Spatial organization of bacterial flora in normal and inflamed intestine: a fluorescence in situ hybridization study in mice. World J Gastroenterol. 2005;11:1131–40.
62. Dejea CM, Wick EC, Hechenbleikner EM, et al. Microbiota organization is a distinct feature of proximal colorectal cancers. Proc Natl Acad Sci U S A. 2014;111:18321–6.
63. Johnson CH, Dejea CM, Edler D, et al. Metabolism links bacterial biofilms and colon carcinogenesis. Cell Metab. 2015;21:891–7.
64. Tomkovich S, Dejea CM, Winglee K, et al. Human colon mucosal biofilms from healthy or colon cancer hosts are carcinogenic. J Clin Invest. 2019;129:1699–712.
65. Flemming HC, Wingender J, Szewzyk U, et al. Biofilms: an emergent form of bacterial life. Nat Rev Microbiol. 2016;14:563–75.
66. Mehta RS, Nishihara R, Cao Y, et al. Association of Dietary Patterns with Risk of colorectal cancer subtypes classified by Fusobacterium nucleatum in tumor tissue. JAMA Oncol. 2017;3:921–7.
67. Orlich MJ, Singh PN, Sabate J, et al. Vegetarian dietary patterns and the risk of colorectal cancers. JAMA Intern Med. 2015;175:767–76.
68. Ciorba MA, Riehl TE, Rao MS, et al. Lactobacillus probiotic protects intestinal epithelium from radiation injury in a TLR-2/cyclo-oxygenase-2-dependent manner. Gut. 2012;61:829–38.
69. David LA, Maurice CF, Carmody RN, et al. Diet rapidly and reproducibly alters the human gut microbiome. Nature. 2014;505:559–63.
70. Turnbaugh PJ, Hamady M, Yatsunenko T, et al. A core gut microbiome in obese and lean twins. Nature. 2009;457:480–4.
71. Wu GD, Chen J, Hoffmann C, et al. Linking long-term dietary patterns with gut microbial enterotypes. Science (New York, N Y). 2011;334:105–8.
72. Muegge BD, Kuczynski J, Knights D, et al. Diet drives convergence in gut microbiome functions across mammalian phylogeny and within humans. Science. 2011;332:970–4.
73. Sonnenburg ED, Smits SA, Tikhonov M, et al. Diet-induced extinctions in the gut microbiota compound over generations. Nature. 2016;529:212–5.
74. Thaiss CA, Itav S, Rothschild D, et al. Persistent microbiome alterations modulate the rate of post-dieting weight regain. Nature. 2016;540:544–51.
75. Begley M, Gahan CG, Hill C. The interaction between bacteria and bile. FEMS Microbiol Rev. 2005;29:625–51.
76. Islam KB, Fukiya S, Hagio M, et al. Bile acid is a host factor that regulates the composition of the cecal microbiota in rats. Gastroenterology. 2011;141:1773–81.
77. Lechner S, Muller-Ladner U, Schlottmann K, et al. Bile acids mimic oxidative stress induced upregulation of thioredoxin reductase in colon cancer cell lines. Carcinogenesis. 2002;23:1281–8.
78. Dvorak K, Payne CM, Chavarria M, et al. Bile acids in combination with low pH induce oxidative stress and oxidative DNA damage: relevance to the pathogenesis of Barrett's oesophagus. Gut. 2007;56:763–71.
79. Bernstein H, Payne CM, Bernstein C, et al. Activation of the promoters of genes associated with DNA damage, oxidative stress, ER stress and protein malfolding by the bile salt, deoxycholate. Toxicol Lett. 1999;108:37–46.
80. Washo-Stultz D, Hoglen N, Bernstein H, et al. Role of nitric oxide and peroxynitrite in bile salt-induced apoptosis: relevance to colon carcinogenesis. Nutr Cancer. 1999;35:180–8.
81. Booth LA, Gilmore IT, Bilton RF. Secondary bile acid induced DNA damage in HT29 cells: are free radicals involved? Free Radic Res. 1997;26:135–44.
82. Rath S, Rud T, Karch A, et al. Pathogenic functions of host microbiota. Microbiome. 2018;6:174.
83. O'Keefe SJ, Li JV, Lahti L, et al. Fat, fibre and cancer risk in African Americans and rural Africans. Nat Commun. 2015;6:6342.
84. Reddy BS, Simi B, Patel N, et al. Effect of amount and types of dietary fat on intestinal bacterial 7 alpha-dehydroxylase and phosphatidylinositol-specific phospholipase C and colonic mucosal diacylglycerol kinase and PKC activities during stages of colon tumor promotion. Cancer Res. 1996;56:2314–20.
85. Bayerdorffer E, Mannes GA, Ochsenkuhn T, et al. Variation of serum bile acids in patients with colorectal adenomas during a one-year follow-up. Digestion. 1994;55:121–9.
86. Bayerdorffer E, Mannes GA, Ochsenkuhn T, et al. Unconjugated secondary bile acids in the serum of patients with colorectal adenomas. Gut. 1995;36:268–73.
87. Reddy BS, Wynder EL. Metabolic epidemiology of colon cancer. Fecal bile acids and neutral sterols in colon cancer patients and patients with adenomatous polyps. Cancer. 1977;39:2533–9.
88. O'Keefe SJ, Kidd M, Espitalier-Noel G, et al. Rarity of colon cancer in Africans is associated with low animal product consumption, not fiber. Am J Gastroenterol. 1999;94:1373–80.
89. Prentice RL, Thomson CA, Caan B, et al. Low-fat dietary pattern and cancer incidence in the Women's Health Initiative dietary modification randomized controlled trial. J Natl Cancer Inst. 2007;99:1534–43.
90. Prentice RL, Aragaki AK, Howard BV, et al. Low-fat dietary pattern among postmenopausal women influences long-term cancer, cardiovascular disease, and diabetes outcomes. J Nutr. 2019;149:1565–74.
91. Flynn C, Montrose DC, Swank DL, et al. Deoxycholic acid promotes the growth of colonic aberrant crypt foci. Mol Carcinog. 2007;46:60–70.
92. Bernstein C, Holubec H, Bhattacharyya AK, et al. Carcinogenicity of deoxycholate, a secondary bile acid. Arch Toxicol. 2011;85:863–71.
93. Liu L, Dong W, Wang S, et al. Deoxycholic acid disrupts the intestinal mucosal barrier and promotes intestinal tumorigenesis. Food Funct. 2018;9:5588–97.
94. Raufman JP, Dawson PA, Rao A, et al. Slc10a2-null mice uncover colon cancer-promoting actions of endogenous fecal bile acids. Carcinogenesis. 2015;36:1193–200.
95. Dermadi D, Valo S, Ollila S, et al. Western diet deregulates bile acid homeostasis, cell proliferation, and tumorigenesis in colon. Cancer Res. 2017;77:3352–63.
96. Li R, Grimm SA, Mav D, et al. Transcriptome and DNA methylome analysis in a mouse model of diet-induced obesity predicts increased risk of colorectal cancer. Cell Rep. 2018;22:624–37.
97. Qin Y, Roberts JD, Grimm SA, et al. An obesity-associated gut microbiome reprograms the intestinal epigenome and leads to altered colonic gene expression. Genome Biol. 2018;19:7.
98. Patterson E, O'doherty RM, Murphy EF, et al. Impact of dietary fatty acids on metabolic activity and host intestinal microbiota composition in C57BL/6J mice. Br J Nutr. 2014;111:1905–17.
99. Piazzi G, D'Argenio G, Prossomariti A, et al. Eicosapentaenoic acid free fatty acid prevents and suppresses colonic neoplasia in colitis-associated colorectal cancer acting on notch signaling and gut microbiota. Int J Cancer. 2014;135:2004–13.
100. Watson H, Mitra S, Croden FC, et al. A randomised trial of the effect of omega-3 polyunsaturated fatty acid supplements on the human intestinal microbiota. Gut. 2018;67:1974–83.
101. IARC Working Group on the Evaluation of Carcinogenic Risks to Humans. Red meat and processed meat. Lyon, FR: International Agency for Research on Cancer; 2018.
102. Lewin MH, Bailey N, Bandaletova T, et al. Red meat enhances the colonic formation of the DNA adduct O6-carboxymethyl guanine: implications for colorectal cancer risk. Cancer Res. 2006;66:1859–65.

103. Cross AJ, Pollock JR, Bingham SA. Haem, not protein or inorganic iron, is responsible for endogenous intestinal N-nitrosation arising from red meat. Cancer Res. 2003;63:2358–60.
104. Bingham SA, Pignatelli B, Pollock JR, et al. Does increased endogenous formation of N-nitroso compounds in the human colon explain the association between red meat and colon cancer? Carcinogenesis. 1996;17:515–23.
105. Ijssennagger N, Rijnierse A, de Wit NJ, et al. Dietary heme induces acute oxidative stress, but delayed cytotoxicity and compensatory hyperproliferation in mouse colon. Carcinogenesis. 2013;34:1628–35.
106. Ijssennagger N, Belzer C, Hooiveld GJ, et al. Gut microbiota facilitates dietary heme-induced epithelial hyperproliferation by opening the mucus barrier in colon. Proc Natl Acad Sci U S A. 2015;112:10038–43.
107. Abu-Ghazaleh N, Chua WJ, Gopalan V. Intestinal microbiota and its association with colon cancer and red/processed meat consumption. J Gastroenterol Hepatol. 2021;36:75–88.
108. Davila AM, Blachier F, Gotteland M, et al. Intestinal luminal nitrogen metabolism: role of the gut microbiota and consequences for the host. Pharmacol Res. 2013;68:95–107.
109. Humblot C, Murkovic M, Rigottier-Gois L, et al. Beta-glucuronidase in human intestinal microbiota is necessary for the colonic genotoxicity of the food-borne carcinogen 2-amino-3-methylimidazo[4,5-f]quinoline in rats. Carcinogenesis. 2007;28:2419–25.
110. Zhang J, Lacroix C, Wortmann E, et al. Gut microbial beta-glucuronidase and glycerol/diol dehydratase activity contribute to dietary heterocyclic amine biotransformation. BMC Microbiol. 2019;19:99.
111. Parada Venegas D, De la Fuente MK, Landskron G, et al. Short chain fatty acids (SCFAs)-mediated gut epithelial and immune regulation and its relevance for inflammatory bowel diseases. Front Immunol. 2019;10:277.
112. Bultman SJ. Molecular pathways: gene-environment interactions regulating dietary fiber induction of proliferation and apoptosis via butyrate for cancer prevention. Clin Cancer Res. 2014;20:799–803.
113. Bultman SJ. Interplay between diet, gut microbiota, epigenetic events, and colorectal cancer. Mol Nutr Food Res 2017;61.
114. Willemsen LE, Koetsier MA, van Deventer SJ, et al. Short chain fatty acids stimulate epithelial mucin 2 expression through differential effects on prostaglandin E(1) and E(2) production by intestinal myofibroblasts. Gut. 2003;52:1442–7.
115. Louis P, Flint HJ. Formation of propionate and butyrate by the human colonic microbiota. Environ Microbiol. 2017;19:29–41.
116. Singh N, Gurav A, Sivaprakasam S, et al. Activation of Gpr109a, receptor for niacin and the commensal metabolite butyrate, suppresses colonic inflammation and carcinogenesis. Immunity. 2014;40:128–39.
117. Roediger WE. Role of anaerobic bacteria in the metabolic welfare of the colonic mucosa in man. Gut. 1980;21:793–8.
118. Kelly CJ, Zheng L, Campbell EL, et al. Crosstalk between microbiota-derived short-chain fatty acids and intestinal epithelial HIF augments tissue barrier function. Cell Host Microbe. 2015;17:662–71.
119. Donohoe DR, Collins LB, Wali A, et al. The Warburg effect dictates the mechanism of butyrate-mediated histone acetylation and cell proliferation. Mol Cell. 2012;48:612–26.
120. Ploger S, Stumpff F, Penner GB, et al. Microbial butyrate and its role for barrier function in the gastrointestinal tract. Ann N Y Acad Sci. 2012;1258:52–9.
121. Desai MS, Seekatz AM, Koropatkin NM, et al. A dietary fiber-deprived gut microbiota degrades the colonic mucus barrier and enhances pathogen susceptibility. Cell. 2016;167:1339–1353.e21.
122. Bishehsari F, Engen PA, Preite NZ, et al. Dietary fiber treatment corrects the composition of gut microbiota, promotes SCFA production, and suppresses colon carcinogenesis. Genes (Basel). 2018;9:9.
123. Donohoe DR, Holley D, Collins LB, et al. A gnotobiotic mouse model demonstrates that dietary fiber protects against colorectal tumorigenesis in a microbiota- and butyrate-dependent manner. Cancer Discov. 2014;4:1387–97.
124. Mehta RS, Song M, Nishihara R, et al. Dietary patterns and risk of colorectal cancer: analysis by tumor location and molecular subtypes. Gastroenterology. 2017;152:1944–1953e1.
125. Kunzmann AT, Coleman HG, Huang WY, et al. Dietary fiber intake and risk of colorectal cancer and incident and recurrent adenoma in the prostate, lung, colorectal, and ovarian cancer screening trial. Am J Clin Nutr. 2015;102:881–90.
126. Fuchs CS, Giovannucci EL, Colditz GA, et al. Dietary fiber and the risk of colorectal cancer and adenoma in women. N Engl J Med. 1999;340:169–76.
127. Ananthakrishnan AN, Khalili H, Konijeti GG, et al. A prospective study of long-term intake of dietary fiber and risk of Crohn's disease and ulcerative colitis. Gastroenterology. 2013;145:970–7.
128. Chen HM, Yu YN, Wang JL, et al. Decreased dietary fiber intake and structural alteration of gut microbiota in patients with advanced colorectal adenoma. Am J Clin Nutr. 2013;97:1044–52.
129. DeCosse JJ, Miller HH, Lesser ML. Effect of wheat fiber and vitamins C and E on rectal polyps in patients with familial adenomatous polyposis. J Natl Cancer Inst. 1989;81:1290–7.
130. MacLennan R, Macrae F, Bain C, et al. Randomized trial of intake of fat, fiber, and beta carotene to prevent colorectal adenomas. J Natl Cancer Inst. 1995;87:1760–6.
131. Vernia F, Longo S, Stefanelli G, et al. Dietary factors modulating colorectal carcinogenesis. Nutrients. 2021;13:143.
132. Michels KB, Willett WC, Vaidya R, et al. Yogurt consumption and colorectal cancer incidence and mortality in the nurses' health study and the health professionals follow-up study. Am J Clin Nutr. 2020;112:1566–75.
133. Rifkin SB, Giardiello FM, Zhu X, et al. Yogurt consumption and colorectal polyps. Br J Nutr. 2020;124:1–12.
134. Bouvard V, Loomis D, Guyton KZ, et al. Carcinogenicity of consumption of red and processed meat. Lancet Oncol. 2015;16:1599–600.
135. Andreux PA, Blanco-Bose W, Ryu D, et al. The mitophagy activator urolithin a is safe and induces a molecular signature of improved mitochondrial and cellular health in humans. Nat Metab. 2019;1:595–603.
136. Collaborators GBDIBD. The global, regional, and national burden of inflammatory bowel disease in 195 countries and territories, 1990-2017: a systematic analysis for the global burden of disease study 2017. Lancet Gastroenterol Hepatol. 2020;5:17–30.
137. Jess T, Rungoe C, Peyrin-Biroulet L. Risk of colorectal cancer in patients with ulcerative colitis: a meta-analysis of population-based cohort studies. Clin Gastroenterol Hepatol. 2012;10:639–45.
138. Fumery M, Dulai PS, Gupta S, et al. Incidence, risk factors, and outcomes of colorectal cancer in patients with ulcerative colitis with low-grade dysplasia: a systematic review and meta-analysis. Clin Gastroenterol Hepatol. 2017;15:665–674.e5.
139. Ullman TA, Itzkowitz SH. Intestinal inflammation and cancer. Gastroenterology. 2011;140:1807–16.
140. Waldner MJ, Neurath MF. Mechanisms of immune signaling in colitis-associated cancer. Cell Mol Gastroenterol Hepatol. 2015;1:6–16.
141. Hanahan D, Weinberg RA. Hallmarks of cancer: the next generation. Cell. 2011;144:646–74.
142. Schultsz C, Van Den Berg FM, Ten Kate FW, et al. The intestinal mucus layer from patients with inflammatory bowel dis-

ease harbors high numbers of bacteria compared with controls. Gastroenterology. 1999;117:1089–97.
143. Fyderek K, Strus M, Kowalska-Duplaga K, et al. Mucosal bacterial microflora and mucus layer thickness in adolescents with inflammatory bowel disease. World J Gastroenterol. 2009;15:5287–94.
144. Bhatia R, Gautam SK, Cannon A, et al. Cancer-associated mucins: role in immune modulation and metastasis. Cancer Metastasis Rev. 2019;38:223–36.
145. Van der Sluis M, De Koning BA, De Bruijn AC, et al. Muc2-deficient mice spontaneously develop colitis, indicating that MUC2 is critical for colonic protection. Gastroenterology. 2006;131:117–29.
146. Velcich A, Yang W, Heyer J, et al. Colorectal cancer in mice genetically deficient in the mucin Muc2. Science. 2002;295:1726–9.
147. Keubler LM, Buettner M, Hager C, et al. A multihit model: colitis lessons from the Interleukin-10-deficient mouse. Inflamm Bowel Dis. 2015;21:1967–75.
148. Kuhn R, Lohler J, Rennick D, et al. Interleukin-10-deficient mice develop chronic enterocolitis. Cell. 1993;75:263–74.
149. Schwerbrock NM, Makkink MK, van der Sluis M, et al. Interleukin 10-deficient mice exhibit defective colonic Muc2 synthesis before and after induction of colitis by commensal bacteria. Inflamm Bowel Dis. 2004;10:811–23.
150. Liso M, De Santis S, Verna G, et al. A specific mutation in Muc2 determines early Dysbiosis in colitis-prone Winnie mice. Inflamm Bowel Dis. 2020;26:546–56.
151. Das S, Rachagani S, Sheinin Y, et al. Mice deficient in Muc4 are resistant to experimental colitis and colitis-associated colorectal cancer. Oncogene. 2016;35:2645–54.
152. De Arcangelis A, Hamade H, Alpy F, et al. Hemidesmosome integrity protects the colon against colitis and colorectal cancer. Gut. 2017;66:1748–60.
153. Low END, Mokhtar NM, Wong Z, et al. Colonic mucosal transcriptomic changes in patients with long-duration ulcerative colitis revealed colitis-associated cancer pathways. J Crohns Colitis. 2019;13:755–63.
154. Okumura R, Kurakawa T, Nakano T, et al. Lypd8 promotes the segregation of flagellated microbiota and colonic epithelia. Nature. 2016;532:117–21.
155. Xu J, Qian J, Zhang W, et al. LYPD8 regulates the proliferation and migration of colorectal cancer cells through inhibiting the secretion of IL6 and TNFalpha. Oncol Rep. 2019;41:2389–95.
156. Singh V, Yeoh BS, Chassaing B, et al. Microbiota-inducible innate immune, siderophore binding protein Lipocalin 2 is critical for intestinal homeostasis. Cell Mol Gastroenterol Hepatol. 2016;2:482–498.e6.
157. Lloyd-Price J, Arze C, Ananthakrishnan AN, et al. Multi-omics of the gut microbial ecosystem in inflammatory bowel diseases. Nature. 2019;569:655–62.
158. Moschen AR, Gerner RR, Wang J, et al. Lipocalin 2 protects from inflammation and tumorigenesis associated with gut microbiota alterations. Cell Host Microbe. 2016;19:455–69.
159. Vinson KE, George DC, Fender AW, et al. The notch pathway in colorectal cancer. Int J Cancer. 2016;138:1835–42.
160. Pinho SS, Reis CA. Glycosylation in cancer: mechanisms and clinical implications. Nat Rev Cancer. 2015;15:540–55.
161. Wang Y, Huang D, Chen KY, et al. Fucosylation deficiency in mice leads to colitis and adenocarcinoma. Gastroenterology. 2017;152:193–205.e10.
162. Misiorek JO, Lahdeniemi IAK, Nystrom JH, et al. Keratin 8-deletion induced colitis predisposes to murine colorectal cancer enforced by the inflammasome and IL-22 pathway. Carcinogenesis. 2016;37:777–86.
163. Liu C, Liu ED, Meng YX, et al. Keratin 8 reduces colonic permeability and maintains gut microbiota homeostasis, protecting against colitis and colitis-associated tumorigenesis. Oncotarget. 2017;8:96774–90.
164. Peuker K, Muff S, Wang J, et al. Epithelial calcineurin controls microbiota-dependent intestinal tumor development. Nat Med. 2016;22:506–15.
165. Gehart H, Clevers H. Tales from the crypt: new insights into intestinal stem cells. Nat Rev Gastroenterol Hepatol. 2019;16:19–34.
166. Fukata M, Chen A, Vamadevan AS, et al. Toll-like receptor-4 promotes the development of colitis-associated colorectal tumors. Gastroenterology. 2007;133:1869–81.
167. Chen L, Wilson JE, Koenigsknecht MJ, et al. NLRP12 attenuates colon inflammation by maintaining colonic microbial diversity and promoting protective commensal bacterial growth. Nat Immunol. 2017;18:541–51.
168. Zaki MH, Vogel P, Malireddi RK, et al. The NOD-like receptor NLRP12 attenuates colon inflammation and tumorigenesis. Cancer Cell. 2011;20:649–60.
169. Rakoff-Nahoum S, Medzhitov R. Regulation of spontaneous intestinal tumorigenesis through the adaptor protein MyD88. Science. 2007;317:124–7.
170. Li Y, Kundu P, Seow SW, et al. Gut microbiota accelerate tumor growth via c-Jun and STAT3 phosphorylation in APCMin/+ mice. Carcinogenesis. 2012;33:1231–8.
171. Grivennikov SI, Wang K, Mucida D, et al. Adenoma-linked barrier defects and microbial products drive IL-23/IL-17-mediated tumour growth. Nature. 2012;491:254–8.
172. Santaolalla R, Sussman DA, Ruiz JR, et al. TLR4 activates the beta-catenin pathway to cause intestinal neoplasia. PLoS One. 2013;8:e63298.
173. Kesselring R, Glaesner J, Hiergeist A, et al. Irak-M expression in tumor cells supports colorectal cancer progression through reduction of antimicrobial defense and stabilization of STAT3. Cancer Cell. 2016;29:684–96.
174. Coleman OI, Lobner EM, Bierwirth S, et al. Activated ATF6 induces intestinal Dysbiosis and innate immune response to promote colorectal tumorigenesis. Gastroenterology. 2018;155:1539–1552.e12.
175. Tian M, Wang X, Sun J, et al. IRF3 prevents colorectal tumorigenesis via inhibiting the nuclear translocation of beta-catenin. Nat Commun. 2020;11:5762.
176. Andoh A, Zhang Z, Inatomi O, et al. Interleukin-22, a member of the IL-10 subfamily, induces inflammatory responses in colonic subepithelial myofibroblasts. Gastroenterology. 2005;129:969–84.
177. Gronke K, Hernandez PP, Zimmermann J, et al. Interleukin-22 protects intestinal stem cells against genotoxic stress. Nature. 2019;566:249–53.
178. Jiang R, Wang H, Deng L, et al. IL-22 is related to development of human colon cancer by activation of STAT3. BMC Cancer. 2013;13:59.
179. Wu T, Wang Z, Liu Y, et al. Interleukin 22 protects colorectal cancer cells from chemotherapy by activating the STAT3 pathway and inducing autocrine expression of interleukin 8. Clin Immunol. 2014;154:116–26.
180. Cash HL, Whitham CV, Behrendt CL, et al. Symbiotic bacteria direct expression of an intestinal bactericidal lectin. Science. 2006;313:1126–30.
181. Mukherjee S, Zheng H, Derebe MG, et al. Antibacterial membrane attack by a pore-forming intestinal C-type lectin. Nature. 2014;505:103–7.
182. Kempski J, Giannou AD, Riecken K, et al. IL22BP mediates the antitumor effects of Lymphotoxin against colorectal tumors in mice and humans. Gastroenterology. 2020;159:1417–1430.e3.
183. Bouskra D, Brezillon C, Berard M, et al. Lymphoid tissue genesis induced by commensals through NOD1 regulates intestinal homeostasis. Nature. 2008;456:507–10.

184. Ervin SM, Ramanan SV, Bhatt AP. Relationship between the gut microbiome and systemic chemotherapy. Dig Dis Sci. 2020;65:874–84.
185. Koppel N, Maini Rekdal V, Balskus EP. Chemical transformation of xenobiotics by the human gut microbiota. Science. 2017;356:eaag2770.
186. Haiser HJ, Turnbaugh PJ. Developing a metagenomic view of xenobiotic metabolism. Pharmacol Res. 2013;69:21–31.
187. Spanogiannopoulos P, Bess EN, Carmody RN, et al. The microbial pharmacists within us: a metagenomic view of xenobiotic metabolism. Nat Rev Microbiol. 2016;14:273–87.
188. Wilson ID, Nicholson JK. Gut microbiome interactions with drug metabolism, efficacy, and toxicity. Transl Res. 2017;179:204–22.
189. Alexander JL, Wilson ID, Teare J, et al. Gut microbiota modulation of chemotherapy efficacy and toxicity. Nat Rev Gastroenterol Hepatol. 2017;14:356–65.
190. Zitvogel L, Ma Y, Raoult D, et al. The microbiome in cancer immunotherapy: diagnostic tools and therapeutic strategies. Science. 2018;359:1366–70.
191. Fidelle M, Yonekura S, Picard M, et al. Resolving the paradox of colon cancer through the integration of genetics, immunology, and the microbiota. Front Immunol. 2020;11:600886.
192. Petroni G, Buque A, Zitvogel L, et al. Immunomodulation by targeted anticancer agents. Cancer Cell. 2020;39:310–45.

The Role of the Microbiome in Inflammatory Bowel Disease

Will Takakura, Ali Rezaie, and Mark Pimentel

Highlights

- Evidence that changes in the microbiome contribute to inflammatory bowel disease (IBD) include the high prevalence of *Mycobacterium avium* subspecies *paratuberculosis* (MAP) in Crohn's disease (CD), increases in IBD rates after bacterial, parasitic, and viral infections, and some successes in using antibiotics to treat IBD.
- Fungal species including *Saccharomyces cerevisiae*, *Clavispora lusitaniae*, *Cyberlindnera jadinii*, *Candida albicans*, and *Kluy-veromyces marxianus*, are overrepresented in CD. The fungus *Malassezia restricta* is associated with a polymorphism in the gene for CARD9, which is important in sensing fungal Dectin-1, Dectin-2, and Mincle, and a known risk factor for IBD.
- Symptomatic CD and ulcerative colitis (UC) patients have more flatline breath tests than patients without IBD, which may indicate increased colonization with hydrogen sulfide (H_2S) producers such as *Desulfovibrio.* H_2S may inhibit butyrate oxidation in colonocytes and reduce disulfide bonds in the mucosal biofilm, increasing colonic permeability to enteric pathogens.
- An elemental diet can induce remission in pediatric and adult CD, although the positive effects may be due to improvements in concomitant small intestinal bacterial overgrowth (SIBO). Hydrogen-predominant SIBO is the most common type of SIBO in CD and UC patients, and the successes of antibiotics in IBD patients may be due, at least in part, to treatment of concomitant SIBO.

1 Historical Evidence for a Microbial Cause of IBD

The concept of an infectious agent in IBD originated from finding granulomas in the tissue of patients with Crohn's disease, resembling the histology of tuberculosis. However, no pathogen was ever found to account for ulcerative colitis (UC) or Crohn's disease (CD), and treatments focused on treating the inflammation without a clear understanding of its origins or basis. In the 1980's several studies attempted to elucidate the importance of the gut microbiome in IBD. One interesting approach was the use of total parenteral nutrition (TPN). The postulate was that if the stream of nutrients to the gut was cut off, there would be less nutrients for resident flora. Less flora meant less drivers of inflammation. Studies were performed as a treatment for severe Crohn's disease and while they showed a modest effect, recurrence was high [1]. Subsequent small randomized control trials in either Crohn's disease [2] or ulcerative colitis [3, 4] failed to show any significant difference in subjects with acute IBD flares, although these studies may have been underpowered to detect small differences. The challenge with this approach is obvious as there are multiple potential consequences to TPN, including high cost, line infections, liver complications, and deciding on when to resume oral feeding. However, to this day, the concept of reducing luminal bacterial and fecal material in the gut is still used. For example, an intestinal diversion of chyme is still a treatment for severe refractory colitis [5].

Another more direct approach to reducing the microbiome in IBD which garnered more attention, and was more

W. Takakura
Medically Associated Science and Technology (MAST) Program, Cedars-Sinai Medical Center, West Hollywood, CA, USA

Department of Medicine, Cedars-Sinai Medical Center, Los Angeles, CA, USA

A. Rezaie · M. Pimentel (✉)
Medically Associated Science and Technology (MAST) Program, Cedars-Sinai Medical Center, West Hollywood, CA, USA

Karsh Division of Gastroenterology and Hepatology, Department of Medicine, Cedars-Sinai Medical Center, Los Angeles, CA, USA
e-mail: Ali.Rezaie@cshs.org; Mark.Pimentel@cshs.org

M. Pimentel et al. (eds.), *Clinical Understanding of the Human Gut Microbiome*, https://doi.org/10.1007/978-3-031-46712-7_8

successful, was the use of antibiotics. Initially, antibiotics in a large case-crossover study showed that this approach may be protective against IBD flares, especially in Crohn's disease [6]. Similar to TPN, this work suffered from a number of small randomized control trials in active IBD which were underpowered and thus did not show any significant difference in outcome for those on antibiotics versus those who were not [7–11]. An interesting observation in a study by Sutherland et al. was that the beneficial effect of the antibiotic was seen more in those with colonic Crohn's disease than in those with small bowel Crohn's disease. It is well known that the colon has the highest microbial load within the gastrointestinal (GI) tract. Regardless, current guidelines do not recommend the use of antibiotics to treat acute flares of IBD [12, 13], although early data pose the question of whether modulation of the gut microbiome may have a more subtle effect on IBD.

During the same period, another interesting finding reignited interest in the microbiome specifically in Crohn's disease. Chiodini et al. cultured the *Mycobacterium avium* subspecies *paratuberculosis* (MAP) in patients with Crohn's disease [14]. MAP is a causative agent in Johne's disease, a similar disease that affects the small intestine of cattle, and is found in food sources such as milk [15]. There seemed to be a relatively high prevalence of MAP in Crohn's disease, where in one meta-analysis the OR of having MAP in Crohn's disease was 7.01 compared to non-IBD patients, and 4.13 compared to ulcerative colitis patients [16]. To support the hypothesis that MAP plays a causal role in the pathophysiology of Crohn's disease, two small trials have shown clinical improvement in Crohn's disease after treatment with macrolide antibiotics [17, 18]. Indeed, it is feasible that Crohn's disease as a granulomatosis disease may share some common pathophysiological features with other granulomatous diseases such as those caused by mycobacteria.

To this day, the causative role of MAP in a subset of patients with Crohn's disease is a matter of debate [19], although current guidelines do not recommend the use of anti-mycobacterials for the treatment of Crohn's disease, given the lack of overall efficacy [12], and the lack of universally finding MAP in patients with Crohn's disease. One common argument against MAP being a causative agent is the lack of worsening of IBD with active immunosuppression used to treat IBD. Second is the lack of transmission to humans from animals affected by MAP, and the lack of a verified epidemiological model of human transmission [19]. Regardless, the high prevalence of MAP in Crohn's disease continues to suggest a link between the pathogen and the disease, although it likely plays a much smaller role in the background of other factors such as the host's genetics, epigenetics, and environmental factors like diet.

Lastly, intestinal infection may increase the risk of developing IBD, as epidemiological studies have shown an increase in rates of IBD after bacterial, parasitic, and viral infections [20, 21]. There appears to be quite a degree of heterogeneity in these studies and detection and misclassification bias may prohibit us from making any significant conclusion [22, 23].

These studies provide evidence for microbial-host interactions that likely play a role in IBD. Like many other relationships surrounding the microbiome, the connection between IBD and the gut flora is part of a complex network influenced by numerous other host and environmental factors.

2 Intestinal Gases and IBD

Of particular interest, gases produced by the gut microbiome can be measured via breath tests [24]. These breath tests provide a simple way to study the microbial load of patients without performing expensive and sometimes invasive studies. The premise of the breath test is that under normal physiology, hydrogen and methane are exclusively produced by the intestinal flora [25, 26]. More recently hydrogen sulfide has been added to the armamentarium for clinical use [27]; however, the concept that these gas-producing microbes, and possibly these gases themselves, are linked to IBD is not new.

2.1 Methane

Methane appears to play an important role in IBD. Methane is a gas produced predominantly by another microbial domain of life, the archaea. The predominant methanogen in humans is believed to be the archaeon *Methanobrevibacter smithii* [28, 29]. Since the 1990's and early 2000's, observational studies have described a lower prevalence of methane positive breath tests in patients with IBD [30–32]. Methane appears to be strongly tied to the bowel pattern of the patients and has been shown to be decreased in especially those with IBD and diarrhea. Interestingly, animal models of IBD have shown anti-inflammatory effects of methane [33]. The severity of dextran sulfate sodium-induced inflammatory bowel disease was attenuated with intraperitoneal injection of methane-rich saline. Interestingly this decrease in inflammation was seen with a decrease in toll-like receptor 4, a receptor for lipopolysaccharide, as well as an increase in IL-10, an anti-inflammatory cytokine. Lastly, this anti-inflammatory effect was seen with an increase in glutathione and superoxide dismutase as well as a decrease in malondialdehyde, suggesting an antioxidant property of methane. Furthermore, an ischemia-reperfusion model in canine intestine has found a similar protective effect with methane with protection of intestinal architecture associated with a decrease in superoxide production [34]. This remedial and protective effect of methane has been shown in other disease models such as myocardial infarction [35], concanavalin A-induced autoimmune hepatitis [36], ischemia/reperfusion injury model in surgical skin flaps [37], and the liver [38]. Although this level of anti-inflammatory effect seen in these experiments may not be seen in normal physiologic condition given the extremely high concentrations used in these experimental settings (Boros et al. used 2.5% methane) [34], it is still notable that lower rates of methane are found in IBD human patients. Perhaps methanogens play an important role in the pathophysiology of IBD and their use may be therapeutic. Future studies should focus on elucidating the pathophysiological role of methane are more physiologic concentrations, as well as clinical trials utilizing methane to treat IBD.

2.2 Hydrogen Sulfide

Another gas that was discovered in the 90's to be highly prevalent in IBD is hydrogen sulfide. Sulfate-reducing bacteria such as *Desulfovibrio* have been implicated in the pathogenesis of IBD and, in particular, ulcerative colitis [39]. Culture based techniques have shown an increase in *Desulfovibrio* [40, 41] with one study showing 92% of patients with ulcerative colitis being positive for these organisms compared to 52% in controls [41]. Interestingly PCR assays have shown that these sulfate-reducing bacteria are found ubiquitously in both patients with ulcerative colitis and healthy controls [41, 42].

The mechanistic underpinning of hydrogen sulfide is interesting. Hydrogen sulfide is thought to inhibit butyrate oxidation [43] in colonocytes and exhibits signs of toxicity in colonic epithelium which shares similar metabolic profiles seen in ulcerative colitis [44]. In fact, a reduction in *Desulfovibrio* by Glycomacropeptide was associated with a decrease in proinflammatory cytokines such as interferon (IFN)-γ, TNF-α, IL-1β and IL-2. Also, in isolated mitochondria, 10 μM of hydrogen sulfide can inhibit respiration by binding to cytochrome oxidase [45], and 0.5–1 mM of sodium hydrosulfide (a donor of hydrogen sulfide) appear to increase hydrogen peroxide induced damage. However, at higher concentrations (1.5 mM) sodium hydrosulfide appears to protect cells against hydrogen peroxide induced toxicity [46]. To further add to the controversy, 0.2–0.5 mM appear to induce proliferation in intestinal epithelial cells (IECs), whereas higher concentrations (0.4–1.0 mM) appear to inhibit cell proliferation [47]. Overall, the effects of hydrogen sulfide and the bacteria responsible for their production appear to be dependent on its concentration and environment. This complex nature of hydrogen sulfide may be the reason why a simple experiment of eliminating hydrogen sulfide by binding bismuth did not remediate the effect of dextran sulfate-induced colitis in one study [48].

In terms of motility, hydrogen sulfide appears to have a role in relaxing smooth muscles. In the ileum of guinea pigs, hydrogen sulfide appear to augment the relaxation induced by nitrogenous compounds [49]. A similar effect was seen in rabbit ileum [50]. Hydrogen sulfide also appears to inhibit the amplitude and pacemaker frequency of ex-vivo mouse intestinal cells of Cajal working conjointly with nitic oxide [51, 52]. Also, inflammatory conditions appear to influence ion channel activity such at K_{ATP}, which is hypothesized to affect smooth muscle regulation of motility in the inflamed bowel [53]. Hydrogen sulfide appears to play an important role in regulating these channels through post-translational modification [54]. This evidence seems to contradict a recent epidemiological study which showed that increased levels of hydrogen sulfide is associated with diarrhea in humans [27]. This may be partially explained by hydrogen sulfide's direct effect on intestinal smooth muscle and the prosecretory effects on intestinal epithelium [55].

Of note, hydrogen sulfide is produced by human cells and is used in regulation within the gut and the nervous system [56]. There appears to be variable effects depending on the origin and location of this gas, which further complicates the study of hydrogen sulfide.

3 The Modern Era of Microbiome Investigation in IBD

The relationship between IBD and the gut microbiome is a longstanding one that extends well before the term "microbiome" was established. Early papers postulated on the possibility of a "transmissible agent" for Crohn's disease but failed to demonstrate one [57]. However, there is circumstantial evidence to suggest that gut bacteria may be important in IBD. These points of evidence include the belief that inflammation is a reaction to a luminal factor, the benefits of antibiotics in relieving symptoms, and the finding of granulomas in the case of CD.

The most provocative circumstantial evidence is that antibiotics appear to benefit subjects with IBD, particularly CD. Randomized trials in the early 1980's even suggested that drugs like metronidazole were of greater benefit to patients with CD than even sulfasalazine [58]. This led to clinicians of the day using chronic antibiotics for CD. Despite the successful use of antibiotics, an active microbial agent was not discovered. In fact, the momentum of looking for a luminal infectious agent or microbial factor was dampened by the development of biologic agents. These agents were so successful that the field of investigation in IBD completely shifted back towards blocking immune pathways as a means of treating these conditions. In the new age of the microbiome, there has been a renewed interest in the benefits of antibiotics as demonstrated by a 2011 meta-analysis suggesting an overall relative risk of 0.85 (CI = 0.73–0.99) inclusive [59]. This included benefits of fluoroquinolones, rifaximin, anti-mycobacterials, 5-nitroimidazoles, and macrolides. Interestingly, the benefit was present regardless of antibiotic category used. This provided little clarity regarding the mechanism underlying this benefit.

Another important factor suggesting an infectious agent is the presence of granulomas specifically seen in some patients with Crohn's disease. A genus known to produce granulomas during human infection is *Mycobacterium*. In fact, there was specific interest in *M. paratuberculosis* in the 1980's based on the similarities between the lesions created by this organism and Crohn's disease [60]. As already discussed, these findings led to the use of anti-mycobacterial therapies as shown in the above-described meta-analysis [59].

4 Molecular Biology Clues to Suggest a Microbial Agent in IBD

Understanding the inflammatory cascade in CD and UC has provided some strong (albeit still indirect) support for the importance of the microbiome these conditions (see Table 1). In this section, we will focus on four of these lines of research.

An important factor in the control of the inflammatory response is interleukin-10 (IL-10). This agent has an anti-inflammatory effect which offsets aggressive pro-inflammatory factors such as tumor necrosis factor-α (TNF-α). The main reason for mentioning IL-10 here is that a key lesson has been discovered by studying the IL-10 knockout mouse model. The IL-10 knockout mouse has been found to develop colitis more easily and is used as a model of ulcerative colitis [61]. In this model, mice develop mucosal inflammation more easily, allowing the study of provocation and mitigation of IBD. In a now classic study, the development of colitis in this model was shown to be dependent on the presence of a microbiome, since colitis fails to develop if IL-10 knockout mice are raised under germ-free conditions [62]. In this study, IL-10 deficient mice colonized with specific pathogen free microbes developed colitis in multiple segments, as well as an upregulated mucosal immune response. If the same mice were raised under germ-free conditions, no inflammation was seen. In this same model, certain bacteria may induce an even more vigorous response. Balish et al. divided germ free IL-10 knockout mice into three groups [63]. One group remained germ-free and developed no colitis as expected. The second group was colonized in adulthood with a mixture of yeast, *Escherichia coli*, and various *Lactobacillus* spp. The third germ-free group was colonized with *Enterococcus faecalis*.

Table 1 Evidence to support the importance of the gut microbiome in IBD

Category	Factor	Findings
Host factors	Toll receptors	Respond to bacterial factors such as flagellin
	NOD2 gene mutations	NOD2 important for bacterial recognition
	CARD9 polymorphisms	Important protein in the detection of fungal dectin
	Defensins	Lower production of defensins (which are natural antibiotics)
Gut factors	Bacterial	Changes in diversity including *Bacteroides*
	Fungal	Increased *Candida* spp. *Malassezia restricta* and CARD9 connection
	Viral	
	Mycobacteria	Possible infection with *Paratuberculosis*
	Germ-free animals	Animal models of IBD are less affected when germ free
Killing bacteria	Antibiotics	
Starving bacteria	Diverting gut contents	Diverty ostomies help colitis
	Elemental diet	Elemental diets reduce residua to distal bacteria
	Parenteral nutrition	
Correcting dysbiosis	Probiotics	Not recommended but data are proactive in treating pouchitis
	FMT	See Chapter "Fecal Transplant: The Benefits and Harms of Fecal Microbiota Transplantation"

Colitis only developed in the group with *E. faecalis* but moreover, this group also developed colonic dysplasia and even adenocarcinoma [63].

An important set of cell membrane proteins in IBD, many at the interface between IECs and the microbial environment in the lumen, are toll-like receptors (TLRs). These proteins provide further evidence of a role for luminal bacteria in IBD. There are at least 9 of these receptors with a variety of demonstrated roles in IBD [64], and a full discussion of these would require an entire chapter on its own. Two of the more important of these receptors in connecting the microbiome to IBD are TLR4 and TLR5. Both of these are external membrane receptors on IECs and have different roles in their interaction with luminal bacteria. The TLR4 receptor appears to detect lipopolysaccharide typically found on Gram negative bacteria. This interaction has the effect of inducing inflammation to defend against the possibility of barrier disruption and prevent invasion. It is believed that in IBD, there may be an augmented TLR4 response leading to an aggressive inflammatory response [65]. TLR5 on the other hand appears to specifically detect flagellin. As such, this receptor can detect organisms with flagellae, such as *E. coli* and

Salmonella. In IBD, studies suggest the possibility that polymorphisms in the TLR5 gene in specific populations are associated with the development of UC [66]. The connection between these two toll-like receptors and IBD suggest important roles for the microbiome and for host response.

Defensins are a group of peptides that are part of an important response to changes in gut microbial composition. There are many defensins which are secreted by different cells in the mucosa, but mostly by Paneth cells (while some are secreted by epithelial cells and immune cells). An important defensin that may have an important role in IBD is α-defensin from Paneth cells. Since Paneth cells are only found in the small intestine in humans, this peptide is only secreted in that region of the GI tract [67]. This peptide can be anti-bacterial and an important part of host response to luminal microbes. For example, there is evidence that this α-defensin may be important in the protection from certain pathogens such as *Salmonella* [68]. In CD, another study from the same group demonstrated a reduced expression of α-defensins compared to controls in the ileum [69]. The implication is that subjects with CD may have an innately poor response to luminal microbes leading to poor control of the microbiome, although this remains to be determined. Interestingly, the support for a genetic component to this comes from a study showing that NOD2 mutations are associated with both CD and lower α-defensin expression [70]. However, this is not consistent among studies [71].

Another defensin, β-defensin, is expressed (in contrast to α-defensin) more diffusely, including both the colon and the ileum [72]. In contrast to α-defensin, β-defensin expression is augmented in IBD, predominantly human β-defensin 2 (HD2) [73]. Interestingly, pathogenic bacteria and even probiotics might induce the production of HD2, and interestingly, part of this may be in response again to the bacterial protein flagellin [74]. While these results add to the story of defensins, the exact role defensins play and whether they are reactionary or pre-determinants of the microbiome-host interplay in IBD needs further study.

A very important component of gut immunity and important determinant in the sensing of the microbiome is the NOD2 protein. The gene encoding this cytosolic protein has become one of the first and most studied genes whose polymorphisms are a major risk factor for Crohn's disease specifically. Like other factors discussed above, the NOD2 protein is highly expressed in ileal Paneth cells, but unlike other factors may also be expressed by the gastrointestinal tract epithelium [75, 76]. Importantly, the expression of NOD2 is tightly linked to the luminal microbiome. Its expression is brought about by the presence of lipopolysaccharides and bacteria [77, 78]. In addition, bacterial byproducts such as butyrate also have a role in NOD2 function [79]. Tying the above story together, the activation of NOD2 also plays into the production of antibacterial factors such as α-defensins [80] and cytokines [81]. In addition, and as already discussed, this effect is also activated through toll-like receptors. Thus, the response to the microbiome is a complex and integrated system of factors, some which are known. It is now recognized for almost 20 years that there is an association between specific mutations in the NOD2 gene and the risk of CD [82, 83].

5 Novel Pathogens in IBD

Despite the breadth of research in the gut's own immune response to the microbiome and its components, there has been no smoking gun in the microbiome; no single pathogen known to cause IBD. Nevertheless, efforts continue to try to determine if a potential pathogen or pathogens may trigger these chronic diseases.

While the initial effort to understand the gut microbiome has focused on bacteria, there is an increasing interest in the role of fungi. Ironically, one of the biomarkers used over the last 20 years in the diagnosis of Crohn's disease is an antibody to a yeast (anti-*Saccharomyces cerevisiae* [ASCA] antibodies) [84]. Specifically, this is an antibody to mannan from this yeast (also known as Brewer's yeast). The presence of this antibody in the initial paper demonstrated a sensitivity and specificity of 49% and 97% respectively for CD, although follow on studies were less optimistic. Nevertheless, the concept of antibodies to a fungus as a biomarker in CD at least made note of yeast as important in Crohn's disease.

Recently, the interest in the fungome has discovered new and exciting correlations with IBD beyond the discovery of ASCA. In particular, studies suggest that *Candida* spp. may be associated with CD. In a study of pediatric CD compared to healthy control stool, a number of fungi were overrepresented in CD subjects. These included *Saccharomyces cerevisiae* (the basis for the ASCA serum marker), *Clavispora lusitaniae*, *Cyberlindnera jadinii*, *Candida albicans*, and *Kluy-veromyces marxianus* [85]. There was also a high co-occurrence rate and predictive ability for CD over healthy controls for the presence of these organisms. The strongest correlation to CD was with *C. lusitaniae*. Another study suggested, again in a pediatric population, that a reduction in fungal diversity is associated with a rise or overabundance of *Candida* spp [86]. Extending into the adult CD population, Hoarau et al. made a number of interesting observations [87]. By comparing familial clustered CD patients with non-disease affected relatives, they found a higher abundance of *Candida tropicalis* in stool of CD affected individuals, and this fungus was also associated with the presence of serum ASCA. They also found positive interactions between elevations in *C. tropicalis* and bacterial dysbiosis (including elevated *E. coli* and *Serratia marcescens*). The increased abundance of *Candida* spp. is also supported by other work [88, 89].

Another major finding is that a novel fungal element (*Malassezia restricta*), which is part of the skin microbiome, appears to be associated with major genetic alterations in IBD. An important genetic polymorphism that is one of the most important risk factors for IBD is located in the CARD9 gene [90, 91]. The CARD9 protein is an important protein for signaling inflammatory response and sensing fungal Dectin-1, Dectin-2, and Mincle [92]. Based on this background, a recent study has elegantly confirmed a number of prior observations, but pointed to *Malassezia* spp. (*M. restricta* and modestly *M. globosa*) as associated with CD but more importantly most associated with the presence of a single nucleotide polymorphism in CARD9 [93]. In addition, the group found patients with high ASCA levels to be reactive to *M. restricta*.

The final area of the microbiome that warrants exploration in IBD is the virome. While a common viral diagnosis in IBD patients is cytomegalovirus (CMV), from the Herpesviridae family of viruses, this virus is common in humans and reactivation in inflamed tissue such as in patients with ulcerative colitis is common. In fact, a systematic review from 2006 suggests that presence of CMV ranges between studies [94] but using PCR in prospective patients, from 16–60% of UC subjects have detectable CMV virus in colonic samples. In this instance, it is believed that CMV is not the cause of UC but rather activated by the damaged tissue and inflammation in the colon. Another eukaryotic virus found in sequencing stool from UC and CD compared to healthy controls was Anellovirus, but this was not detected in all samples and did not correlate with disease activity [95].

Another more intriguing virome in IBD is that of the bacteriophage. Bacteriophages are viruses that infect prokaryotic rather than eukaryotic cells. The challenge in any study of bacteriophages is that bacteriophages can be specific to the bacterial composition, as there can be specific phages that target specific bacteria, so if there is a change in the microbiome and virome, it is unclear which came first. Phages can also have a substantive impact on the microbiome [96]. Bearing this in mind, there are notable discoveries in the virome related to IBD. In an important study attempting classify changes in the virome in IBD, Norman, et al. found an increase in the Caudovirales order of bacteriophages in CD and UC both in the UK and an American validation set [95]. While the changes in this order were associated with reductions in the Bacteroidaceae family of bacteria in these subjects, and increases in other bacteria, these bacterial correlates were not confirmed in the US [95]. Similarly, in a pediatric IBD study, Fernandes, et al. demonstrated a greater relative abundance of Caudovirales than Microviridae [97]. Using more modern techniques, it was found that re-analysis of the Norman, et al. data revealed that combining viral and bacterial architecture simultaneously might separate IBD from healthy more reliably, based on complex observed changes in both the virome and prokaryotic profiles [98]. In a specific study with UC patients and more particularly using rectal mucosal tissue, there was again the common theme of an increase in the Caudovirales bacteriophages pool but with lower diversity and richness [99]. Interestingly, they also showed a greater abundance of *Escherichia* and *Enterobacteria* phages in UC. Taken one step further, Nishiyama, et al. through an analysis of whole-metagenome shotgun sequencing data from the IBD Multi-omics Database that was included in the Human Microbiome Project 2 [100], Nishiyawa, et al. found an increase in temperate bacteriophages infecting *Bacteroides uniformis* and *Bacteroides thetaiotamicron* [101]. Thus, these two beneficial bacteria were reduced in abundance, suggesting cause-and-effect.

While these data on bacteriophages are a promising start to a new line of discovery in the microbiome, more data is needed. Nevertheless, the accruing pieces of evidence attempting to identify microbial signatures of IBD are not ready to prove a cause-and-effect relationship to date. However, Table 1 is a convincing list of the evidence that the microbiome has a role in the development and/or maintenance of IBD. What is important to take away from this section is that the study of the microbiome in IBD may require a broad approach to microbial changes to include host-gut interactions, bacteria, viruses, and fungi. The disharmony of this system may eventually yield a better understanding of the pathophysiology of IBD and microbiome directed treatments.

6 Treating the Microbiome as a Means of Improving IBD

While there is mounting evidence on the crucial role of the gut microbiome in IBD, we are yet to see a microbiome-modulating drug to drastically alter the course of disease. The only exception is elemental diet, which has been shown to be effective in induction of remission of IBD specially in the pediatric population [102]. As will be discussed below, one potential mechanism of action of elemental diet is via modification of the gut microbiome. Furthermore, there are no microbiome-modulating drugs to show a promise in prevention of IBD in high risk populations. Studies to assess prevention of IBD are expensive, with long periods of follow-up. At least for the near future, we are unlikely to see such studies conducted. Below we summarize the role of drugs which modify the gut microbiome in IBD:

6.1 Antibiotics

Several trials have been performed to assess the efficacy and safety of antibiotics in induction of remission, and even maintenance of remission, in IBD patients [103, 104]. However, generally these studies were not performed with the currently acceptable endpoints for IBD, such as mucosal or histologic remission. Hence, it is hard to estimate the true beneficial effects of antibiotics in IBD.

Pooling the data from ten randomized placebo-controlled trials assessing antibiotics in UC shows that antibiotics are superior to placebo, with an odds ratio (OR) of 2.14 (95% confidence interval [CI], 1.48–3.09; $P < 0.01$) in induction of remission [104]. Interestingly, pooling of studies assessing the role of antibiotics in CD shows a magnitude of response similar to that observed in UC patients. Meta-analysis of six randomized placebo-controlled trials showed an OR of 2.26 (95% CI, 1.68–3.0; $P < 0.001$).

Rifaximin (which is a non-systemic antibiotic) has been shown to have a modest effect in induction of remission in both UC and CD [103, 104]. This suggests that the mechanism of action of antibiotics in IBD is an intraluminal effect and most likely involves modulation of the gut microbiome. Studies exploring changes in the gut microbiome in IBD patients who do respond to antibiotic therapy are lacking. As the above-noted trials used clinical activity indexes that are heavily affected by clinical symptoms, one alternative theory is that the response observed in IBD patients with antibiotics is due to improvement of concomitant SIBO and/or IBS which are known to be responsive to antibiotics [105, 106]. In the largest study to date to assess SIBO and IMO in symptomatic IBD patients, SIBO was observed in a significant proportion of CD ($n = 163$, 56.2%) and UC ($n = 101$, 57.7%) patients. Hydrogen-predominant SIBO was the most common type of SIBO among CD (46.2%) and UC (45.1%) patients. CD (OR 0.21, 95% CI 0.11–0.38, $p < 0.0001$) and UC (OR 0.39, 95% CI 0.22–0.70, $p = 0.01$) patients were less likely to be excessive CH_4 producers than non-IBD patients. IBD patients with SIBO who were treated with antibiotics showed a 57.3% symptomatic improvement. The rate of response to antibiotics was statistically similar between CD (52.1%) and UC (65.9%) ($p = 0.18$). Therefore, it is quite feasible that the observed clinical response seen in the studies of antibiotics in IBD patients is at least in part due to treatment of concomitant SIBO.

Interestingly, symptomatic CD (5.5%) and UC (5.1%) patients have more flatline breath tests (BT) (non-CH_4, fixed-H_2 production) than patients without IBD (3.0%), potentially indicating excessive intraluminal hydrogenotrophic bacteria (e.g. hydrogen sulfide (H_2S) producing bacteria). Recent evidence suggests that a flatline BT may in part reflect excessive H_2S producing bacteria, which is not detected by conventional gas chromatography machines [107]. Hydrogenotrophic bacteria out-compete methanogenic archaea for H_2 to produce H_2S, resulting in lack of CH_4 or H_2 detected during breath testing [108]. H_2S may have a role in the pathophysiology of IBD, particularly UC [43]. H_2S likely contributes to the pathogenesis of UC by suppressing butyrate oxidation in the colonic cells and reducing the disulfide bonds of the mucosal biofilm, leading to increased colonic permeability, allowing enteric pathogens to penetrate deeper layers [43]. It has been shown that 56% of symptomatic patients with a flat-line pattern on breath testing respond to antibiotic therapy [103]. Therefore, another potential hypothesis for the modest effect of antibiotics in IBD is eradication of H_2S-predominant bacterial overgrowth.

Sulfasalzine, which is also used in treatment of UC, breaks into Sulfapyridine and 5-Aminosalicylates (5-ASA) [109]. Sulfapyridine has antibiotic properties which likely contributes to the efficacy of sulfasalazine in UC.

Patients with perianal CD fistula have also been shown to modestly benefit from antibiotic therapy. Pooling of three trials ($n = 123$) showed a small but statistically significant effect in reducing drainage of fistula in CD patients ($RR = 0.8$; 95% $CI = 0.66$–0.98) [59]. A distinct microbiome profile has been found in the fistula of CD patients as compared to their stool [110]. *Bifidobacterium* is higher in stool than in fistula samples, while *Achromobacter* and *Corynebacterium* are more prevalent in fistula samples. The same study showed that antibiotics do affect the microbiome composition of CD-related fistula.

A role for the gut microbiome has also been elucidated in patients with pouchitis after ileal pouch-anal anastomosis for treatment of UC, with a distinct taxonomic profile [111]. Patients with pouchitis have decreased diversity of the microbiome when compared to patients with normal pouch. The genera *Ruminococcus, Lachnospira,* and *Coprococcus* were significantly decreased in pouchitis subjects [111]. Acute and chronic pouchitis have both been shown to be responsive to antibiotics to some degree. Metronidazole, as the most commonly used treatment for acute pouchitis, is associated with remission with RR of 2.68 (95% CI 1.13 to 6.35). While not assessed in controlled trials, alternating antibiotics are commonly used in patients with chronic pouchitis. Antibiotics have even been used to prevent pouchitis. Tinidazole in small studies has shown superiority over placebo in prevention of pouchitis after ileal pouch-anal anastomosis for the treatment of UC, with RR of 1.38 (95% CI 0.83 to 2.31) [112].

While robust data on the true efficacy of antibiotics using the current acceptable endoscopic and histologic endpoints in IBD are lacking, in clinical practice it is not uncommon to use antibiotics in induction of remission, particularly in severe cases. Larger blinded studies with non-systemic antibiotics are needed to elucidate the stand-alone or objective efficacy of antibiotics in IBD.

6.2 Probiotics

Given the role of microbiome in pathogenesis of IBD, it seems plausible for probiotics to have beneficial effects; however, the results have not been as promising as expected. As per 2020 American Gastroenterological Association (AGA) guidelines, probiotics are not recommended for the induction or maintenance of remission of UC or CD [113]. Probiotics trials in IBD are mired with methodological issues such as heterogenous patient populations and designs, small sample sizes and variable probiotic formulation, preparation, dose, and treatment duration. Furthermore, remission is variably defined in these trials.

A comprehensive review by the AGA found 12 studies in CD which assessed the following probiotics: *E. coli* Nissle 1917, *Saccharomyces boulardii, Lactobaccilus rhamnosus* GG, *Lactobacillus johnsonii* LA1, and an 8-strain probiotic mix of *Lactobacillus* and *Bifidobacterium*. Overall, no robust benefit was observed for any of the probiotics in the induction or maintenance of remission of CD.

Similarly, 17 studies were found on the use of probiotics in the induction or maintenance of remission in UC. The strain(s) of probiotics included: *Bifidobacterium breve*, *B. bifidum* Yakult, *L. acidophilus*; *B. longum; E coli* Nissle 1917; *L. reuteri* and *L. rhammosus*. None of these studies provided convincing evidence of efficacy of probiotics in UC.

Patients with refractory ulcerative colitis or dysplasia may undergo total proctocolectomy and ileal pouch-anal anastomosis (IPAA). Commonly, IPAA is complicated with pouchitis which is generally treated with antibiotics [114]. An 8-strain combination of *Lactobacillus paracasei* subsp. *paracasei, L. plantarum, L. acidophilus, L. delbrueckii* subsp. *bulgaricus, B. longum* subsp. *longum, B. breve, B. longum* subsp. *infantis*, and *Streptococcus salivarius* subsp. *thermophilus* has been shown in several studies to have beneficial effects in the prevention and treatment of pouchitis. Nevertheless, the quality of evidence remains low.

6.3 Elemental Diet

An elemental diet is a liquid formula containing amino acids along with easily digestible carbohydrates (generally in the form of mono- or disaccharides), minerals and vitamins. Fat is present in very small quantities. Systematic reviews have demonstrated the effects of an elemental diet in inducing remission in pediatric and adult CD [115]. Interestingly, an elemental diet is even effective in pneumatosis coli which further attests that its effectiveness is mediated via modulation of the microbiome, resulting in decreased gas production and improvement of pneumatosis [116]. An elemental diet has even been used in conjunction with antibiotics to induce remission in CD as effectively as steroids [117].

In mice, an elemental diet has been shown to decrease the diversity of the stool microbiota [118] and to decrease inflammation with decreased IFN-γ and IL-6 secretion [119]. However, in humans, an elemental diet was not associated with decreased diversity of the fecal microbiota [120].

A four-week exclusive elemental diet has also been shown to be effective in a study of 7 patients with pouchitis [121]. Five of the 7 patients (71%) reported a reduction in stool frequency and pouch activity index score. Interestingly, clinical improvement did not correlate with endoscopic and histologic improvement, and the elemental diet did not change the concentration of bacteria in IPAA patients. All but one patient had a reduction in weight following treatment with elemental diet, which is common with exclusive elemental diet. While reduction in weight and body fat mass is common with an elemental diet, this is not associated with lean body mass reduction or sarcopenia [122].

An elemental diet has also been shown to prevent postoperative recurrence of Crohn's disease in an uncontrolled study [123].

Notably, the positive effects of an elemental diet on the symptoms of IBD may also be due to improvements in concomitant SIBO, which is prevalent in IBD. The efficacy of elemental diet has been shown in an uncontrolled trial, with normalization of breath test in 85% of subjects after 3 weeks of an exclusive elemental diet [124].

Despite these encouraging results, the use of elemental diets in clinical practice remains limited. This is due to poor palatability and the tediousness of being on an exclusive liquid diet, especially in adults.

6.4 FMT

There has been great interest in the use of FMT to treat refractory IBD, and in particular UC. Study outcomes have been mixed and are discussed at length in Chapter "Fecal Transplant: The Benefits and Harms of Fecal Microbiota Transplantation".

7 Future Directions

Based on a growing body of evidence, it is clear that the microbiome has a role in either the pathophysiology or the exacerbation of IBD. While there are differences in these findings between CD and UC, there are also consistencies. It is clear that the answer lies somewhere in the culmination of the host (genetics, diet, and physical environment) and the luminal microbes (bacteria, fungi, viruses or all the collective). Unravelling this mystery may unleash a new series of diagnostics and therapeutics in this area.

References

1. Muller JM, Keller HW, Erasmi H, et al. Total parenteral nutrition as the sole therapy in Crohn's disease—a prospective study. Br J Surg. 1983;70:40–3.
2. Greenberg GR, Fleming CR, Jeejeebhoy KN, et al. Controlled trial of bowel rest and nutritional support in the management of Crohn's disease. Gut. 1988;29:1309–15.
3. McIntyre PB, Powell-Tuck J, Wood SR, et al. Controlled trial of bowel rest in the treatment of severe acute colitis. Gut. 1986;27:481–5.
4. Dickinson RJ, Ashton MG, Axon AT, et al. Controlled trial of intravenous hyperalimentation and total bowel rest as an adjunct to the routine therapy of acute colitis. Gastroenterology. 1980;79:1199–204.
5. Russell TA, Dawes AJ, Graham DS, et al. Rescue diverting loop ileostomy: an alternative to emergent colectomy in the setting of severe acute refractory IBD-colitis. Dis Colon Rectum. 2018;61:214–20.
6. Aberra FN, Brensinger CM, Bilker WB, et al. Antibiotic use and the risk of flare of inflammatory bowel disease. Clin Gastroenterol Hepatol. 2005;3:459–65.
7. Sutherland L, Singleton J, Sessions J, et al. Double blind, placebo controlled trial of metronidazole in Crohn's disease. Gut. 1991;32:1071–5.
8. Chapman RW, Selby WS, Jewell DP. Controlled trial of intravenous metronidazole as an adjunct to corticosteroids in severe ulcerative colitis. Gut. 1986;27:1210–2.
9. Mantzaris GJ, Hatzis A, Kontogiannis P, et al. Intravenous tobramycin and metronidazole as an adjunct to corticosteroids in acute, severe ulcerative colitis. Am J Gastroenterol. 1994;89:43–6.
10. Mantzaris GJ, Petraki K, Archavlis E, et al. A prospective randomized controlled trial of intravenous ciprofloxacin as an adjunct to corticosteroids in acute, severe ulcerative colitis. Scand J Gastroenterol. 2001;36:971–4.
11. Dickinson RJ, O'Connor HJ, Pinder I, et al. Double blind controlled trial of oral vancomycin as adjunctive treatment in acute exacerbations of idiopathic colitis. Gut. 1985;26:1380–4.
12. Lichtenstein GR, Loftus EV, Isaacs KL, et al. ACG clinical guideline: management of Crohn's disease in adults. Am J Gastroenterol. 2018;113:481–517.
13. Rubin DT, Ananthakrishnan AN, Siegel CA, et al. ACG clinical guideline: ulcerative colitis in adults. Am J Gastroenterol. 2019;114:384–413.
14. Chiodini RJ, Van Kruiningen HJ, Thayer WR, et al. Possible role of mycobacteria in inflammatory bowel disease. I. An unclassified Mycobacterium species isolated from patients with Crohn's disease. Dig Dis Sci. 1984;29:1073–9.
15. Millar D, Ford J, Sanderson J, et al. IS900 PCR to detect Mycobacterium paratuberculosis in retail supplies of whole pasteurized cows' milk in England and Wales. Appl Environ Microbiol. 1996;62:3446–52.
16. Feller M, Huwiler K, Stephan R, et al. Mycobacterium avium subspecies paratuberculosis and Crohn's disease: a systematic review and meta-analysis. Lancet Infect Dis. 2007;7:607–13.
17. Gui GP, Thomas PR, Tizard ML, et al. Two-year-outcomes analysis of Crohn's disease treated with rifabutin and macrolide antibiotics. J Antimicrob Chemother. 1997;39:393–400.
18. Shafran I, Kugler L, El-Zaatari FA, et al. Open clinical trial of rifabutin and clarithromycin therapy in Crohn's disease. Dig Liver Dis. 2002;34:22–8.
19. Sartor RB. Does Mycobacterium avium subspecies paratuberculosis cause Crohn's disease? Gut. 2005;54:896–8.
20. Axelrad JE, Olen O, Askling J, et al. Gastrointestinal infection increases odds of inflammatory bowel disease in a Nationwide case-control study. Clin Gastroenterol Hepatol. 2019;17:1311–1322.e7.
21. Gradel KO, Nielsen HL, Schonheyder HC, et al. Increased short- and long-term risk of inflammatory bowel disease after salmonella or campylobacter gastroenteritis. Gastroenterology. 2009;137:495–501.
22. Nielsen HL, Dalager-Pedersen M, Nielsen H. Risk of inflammatory bowel disease after campylobacter jejuni and campylobacter concisus infection: a population-based cohort study. Scand J Gastroenterol. 2019;54:265–72.
23. Jess T, Simonsen J, Nielsen NM, et al. Enteric salmonella or campylobacter infections and the risk of inflammatory bowel disease. Gut. 2011;60:318–24.
24. Pimentel M, Saad RJ, Long MD, et al. ACG clinical guideline: small intestinal bacterial overgrowth. Am J Gastroenterol. 2020;115:165.
25. Bond JH, Engel RR, Levitt MD. Factors influencing pulmonary methane excretion in man. An indirect method of studying the in situ metabolism of the methane-producing colonic bacteria. J Exp Med. 1971;133:572–88.
26. Christl SU, Murgatroyd PR, Gibson GR, et al. Production, metabolism, and excretion of hydrogen in the large intestine. Gastroenterology. 1992;102:1269–77.
27. Singer-Englar T, Rezaie A, Gupta K, et al. 182 - competitive hydrogen gas utilization by methane- and hydrogen sulfide-producing microorganisms and associated symptoms: results of a novel 4-gas breath test machine. Gastroenterology. 2018;154:S–47.
28. Miller TL, Wolin MJ, de Macario EC, et al. Isolation of Methanobrevibacter smithii from human feces. Appl Environ Microbiol. 1982;43:227–32.
29. Weaver GA, Krause JA, Miller TL, et al. Incidence of methanogenic bacteria in a sigmoidoscopy population: an association of methanogenic bacteria and diverticulosis. Gut. 1986;27:698–704.
30. Rana SV, Sharma S, Malik A, et al. Small intestinal bacterial overgrowth and orocecal transit time in patients of inflammatory bowel disease. Dig Dis Sci. 2013;58:2594–8.
31. Gu P, Patel D, Lakhoo K, et al. Breath test gas patterns in inflammatory bowel disease with concomitant irritable bowel syndrome-like symptoms: a controlled large-scale database linkage analysis. Dig Dis Sci. 2020;65:2388–96.
32. Pimentel M, Mayer AG, Park S, et al. Methane production during lactulose breath test is associated with gastrointestinal disease presentation. Dig Dis Sci. 2003;48:86–92.
33. Shen N, Wang Z, Wang C, et al. Methane alleviates inflammation and apoptosis of dextran sulfate sodium-induced inflammatory bowel diseases by inhibiting toll-like receptor 4 (TLR4)/myeloid differentiation factor 88 (MyD88)/nuclear translocation of nuclear factor-kappaB (NF-kappaB) and endoplasmic reticulum stress pathways in mice. Med Sci Monit. 2020;26:e922248.
34. Boros M, Ghyczy M, Erces D, et al. The anti-inflammatory effects of methane. Crit Care Med. 2012;40:1269–78.
35. Chen O, Ye Z, Cao Z, et al. Methane attenuates myocardial ischemia injury in rats through anti-oxidative, anti-apoptotic and anti-inflammatory actions. Free Radic Biol Med. 2016;90:1–11.
36. He R, Wang L, Zhu J, et al. Methane-rich saline protects against concanavalin A-induced autoimmune hepatitis in mice through anti-inflammatory and anti-oxidative pathways. Biochem Biophys Res Commun. 2016;470:22–8.
37. Song K, Zhang M, Hu J, et al. Methane-rich saline attenuates ischemia/reperfusion injury of abdominal skin flaps in rats via regulating apoptosis level. BMC Surg. 2015;15:92.
38. Ye Z, Chen O, Zhang R, et al. Methane attenuates hepatic ischemia/reperfusion injury in rats through Antiapoptotic, anti-inflammatory, and antioxidative actions. Shock. 2015;44:181–7.
39. Gibson GR, Cummings JH, Macfarlane GT. Growth and activities of sulphate-reducing bacteria in gut contents of healthy sub-

jects and patients with ulcerative colitis. FEMS Microbiol Lett. 1991;86:103–11.
40. Loubinoux J, Bronowicki JP, Pereira IA, et al. Sulfate-reducing bacteria in human feces and their association with inflammatory bowel diseases. FEMS Microbiol Ecol. 2002;40:107–12.
41. Zinkevich VV, Beech IB. Screening of sulfate-reducing bacteria in colonoscopy samples from healthy and colitic human gut mucosa. FEMS Microbiol Ecol. 2000;34:147–55.
42. Fite A, Macfarlane GT, Cummings JH, et al. Identification and quantitation of mucosal and faecal desulfovibrios using real time polymerase chain reaction. Gut. 2004;53:523–9.
43. Roediger WE, Duncan A, Kapaniris O, et al. Reducing sulfur compounds of the colon impair colonocyte nutrition: implications for ulcerative colitis. Gastroenterology. 1993;104:802–9.
44. Chapman MA, Grahn MF, Boyle MA, et al. Butyrate oxidation is impaired in the colonic mucosa of sufferers of quiescent ulcerative colitis. Gut. 1994;35:73–6.
45. Cooper CE, Brown GC. The inhibition of mitochondrial cytochrome oxidase by the gases carbon monoxide, nitric oxide, hydrogen cyanide and hydrogen sulfide: chemical mechanism and physiological significance. J Bioenerg Biomembr. 2008;40:533–9.
46. Yonezawa D, Sekiguchi F, Miyamoto M, et al. A protective role of hydrogen sulfide against oxidative stress in rat gastric mucosal epithelium. Toxicology. 2007;241:11–8.
47. Wu YC, Wang XJ, Yu L, et al. Hydrogen sulfide lowers proliferation and induces protective autophagy in colon epithelial cells. PLoS One. 2012;7:e37572.
48. Furne JK, Suarez FL, Ewing SL, et al. Binding of hydrogen sulfide by bismuth does not prevent dextran sulfate-induced colitis in rats. Dig Dis Sci. 2000;45:1439–43.
49. Kruszyna H, Kruszyna R, Smith RP. Cyanide and sulfide interact with nitrogenous compounds to influence the relaxation of various smooth muscles. Proc Soc Exp Biol Med. 1985;179:44–9.
50. Teague B, Asiedu S, Moore PK. The smooth muscle relaxant effect of hydrogen sulphide in vitro: evidence for a physiological role to control intestinal contractility. Br J Pharmacol. 2002;137:139–45.
51. Parajuli SP, Choi S, Lee J, et al. The inhibitory effects of hydrogen sulfide on pacemaker activity of interstitial cells of cajal from mouse small intestine. Korean J Physiol Pharmacol. 2010;14:83–9.
52. Yoon PJ, Parajuli SP, Zuo DC, et al. Interplay of hydrogen sulfide and nitric oxide on the pacemaker activity of interstitial cells of cajal from mouse small intestine. Chonnam Med J. 2011;47:72–9.
53. Jin X, Malykhina AP, Lupu F, et al. Altered gene expression and increased bursting activity of colonic smooth muscle ATP-sensitive K+ channels in experimental colitis. Am J Physiol Gastrointest Liver Physiol. 2004;287:G274–85.
54. Kang M, Hashimoto A, Gade A, et al. Interaction between hydrogen sulfide-induced sulfhydration and tyrosine nitration in the KATP channel complex. Am J Physiol Gastrointest Liver Physiol. 2015;308:G532–9.
55. Schicho R, Krueger D, Zeller F, et al. Hydrogen sulfide is a novel prosecretory neuromodulator in the Guinea-pig and human colon. Gastroenterology. 2006;131:1542–52.
56. Linden DR. Hydrogen sulfide signaling in the gastrointestinal tract. Antioxid Redox Signal. 2014;20:818–30.
57. Bolton PM, Owen E, Heatley RV, et al. Negative findings in laboratory animals for a transmissible agent in Crohn's disease. Lancet. 1973;2:1122–4.
58. Ursing B, Alm T, Barany F, et al. A comparative study of metronidazole and sulfasalazine for active Crohn's disease: the cooperative Crohn's disease study in Sweden. II Result. Gastroenterology. 1982;83:550–62.
59. Khan KJ, Ullman TA, Ford AC, et al. Antibiotic therapy in inflammatory bowel disease: a systematic review and meta-analysis. Am J Gastroenterol. 2011;106:661–73.
60. Chiodini RJ. Crohn's disease and the mycobacterioses: a review and comparison of two disease entities. Clin Microbiol Rev. 1989;2:90–117.
61. Keubler LM, Buettner M, Hager C, et al. A multihit model: colitis lessons from the Interleukin-10-deficient mouse. Inflamm Bowel Dis. 2015;21:1967–75.
62. Sellon RK, Tonkonogy S, Schultz M, et al. Resident enteric bacteria are necessary for development of spontaneous colitis and immune system activation in interleukin-10-deficient mice. Infect Immun. 1998;66:5224–31.
63. Balish E, Warner T. Enterococcus faecalis induces inflammatory bowel disease in interleukin-10 knockout mice. Am J Pathol. 2002;160:2253–7.
64. Lu Y, Li X, Liu S, et al. Toll-like receptors and inflammatory bowel disease. Front Immunol. 2018;9:72.
65. Fukata M, Abreu MT. TLR4 signalling in the intestine in health and disease. Biochem Soc Trans. 2007;35:1473–8.
66. Meena NK, Ahuja V, Meena K, et al. Association of TLR5 gene polymorphisms in ulcerative colitis patients of North India and their role in cytokine homeostasis. PLoS One. 2015;10:e0120697.
67. Cunliffe RN. Alpha-defensins in the gastrointestinal tract. Mol Immunol. 2003;40:463–7.
68. Salzman NH, Ghosh D, Huttner KM, et al. Protection against enteric salmonellosis in transgenic mice expressing a human intestinal defensin. Nature. 2003;422:522–6.
69. Wehkamp J, Salzman NH, Porter E, et al. Reduced Paneth cell alpha-defensins in ileal Crohn's disease. Proc Natl Acad Sci U S A. 2005;102:18129–34.
70. Wehkamp J, Harder J, Weichenthal M, et al. NOD2 (CARD15) mutations in Crohn's disease are associated with diminished mucosal alpha-defensin expression. Gut. 2004;53:1658–64.
71. Simms LA, Doecke JD, Walsh MD, et al. Reduced alpha-defensin expression is associated with inflammation and not NOD2 mutation status in ileal Crohn's disease. Gut. 2008;57:903–10.
72. O'Neil DA, Porter EM, Elewaut D, et al. Expression and regulation of the human beta-defensins hBD-1 and hBD-2 in intestinal epithelium. J Immunol. 1999;163:6718–24.
73. Wehkamp J, Fellermann K, Herrlinger KR, et al. Human beta-defensin 2 but not beta-defensin 1 is expressed preferentially in colonic mucosa of inflammatory bowel disease. Eur J Gastroenterol Hepatol. 2002;14:745–52.
74. Schlee M, Wehkamp J, Altenhoefer A, et al. Induction of human beta-defensin 2 by the probiotic Escherichia coli Nissle 1917 is mediated through flagellin. Infect Immun. 2007;75:2399–407.
75. Lala S, Ogura Y, Osborne C, et al. Crohn's disease and the NOD2 gene: a role for paneth cells. Gastroenterology. 2003;125:47–57.
76. Ogura Y, Lala S, Xin W, et al. Expression of NOD2 in Paneth cells: a possible link to Crohn's ileitis. Gut. 2003;52:1591–7.
77. Ikeuchi T, Nakamura T, Fukumoto S, et al. A vitamin D3 analog augmented interleukin-8 production by human monocytic cells in response to various microbe-related synthetic ligands, especially NOD2 agonistic muramyldipeptide. Int Immunopharmacol. 2013;15:15–22.
78. Hisamatsu T, Suzuki M, Reinecker HC, et al. CARD15/NOD2 functions as an antibacterial factor in human intestinal epithelial cells. Gastroenterology. 2003;124:993–1000.
79. Leung CH, Lam W, Ma DL, et al. Butyrate mediates nucleotide-binding and oligomerisation domain (NOD) 2-dependent mucosal immune responses against peptidoglycan. Eur J Immunol. 2009;39:3529–37.
80. Biswas A, Liu YJ, Hao L, et al. Induction and rescue of Nod2-dependent Th1-driven granulomatous inflammation of the ileum. Proc Natl Acad Sci U S A. 2010;107:14739–44.
81. Magalhaes JG, Fritz JH, Le Bourhis L, et al. Nod2-dependent Th2 polarization of antigen-specific immunity. J Immunol. 2008;181:7925–35.

82. Hugot JP, Chamaillard M, Zouali H, et al. Association of NOD2 leucine-rich repeat variants with susceptibility to Crohn's disease. Nature. 2001;411:599–603.
83. Ogura Y, Bonen DK, Inohara N, et al. A frameshift mutation in NOD2 associated with susceptibility to Crohn's disease. Nature. 2001;411:603–6.
84. Quinton JF, Sendid B, Reumaux D, et al. Anti-Saccharomyces cerevisiae mannan antibodies combined with antineutrophil cytoplasmic autoantibodies in inflammatory bowel disease: prevalence and diagnostic role. Gut. 1998;42:788–91.
85. Lewis JD, Chen EZ, Baldassano RN, et al. Inflammation, antibiotics, and diet as environmental stressors of the gut microbiome in pediatric Crohn's disease. Cell Host Microbe. 2015;18:489–500.
86. Chehoud C, Albenberg LG, Judge C, et al. Fungal signature in the gut microbiota of pediatric patients with inflammatory bowel disease. Inflamm Bowel Dis. 2015;21:1948–56.
87. Hoarau G, Mukherjee PK, Gower-Rousseau C, et al. Bacteriome and mycobiome interactions underscore microbial dysbiosis in familial crohn's disease. mBio. 2016;7:e01250.
88. Sokol H, Leducq V, Aschard H, et al. Fungal microbiota dysbiosis in IBD. Gut. 2017;66:1039–48.
89. Liguori G, Lamas B, Richard ML, et al. Fungal Dysbiosis in mucosa-associated microbiota of Crohn's disease patients. J Crohns Colitis. 2016;10:296–305.
90. Rivas MA, Beaudoin M, Gardet A, et al. Deep resequencing of GWAS loci identifies independent rare variants associated with inflammatory bowel disease. Nat Genet. 2011;43:1066–73.
91. Jostins L, Ripke S, Weersma RK, et al. Host-microbe interactions have shaped the genetic architecture of inflammatory bowel disease. Nature. 2012;491:119–24.
92. Roth S, Ruland J. Caspase recruitment domain-containing protein 9 signaling in innate immunity and inflammation. Trends Immunol. 2013;34:243–50.
93. Limon JJ, Tang J, Li D, et al. Malassezia is associated with Crohn's disease and exacerbates colitis in mouse models. Cell Host Microbe. 2019;25:377–388.e6.
94. Criscuoli V, Rizzuto MR, Cottone M. Cytomegalovirus and inflammatory bowel disease: is there a link? World J Gastroenterol. 2006;12:4813–8.
95. Norman JM, Handley SA, Baldridge MT, et al. Disease-specific alterations in the enteric virome in inflammatory bowel disease. Cell. 2015;160:447–60.
96. Reyes A, Wu M, McNulty NP, et al. Gnotobiotic mouse model of phage-bacterial host dynamics in the human gut. Proc Natl Acad Sci U S A. 2013;110:20236–41.
97. Fernandes MA, Verstraete SG, Phan TG, et al. Enteric Virome and bacterial microbiota in children with ulcerative colitis and Crohn disease. J Pediatr Gastroenterol Nutr. 2019;68:30–6.
98. Clooney AG, Sutton TDS, Shkoporov AN, et al. Whole-virome analysis sheds light on viral dark matter in inflammatory bowel disease. Cell Host Microbe. 2019;26:764–778.e5.
99. Zuo T, Lu XJ, Zhang Y, et al. Gut mucosal virome alterations in ulcerative colitis. Gut. 2019;68:1169–79.
100. Lloyd-Price J, Arze C, Ananthakrishnan AN, et al. Multi-omics of the gut microbial ecosystem in inflammatory bowel diseases. Nature. 2019;569:655–62.
101. Nishiyama H, Endo H, Blanc-Mathieu R, et al. Ecological structuring of temperate bacteriophages in the inflammatory bowel disease-affected gut. Microorganisms. 2020;8:8.
102. Takahashi H, Ando T, Watanabe O, et al. Usefulness of an elemental diet in Crohn's disease. Inflammopharmacology. 2007;15:15–7.
103. Rahimi R, Nikfar S, Rezaie A, et al. A meta-analysis of broad-spectrum antibiotic therapy in patients with active Crohn's disease. Clin Ther. 2006;28:1983–8.
104. Rahimi R, Nikfar S, Rezaie A, et al. A meta-analysis of antibiotic therapy for active ulcerative colitis. Dig Dis Sci. 2007;52:2920–5.
105. Barros LL, Farias AQ, Rezaie A. Gastrointestinal motility and absorptive disorders in patients with inflammatory bowel diseases: prevalence, diagnosis and treatment. World J Gastroenterol. 2019;25:4414–26.
106. Rezaie A, Heimanson Z, McCallum R, et al. Lactulose breath testing as a predictor of response to Rifaximin in patients with irritable bowel syndrome with diarrhea. Am J Gastroenterol. 2019;114:1886–93.
107. Birg A, Hu S, Lin HC. Reevaluating our understanding of lactulose breath tests by incorporating hydrogen sulfide measurements. JGH Open. 2019;3:228–33.
108. Strocchi A, Furne J, Ellis C, et al. Methanogens outcompete sulphate reducing bacteria for H2 in the human colon. Gut. 1994;35:1098–101.
109. Nikfar S, Rahimi R, Rezaie A, et al. A meta-analysis of the efficacy of sulfasalazine in comparison with 5-aminosalicylates in the induction of improvement and maintenance of remission in patients with ulcerative colitis. Dig Dis Sci. 2009;54:1157–70.
110. Haac BE, Palmateer NC, Seaton ME, et al. A distinct gut microbiota exists within Crohn's disease-related perianal fistulae. J Surg Res. 2019;242:118–28.
111. Maharshak N, Cohen NA, Reshef L, et al. Alterations of enteric microbiota in patients with a Normal Ileal pouch are predictive of Pouchitis. J Crohns Colitis. 2017;11:314–20.
112. Nguyen N, Zhang B, Holubar SD, et al. Treatment and prevention of pouchitis after ileal pouch-anal anastomosis for chronic ulcerative colitis. Cochrane Database Syst Rev. 2019;11:CD001176.
113. Su GL, Ko CW, Bercik P, et al. AGA clinical practice guidelines on the role of probiotics in the management of gastrointestinal disorders. Gastroenterology. 2020;159:697–705.
114. Rezaei A. Pouchitis: an empirically treated disease in the era of evidence-based medicine. Int J Pharmacol. 2011;7:550–1.
115. Zachos M, Tondeur M, Griffiths AM. Enteral nutritional therapy for inducing remission of Crohn's disease. Cochrane Database Syst Rev. 2001;4:CD000542.
116. Johnston BT, McFarland RJ. Elemental diet in the treatment of pneumatosis coli. Scand J Gastroenterol. 1995;30:1224–7.
117. Saverymuttu S, Hodgson HJ, Chadwick VS. Controlled trial comparing prednisolone with an elemental diet plus non-absorbable antibiotics in active Crohn's disease. Gut. 1985;26:994–8.
118. Andoh A, Inoue R, Kawada Y, et al. Elemental diet induces alterations of the gut microbial community in mice. J Clin Biochem Nutr. 2019;65:118–24.
119. Menezes JS, Andrade MC, Senra B, et al. Immunological activities are modulated by enteral administration of an elemental diet in mice. Clin Nutr. 2006;25:643–52.
120. Shiga H, Kajiura T, Shinozaki J, et al. Changes of faecal microbiota in patients with Crohn's disease treated with an elemental diet and total parenteral nutrition. Dig Liver Dis. 2012;44:736–42.
121. McLaughlin SD, Culkin A, Cole J, et al. Exclusive elemental diet impacts on the gastrointestinal microbiota and improves symptoms in patients with chronic pouchitis. J Crohns Colitis. 2013;7:460–6.
122. Ishikawa T, Yasuda T, Doi T, et al. The amino acid-rich elemental diet Elental(R) preserves lean body mass during chemo- or chemoradiotherapy for esophageal cancer. Oncol Rep. 2016;36:1093–100.
123. Ohara N, Mizushima T, Iijima H, et al. Adherence to an elemental diet for preventing postoperative recurrence of Crohn's disease. Surg Today. 2017;47:1519–25.
124. Pimentel M, Constantino T, Kong Y, et al. A 14-day elemental diet is highly effective in normalizing the lactulose breath test. Dig Dis Sci. 2004;49:73–7.

Fecal Transplant: The Benefits and Harms of Fecal Microbiota Transplantation

David Goldenberg and Gil Y. Melmed

Highlights

- Fecal microbiota transplant (FMT) has been successfully used to treat recurrent or refractory *Clostridioides difficile* infection (CDI), with 1-month cure rates of up to 90% of subjects, but is not currently recommended for the initial treatment of CDI.
- FMT has also shown positive effects in treating Inflammatory Bowel Disease (IBD), although there is more robust data for ulcerative colitis (UC) than for Crohn's disease. Trials in irritable bowel syndrome (IBS) have shown more heterogenous results, although a recent prospective controlled study using a single healthy 'super donor' showed promising results.
- Trials are ongoing to evaluate the potential of FMT in treating other conditions including immune checkpoint inhibitor colitis, hepatic encephalopathy, Parkinson's disease, and multiple sclerosis.
- Careful donor selection and sample screening is essential, as illustrated by the transmission of extended spectrum beta-lactamase (ESBL) *E. coli* bacteremia to recipients in 2 separate trials utilizing the same donor in 2019. This included immunocompromised recipients, one of whom died. This resulted in the FDA issuing stricter guidelines for donor screening.
- RBX2660 and SER-109 are the first microbiome products to receive FDA approval for the treatment of CDI with additional promising treatment options in development.

D. Goldenberg
Division of Gastroenterology, Department of Medicine, Stanford University, Palo Alto, CA, USA
e-mail: degold@stanford.edu

G. Y. Melmed (✉)
Karsh Division of Gastroenterology and Hepatology, Department of Medicine, Cedars-Sinai Medical Center, Los Angeles, CA, USA
e-mail: Gil.Melmed@cshs.org

1 History of FMT

1.1 History of Early FMT Interventions

The gut microbiome has been therapeutically manipulated for over a millennium. The first descriptions date back to the Dong-jin dynasty in fourth century China where Ge Hong, a traditional Chinese medicine doctor described oral preparation of stool for the treatment of food poisoning and those with severe diarrhea. There are several other descriptions throughout traditional Chinese medical literature. It was then discussed by Italian veterinarian Fabricus Aquapendente in the seventeenth century [1]. The first described case series in the modern medical literature in 1958 is attributed to Ben Eismen – a surgeon who successfully treated antibiotic-associated pseudomembranous colitis with fecal enema [2]. Years later he described his thought process:

> In the early days of oral antibiotics we were plagued by frequent diarrhea in our patients due presumably to killing off intestinal bacteria. I was Chief of Surgery at the VA and simplistically considered merely reintroducing normal organisms to counter such absence. Those were days when if one had an idea, we simply tried. It seemed to work and I wrote it up. It made a small splash…Best Wishes [3].

1.2 FMT in the Modern Age

After the success of the Human Genome Project in 2003, a similar initiative was undertaken by scientists to analyze the genetic material of bacteria residing in 18 body sites of 242 adults. This launched a newfound interest in the human flora, and whole microbiome transplantation was reexplored. The first modern randomized, controlled trial of FMT for the treatment of recurrent *C. difficile* infection (CDI) was conducted in Amsterdam in 2013 and halted early due to overwhelmingly positive results [4]. There now exists a significant body of literature to support FMT as a treatment for recurrent CDI in a safe and effective manner.

M. Pimentel et al. (eds.), *Clinical Understanding of the Human Gut Microbiome*, https://doi.org/10.1007/978-3-031-46712-7_9

Given the success FMT has demonstrated with recurrent CDI, it is being investigated for other indications, both in and outside of the gastrointestinal tract, with more tempered results. There are limited data to suggest a role for FMT in the treatment of ulcerative colitis (UC), irritable bowel syndrome (IBS), hepatic encephalopathy, cancer, medication pharmacokinetics, metabolic syndrome, antibiotic-associated diarrhea, and autism [5]. A registry managed collaboratively between major medical societies and biome repositories (American Gastroenterological Association, Infectious Disease Society of America, Crohn's and Colitis Foundation, Openbiome, American Gut) is currently enrolling patients who have received FMT and tracking their outcomes. In 2022, the first FMT therapy (RBX2660) was FDA approved for recurrent CDI based on phase 3 clinical trial data demonstrating superior recurrence-free survival at 8 weeks relative to placebo after a single dose of RBX22660 [6].

2 Technical Considerations

2.1 Route of FMT Administration

Microbiota transplantation was initially performed for the treatment of refractory CDI in the modern era via nasogastric tube [4]; however, other options may be more effective for the treatment of refractory CDI. Rectal enema, colonoscopy, and oral capsule have all been shown to be effective treatment delivery options. Meta-analysis of FMT route incorporating 21 trials and 1191 patients suggested that colonoscopy may be modestly superior to other modalities [pooled cure rate 95.1% (95% CI 92.4% to 97.1%) via colonoscopy vs 84.7% (95% CI 76.3% to 90.7%) via enema, and 77.8% (95% CI 69.6% to 85.0%) via nasogastric tube]. In 2017, a small randomized noninferiority trial did not show any differences between oral capsule preparation and colonoscopic route in 23 patients [7]. A retrospective multicenter analysis suggests FMT may be more effective delivered in the terminal ileum [8].

Importantly, these analyses were conducted in trials with CDI colitis, which is thought to involve dysbiosis of the colon. Route of administration may factor differently in diseases where small intestinal dysbiosis may be important, such as in IBS and/or small intestinal bacterial overgrowth. Optimizing administration will likely be an important consideration as FMT therapies are further defined and targeted to specific flora in various parts of the gastrointestinal tract.

One advantage to endoscopic FMT delivery for CDI is that it allows for simultaneous examination and biopsy of the intestines which may help assess for alternative diagnoses such as inflammatory bowel disease (IBD), and gauge CDI or concomitant disease severity. Thus, endoscopic delivery might be a cost-saving intervention that combines diagnosis and therapy in a single encounter.

Cost may be a factor to be considered in choosing the route of administration. Oral capsules may become the first-line option for treatment of refractory CDI with endoscopic administration reserved for refractory cases or in instances where it is needed for diagnostic purposes. However, cost-effectiveness analyses are needed.

2.2 Single-Donor Vs Pooled Multi-Donor FMT Preparation

Stool for FMT can be administered from a single donor, or can be pooled from multiple donors; there are benefits to each method. Pooling stool will increase the diversity of the bacterial sample because each person has a unique composition of stool flora based on individual environmental factors. Additionally, pooling stool will make larger amounts of stool with similar composition. Theoretically, a single sample of this batch can be screened for safety and, if negative, render the entire batch safe for use.

However, stool from a single donor may offer theoretical benefits. While the specific factors that predispose a patient to gastrointestinal disorders remain unknown, it is conceivable that a single donor could be selected to maximize chance of a successful transplantation. Stool transplant might theoretically be personalized to match each recipient's unique dysbiosis. Additionally, single donor samples may be easily tracked and cataloged to reduce theoretical spread of infection and isolation of contaminated samples. Donor samples from a trusted spouse, family member, or friend may make the experience more agreeable to the patient.

Unpooled samples from individual donors can result in a phenomenon known as "super donors," from whom more effective results might be demonstrated. This was first described in an ulcerative colitis trial wherein seven stool donors provided samples for 75 patients; however, of the nine patients who went into remission, seven received stool obtained from a single donor [9]. However, this phenomenon requires further clarification, as it is unclear how to *a priori* identify "super donors" for FMT.

2.3 Donor Selection

Several factors should be considered when determining the eligibility of donor stool for clinical FMT, primarily focused on minimizing transmission risk of infectious or metabolic disease from the donor to the recipient. Unpooled donors are often chosen on a personalized basis. A family member or unrelated donors can be chosen based on the comfort of patient and availability. Intimate relationships may have similar flora if they share environmental factors. While the specific environmental factors that influence the fecal microbiome are unknown at this time, diet may be the most significant factor. As determinants of stool microbiome become clarified, specific predictive factors may be increasingly utilized to select individual donors.

Exclusion of donors should be considered in the presence of (1) antibiotics in the prior 3 months, (2) intrinsic gastrointestinal conditions, not limited to IBS, IBD, malignancy, functional disorders, dysmotility, (3) major gastrointestinal surgery, (4) history of extra-intestinal autoimmunity, (5) chronic pain, neurological or fatigue syndromes, (6) metabolic syndrome or obesity (body mass index [BMI] >30 kg/m [2]), (7) intentional manipulation of microbiota with probiotics, and (8) prior gastrointestinal infection, especially with drug resistant organisms or CDI.

2.4 Sample Screening

Infectious testing of stool is an evolving area with increasing scrutiny after transition of multi-drug resistant *E. coli* contributed to bacteremia and a death in 2019. The US Food and Drug Administration (FDA) has mandated testing for multidrug resistant organisms. Openbiome, a non-profit stool bank, performs extensive screening for a number of transmissible infections: methicillin-resistant *Staphylococcus aureus* (MRSA) nasal swab, serological testing (HIV, Hepatitis A, B, & C, *Treponema palladium*, HTLV 1 & 2, Stronglyoides IgG), stool testing (CDI, *Salmonella, Shigella, Vibrio, Campylobacter, E. coli* Shiga toxin, *Helicobacter pylori, Cyclospora/Isopsora, Giardia, Cryptosporidium*, Ova and Parasites, *Microsporidium*, Norovirus, Adenovirus, Rotavirus, antibiotic resistance (carbapenemresistant Enterobacteriaceae (CRE), extended spectrum betalactamase (ESBL), and vancomycin resistant *Enterococcus* (VRE)) [10]. While there was initial concern for transmission of COVID-19 due to viral shedding in stool, there have been no published cases of transmission through FMT due to rigorous testing of donors and fecal material.

2.5 Preparation of Samples

Stool samples are filtered and homogenized mechanically through a variety of filters and centrifuges and then aliquoted into delivery mechanism of choice. The samples can be frozen at −80° C for up to 2 years without compromising efficacy for distribution and storage [11].

2.6 Stool Composition

It is unclear whether and how to determine the effectiveness of donor stool in treating CDI, UC, or other conditions. However, increased bacterial diversity of the donor stool has been associated with increased chance of successful transplantation in UC [12–14]. Isolation or amplification of specific organisms is an area under active research that may tailor stool composition to specific deficiencies in an individual's flora.

2.7 Fecal Filtrate

The component of the FMT responsible for therapeutic effect is currently unknown. In fact, stool transplant without bacteria may confer a similar efficacy. Researchers administered stool that was sterile-filtered to remove bacteria and small particles to five patients with recurrent CDI. Remarkably, each of these patients improved and the post-procedure microbiome, virome, and proteome of the recipients were similar to donors [15]. This suggests there may be factors in FMT that contribute to a healthy microbiome in the recipient rather than seeding the colon with donor bacteria. These compounds, if isolated to be non-living substances, could allow for manipulation of the microbiome without the infectious risk of stool transplant.

2.8 Bile Acid Conjugation

It has been hypothesized that bile acid conjugation is a key mediator of *C. difficile* proliferation and a possible explanation for the effect of FMT. Susceptibility to *C. difficile* infection in mouse models after administration of antibiotics was associated with significant loss of secondary bile acids (deoxycholate, lithocholate, ursodeoxycholate, hyodeoxycholate, and ω-muricholate) [16]. Ursodeoxycolic acid was shown to be effective in a case report for the treatment of recurrent CDI in an ileal pouch [17]. CP101, a therapeutic FMT option discussed below, demonstrated increased secondary bile acid conjugation in patients who received treatment. Further research is needed to determine the therapeutic potential of the metabolic effects of microbiome, and there is growing interest in defining the *gut metabolome* and its role in FMT.

3 Indications under Investigation

3.1 Refractory CDI

A meta-analysis of seven randomized control trials and 30 case series published in 2017 showed resolution of recurrent CDI after FMT in a 92% of cases [11]. This was significantly more effective then oral vancomycin (RR: 0.23, 95%CI 0.07–0.80), the previous gold standard. While colonic administration of FMT appeared more effective than nasogastric FMT, the actual magnitude of difference was relatively small 95% (95%CI 92%–97%) *vs* 88% (95%CI 82%–94%). There did not appear to be a difference between frozen or fresh samples. Autologous transplantation (patient's own stool) was inferior to stool donation. Incremental transplantation with multiple administrations over time was shown to be effective in refractory cases as well. This metanalysis did not reveal any serious adverse effects commonly reported. Most recently, a randomized trial did not show significant differences between oral encapsulated FMT and colonoscopic administration of FMT [7].

In a multicenter observational cohort of 259 patients who underwent FMT for CDI, 1-month cure was observed in 90%, of whom 98% received only a single FMT [18]. In this ongoing, national registry, the rate of CDI recurrence among those followed to 6 months was 4%. Notably, new diagnoses of IBS and IBD were each made in 2 patients at 6 months. Further longitudinal follow-up with larger numbers of patients in this ongoing registry will help to further clarify the real-world safety and effectiveness of FMT for CDI.

FMT has been incorporated into society guidelines from both infectious disease and gastroenterology groups as a standard of care in the treatment of refractory CDI. The American College of Gastroenterology (ACG) and Infectious Disease Society of America (IDSA) reserve fecal microbial transplantation for recurrent episodes of CDI, with ACG recommending consideration for severe/fulminant CDI, or after 2 or more recurrences and IDSA after second recurrence (i.e. on the third incident) [19, 20]. Per these guidelines, treatment options for initial presentation of CDI include antibiotics: metronidazole, vancomycin and fidoxamycin. Additionally, bezlotoxumab, a monoclonal antibody directed against toxin B produced in CDI is approved for patients receiving antibiotic therapy; ACG guidelines and IDSA suggest consideration of bezlotoxumab to prevent CDI recurrence in patients at high risk for recurrence. Guidelines do not recommend FMT for the initial treatment of CDI at this time.

The FMT landscape is evolving with FDA approval of live biotherapeutics for recurrent CDI. RBX2660 (marketed as Rebyota®) is a product consisting of live microbes derived from donor stool samples, and the FDA has approved RBX2660 for the indication of recurrent CDI. In a Phase III randomized control trial, 289 patients with recurrent CDI (defined as at least one recurrence) were randomized between treatment and placebo. Treatment response (absence of CDI diarrhea in 8 weeks) was 70.6% compared to 57.5% in placebo group. This trial led to FDA approval of RBX2660 for the treatment of recurrent CDI. Notably, the 70% overall response rate is somewhat disappointing when FMT registry data show >90% response although differences from placebo were significant. Several factors may have contributed to this (1) RBX2660 is administered rectally via enema as opposed to colonoscopy. Prior metanalysis has suggested decreased response via enema compared to colonoscopy 66.3% vs 87.4% [21]. (2) No bowel prep was administered to flush out colonizing bacteria. (3) The trial included patients after first recurrence whereas current clinical guidelines recommend treatment after second recurrence. (4) Inclusion criteria allowed PCR testing for diagnosis of CDI as opposed to toxin testing which may have included patients colonized as opposed to infected with CDI. Additionally, lower response rates may be due to the rigor of a prospective trial compared to prior studies. Retrospective data with 94 patients from 5 study sites over the course of the prospective trials show an 82.8% response rate at 8 weeks [22].

SER-109, produced by Seres Theraputics is a microbiome derived therapeutic of purified Firmicute spores designed for the treatment of recurrent CDI, and was the second approved treatment for prevention of CDI recurrence on the basis of a positive Phase III trial [23]. Firmicutes are hypothesized to increase secondary bile acid conjugation which may reduce *C. difficile* spore germination and prevent recurrent infection. 182 patients were enrolled and randomized to placebo. There was an 88% treatment response rate (defined as sustained clinical response at 8 weeks) compared to 60% of placebo recipients. Several strengths of the trial include confirmation of infection with toxin testing as opposed to PCR, standardization of prior antibiotic use, and inclusion of patients with at least two relapses of CDI which is in line with current medical society guidelines. One advantage to this treatment is the oral route of administration which does not require endoscopy or complete bowel preparation. Magnesium citrate was used to flush out remaining antibiotic which may be more tolerable than bowel preparation for colonoscopy. With the success and safety of oral FMT derived products, incorporation into treatment algorithms for CDI is sure to expand.

Finch Therapeutics illustrates the challenges of microbiome directed therapeutic development. After successful completion of Phase II clinical trial for CP101, an oral live biotherapeutic treatment for CDI [24], the company faced FDA concerns regarding Covid transmission in transplanted materials and there were delays enrolling in Phase III clinical trial. The company discontinued their Phase III trial due to financial pressures; this was a setback to the field with loss of a promising oral microbiome therapeutic.

3.2 Primary *Clostridioides difficile* Infection

Despite the profound effectiveness that FMT has seen for the treatment for recurrent CDI, there have not yet been any large trials comparing FMT to antibiotics as the standard of care for the initial occurrence of CDI. If the effectiveness of

FMT in primary CDI is similar to recurrent CDI, FMT may be more effective than antibiotic therapy. A Cochrane Review in 2017 summarized 22 trials evaluating 12 antibiotics for the primary treatment of CDI. 72% of metronidazole patients achieved cure compared to 79% of vancomycin patients (RR 0.9, 95% CI 0.84,0.97). Fidaxomycin was 71% effective compared to 61% of vancomycin (RR 1.17, 95% CI 1.04, 1.31) [25]. As discussed above, the response rate in recurrent CDI is 92% (95%CI 89%–94%) from metanalysis of 37 publications. None of the antibiotics currently used as *primary* therapy for CDI work as effectively as FMT does for *refractory* therapy. Given that primary CDI is easier to treat then refractory therapy, FMT may therefore be at least as effective in this population.

A Phase III randomized trial comparing metronidazole and FMT is ongoing in Norway (ClinicalTrials.gov number, NCT02301000). In a preliminary analysis of 20 patients, the response rate to FMT group was 78% and metronidazole 45%. While these results were not statistically significant ($p = 0.2$), they do suggest that FMT is more favorable in this very small sample group [26]. Additionally, metronidazole is known to be less effective than vancomycin for primary CDI.

3.3 Non-CDI Gastrointestinal Conditions

Inflammatory Bowel Disease (IBD)

IBD has been associated with alterations of microbiota; however, the role of these findings in pathogenesis remain unclear. It is unknown whether these changes reflect microbial adaptation to a landscape altered by disease or a direct causal relationship. In 2007, rRNA sequencing of bacteria of tissue samples from Crohn's disease (CD) and UC patients first noted low bacterial biodiversity, with reduced Firmicutes (*Clostridium* clusters XIVa and IV), Bacteroidetes, lower butyrate producer *Fecalibacterium prausnitzii*, and higher Proteobacteria and Actinobacteria compared to non-IBD controls [27]. These findings have been followed by several observational studies in both CD and UC, and a number of small randomized trials in UC.

3.4 Ulcerative Colitis (UC)

Four randomized trials have been published with three showing positive effects. In the first published trial 70 patients were randomized to FMT vs placebo in six weekly enemas and demonstrated mucosal healing in 24% vs 5% of controls at 12 weeks. Seven of the nine patients who improved received stool from a single donor suggesting that "super donors" can be identified to have various qualities of stool that make for ideal donors. Remarkably, 39% of patients that received stool from this donor responded. To date, the efficacy of this "super donor's" stool has not been replicated. Another trial compared nasoduodenal infusions of donor stool vs autologous stool in 25 patients [28]. They found that 30% of allogenic vs 20% of autogenic patients met a composite of clinical and endoscopic endpoints, which was not significant. Increasing bacterial diversity was associated with response, but overall the trial was negative. In 2017, 85 patients with mild-moderate UC were randomized to placebo colonoscopic infusion vs colonoscopic transplantation followed by 40 enemas over 8 weeks. Statistically significant steroid free clinical remission with endoscopic response or remission was 27% vs 8% with a risk ratio of 3.6 (95% CI 1.1,11.9). Again, bacterial diversity correlated with response [29]. A fourth trial including 73 adults randomized to autologous vs donor stool via colonoscopy followed by twice weekly enemas for 7 weeks in a composite of endoscopic and clinical outcomes 32% vs 9% response [30].

These randomized trials are suggestive of a treatment effect, but differences between them raise several points for consideration. First, colonoscopic delivery may be more effective. While possibly underpowered, the only trial that utilized upper gastrointestinal tract delivery did not meet statistical significance. Second, there is no validation for maintenance therapy with FMT in UC is at this time. It is unclear if any change in microbiota will be maintained once enemas are completed. While oral FMT agents and enemas could be options for maintenance therapy, patient adherence and tolerability may wane over time if unpalatable. Moreover, UC patients are susceptible to infection because of immunosuppressive medications, malnutrition, frequent hospitalization, and impaired mucosal barrier. Chronic treatment with FMT may pose a cumulative infectious risk over time. Risks of FMT in IBD patients are discussed below.

3.5 Crohn's Disease (CD)

The data on CD are much more limited. There are no randomized control trials published in CD. The largest case series in 2014 of 30 patients in China performed a single FMT via esophagogastroduodenoscopy (EGD) and reported 87% clinical response. However, there was no control group or objective serologic or endoscopic markers of disease severity. An 87% response rate is highly suspect with a low-quality study design. Metanalysis in 2017 of six observational uncontrolled studies showed a pooled response rate of 51.8%; however, these results should be tempered by low-quality study designs.

3.6 Immune Checkpoint Inhibitor Colitis

Immunotherapy has revolutionized the field of oncology through inhibition of immune checkpoints. Unfortunately, these therapies, increase autoimmunity in throughout the gastrointestinal tract, and the colon is particular susceptible. FMT has had early validation with immune checkpoint inhibitor colitis with case series first published in 2018 with two patients that had failed vedolizumab and infliximab. Post FMT, they demonstrated improvement in symptoms and increased microbial biodiversity [31]. This has been replicated in subsequent case reports [32–34].

3.7 Pouchitis

A single case series of eight patients with chronic pouchitis were treated with FMT via nasogastric administration [35]. While there were no changes in clinical response, durable flora changes were noted, and two patients regained sensitivity to ciprofloxacin. Altering microbiota to affect drug resistance may be a potentially novel approach.

3.8 Irritable Bowel Syndrome (IBS)

FMT is being increasingly studied for the treatment of IBS. The microbiome is thought to play a causal role in this disease, and there is data to support the use of both antibiotics and probiotics for this syndrome [36]. There is growing appreciation that gut microbiota can directly affect the gut-brain axis via Brain-Derived Neurotrophic Factor (BDNF). BDNF microRNA expression in the limbic system contributes to anxiety/fear (hippocampus, cingulate cortex, amygdala) and is lower in germ free mouse models compared to pathogen-free control mice [37].

FMT trials for IBS have shown inconsistent results. In one study, 90 patients with severe irritable bowel syndrome-diarrhea (IBS-D) and irritable bowel syndrome-mixed constipation and diarrhea (IBS-M) were randomized to donor vs autologous colonoscopic administration. 65% vs 43% ($p = 0.049$) achieved the endpoint of IBS Symptom Severity Scale (IBS-SSS) increase of 75 points [38]. In another study, 52 patients with moderate/severe unspecified IBS were randomized to placebo vs oral FMT capsules for 12 days. Improvement in IBS-SSS and QoL were actually noted to be *higher* in the placebo group (79% vs FMT group 36%) [39]. In a third study, 48 patients with moderate/severe IBS were randomized to placebo or stool capsules with patient crossover at 12 weeks. There was no effect in primary outcome of IBS-SSS. Another IBS study in which 64 patients were randomized 2:1 autologous to donor nasojejunal administration and 49% vs 29% noted a decrease in abdominal bloating, urgency, and pain as well as increased quality of life; however the validity of the scoring system was unclear [40]. In another study, 16 patients with general IBS were randomized between autologous vs donor FMT delivered via colonoscopy and noted no difference in primary outcome of Gastrointestinal Symptom Rating Scale (GSRS)-IBS, a novel scoring system; however, a secondary endpoint of SF36 (a global measure of quality of life) was positive [41]. A metanalysis of these five studies did not find a significant pooled effect. However, there was significant heterogeneity in study design, intervention, and outcomes that limits the validity of metanalysis (*I^2 75%*) [42].

In a prospective, controlled study, 165 patients with IBS were randomized to FMT from a single healthy donor via gastroscopy [43]. They used two doses of stool (30 g and 60 g) and were evaluated for a 50-point decrease in IBS-SSS. They noted response rates of 23.6%, 76.9%, and 89.1% in placebo, low dose, and high dose stool. In fact, the response rate to this intervention was large with number needed to treat (NNT) of only 1.5 in the high dose group with a durable effect at 3 months. This donor was selected to be a "Super Donor" and his stool was used for each administration. He was screened for known infectious risk factors as well as other factors that may contribute to a healthy microbiome: vaginal birth, breastfed, exercise 1 h five times weekly, nonsmoker, BMI 23. He also ate a diverse diet of 225 food items over the prior year and regularly consumed dietary supplements: Maca root (*Lepidium meyenii*), tribulus (*Tribulus terrestris*), branch chain amino acids, whey protein, Creatine monohydrate, glutamine, Lakerol YUP Mix, vitamin C, vitamin D, Zinc, magnesium, Aerius (an antihistamine), and multivitamins. These are a number of potential factors that may explain these positive data: (1) the FMT was delivered to the duodenum, (2) by using a single donor, each patient is receiving the same stool (it is possible that mixing the material can disrupt a factor that is key in treatment of IBS), and (3) there may be a specific donor factor that mediates the effect. This donor exercised, ate diverse food, took a number of supplements, and was born via vaginal delivery. This study including the identification of potentially desirable donor characteristics deserves replication given the profound effect described.

3.9 Hepatic Encephalopathy

The role of gut flora is well established in the pathophysiology of hepatic encephalopathy, and approved treatments lactulose and rifaximin directly affect the microbiome. In one study, 20 patients with recurrent hepatic encephalopathy were randomized to either standard of care or 5 days of antibiotics followed by a single FMT [44]. None of the 10 patients who received FMT developed recurrent hepatic encephalopathy compared to four of 10 patients in placebo group. Notably, patients in the placebo group did not receive antibiotics prior to the placebo. Therefore, it is not possible to distinguish the effects of 5 days of antibiotics from the effect of FMT. Antibiotics may moderate the efficacy or engraftment of FMT, and this strategy needs further investigation. Interestingly, donors were chosen to highlight Lachnospiraceae and Ruminococcaceae using sequencing and machine learning algorithms to highlight samples rich in these organisms. Targeting specific strains of bacteria may prove to be a unique strategy to boost efficacy of FMT.

3.10 Neurological

Parkinson's Disease

Parkinson's disease is characterized with dysmotility of the entire gastrointestinal tract causing dysphagia, delayed gastric emptying, constipation, and these manifestations are among the most distressing symptoms to patients. IBS correlates with Parkinson's onset, and microbial changes are characteristic and correlate with disease severity [45, 46]. Brain-derived neurotrophic factor (BDNF), which may be moderated by gut flora as discussed with IBS, plays a role in neurodegeneration of Parkinson's disease as well. In mice models, stool transplanted from patients with Parkinson's disease into mice exacerbated α-synuclein mediated deficits, depleted microglia activation via short chain fatty acids alteration, and enhanced motor dysfunction [47]. Trials evaluating FMT in this condition are ongoing.

Multiple Sclerosis

Multiple sclerosis is a debilitating disease characterized by autoimmune T-cell mediated demyelination of neurons, and microbiota appear to play a role in pathogenesis of this disease in animal and *in vitro* models. (Lee, Furusawa) FMT trials are currently underway as potential treatment options of this condition.

Autism

Autism spectrum disorders (ASD) are a group of syndromes characterized by a wide range of impaired social communication and may include restrictive patterns of stereotyped behavior or activities. Gastrointestinal symptoms including bloating, constipation, and diarrhea are often comorbid, and alterations in the stool microbiome have been described. An open-label, non-randomized trial with 18 children with ASD underwent a two-week course of oral vancomycin, followed by a bowel cleanse (Moviprep), acid suppression (Prilosec) and FMT followed by maintenance dose for 7–8 weeks. There was an ~80% response rate in reduction in GI symptoms (as defined by 50% improvement on Gastrointestinal Symptom Rating Scale) including constipation, diarrhea, indigestion, and abdominal pain among patients with improvements durable up to 8 weeks after treatment. Behavioral changes were also seen. A control group of 20 neurotypical children (without ASD) were tested as statistical controls, but did not receive antibiotics, bowel cleanse or autologous FMT. Larger scale randomized trials are needed to replicate these findings [48].

Cancer

The microbiome may theoretically contribute to the pathogenesis of a number of cancers including lymphoma and gastric cancer following *H. pylori* infection, hepatocellular carcinoma with infection of hepatitis B, hepatitis C, EBV, and squamous cell cancer with human papilloma virus [5]. Furthermore, there is increasing awareness that microbiota may contribute to colonic carcinoma [49]. Additionally, microbial signatures are increasingly identified in both pharmacokinetic and pharmacodynamics of chemotherapy agents. Manipulation of fecal microbiota has potential to optimize chemotherapeutics as well as prevent further malignancy. In particular, manipulation of the flora of germ-free mice with melanoma has been shown to modulate the effect of anti-PD1 immunotherapy [50].

3.11 Metabolic Syndrome

Obesity is known to associate with different microbial signatures, and FMT has been considered as a potential intervention. 18 patients were randomized to nasoduodenal infusions of autologous vs lean donor stool. While there were no changes in weight, lipids, or blood pressure over 6 weeks, insulin sensitivity nearly doubled from 26.2 to 45.3 μmol/kg/min [51]. In a repeated experiment with 38 patients over longer term, no improvement in weight or blood pressure was found, but primary endpoints of improvements in insulin

Table 1 Indications for FMT and degree of evidence

Indication	Efficacy	Level of evidence
Recurrent and refractory CDI	Effective	Randomized controlled trials, metanalyses, cost effectiveness trials
Primary CDI	Possibly effective	Randomized controlled trial in process with preliminary data suggest efficacy
Immune checkpoint inhibitor colitis	Possibly effective	Small case series demonstrate efficacy. Further research ongoing.
Ulcerative colitis	Possibly effective	Small randomized controlled trials
Crohn's disease	Unknown	Case series
Pouchitis	Unknown	Case series
Irritable bowel syndrome	Conflicting	Small randomized controlled trials with heterogenous population/interventions
Hepatic encephalopathy	Possibly effective	Small randomized controlled trial
Metabolic syndrome	Possibly effective	Small randomized controlled trial
Autism spectrum disorder	Possibly effective	Non-randomized, non-placebo trial, randomized trials ongoing
Chemotherapy pharmacology	Unknown	Randomized trials ongoing
Parkinson's disease	Unknown	Randomized trials ongoing
Multiple sclerosis	Unknown	Randomized trials ongoing

sensitivity and HgbA1c were noted at 6 weeks. These improvements did not persist at 18 weeks [52].

The indications for FMT and degree of evidence for each indication are summarized in Table 1.

4 Challenges

4.1 Patient Acceptance

Patients are surprisingly amenable to FMT for CDI. A survey of 400 patients (completion rate 48%) *without* CDI noted that 85% would choose to receive FMT compared to antibiotics alone. Patients were more amenable if FMT was offered as a pill (90%) vs endoscopic, NGT, or enema (83%). Patients preferred the intervention in the hospital or physician's office (87%), and 77% were willing to pay out of pocket for FMT [53]. However, completion of FMT protocols can prove undesirable for some patients with potential for high dropout rates [54]. In general, lay media coverage has been largely favorable considering the remarkable efficacy in treatment of life-threatening CDI [55].

4.2 Provider Acceptance

Medical providers have been quick to adopt FMT. After initial seminal descriptions of FMT in 2013, a 2014 survey of 135 physicians in the USA noted already 20% of surveyed physicians had treated patients with FMT and the biggest perceived barriers were patients finding it unappealing (24%), not having a clinical indication (33%) or logistic barriers (23%) [56]. Two years later, a 2016 survey of 120 infectious disease specialists and 84 gastroenterologists in the UK noted 70% were supportive of starting programs even though only seven out of 130 sites performed more than 10 FMT procedures. The biggest barriers to adoption were lack of facilities (45%) and lack of knowledge (38%) [57]. FMT has been incorporated into both gastroenterology and infectious disease society guidelines for treatment of recurrent CDI after broad support, and it has become standard practice for this indication [19, 58].

4.3 Infectious Risk

FMT is a well-tolerated and safe intervention when properly performed; common gastrointestinal symptoms after FMT include abdominal bloating and gas, nausea, abdominal pain, and constipation [20]. However, given the potential for transmission of pathogens, infections remain the primary concern with FMT. There have been reports of gram-negative bacteremia in four patients [59]; however, it was unclear if these infections were result of FMT or underlying conditions of pneumonia, toxic megacolon, and CD. In 2019 two immunosuppressed patients contracted extended spectrum beta-lactamase (ESBL) *E. coli* bacteremia, one of whom died. These patients contracted the infection as part of two separate trials from the same donor. The death was in a patient with myelodysplastic syndrome, and who was enrolled in a phase 2 trial to assess oral FMT capsules pre-and-post-bone marrow transplantation to decrease risk of graft vs host disease. The other patient, affected with hepatitis C cirrhosis, was enrolled in an open label trial of FMT capsules for refractory hepatic encephalopathy. They both had compromised immune systems. Of the 22 patients who received stool from this donor, five developed extended spectrum betalactamase (ESBL)-positive stool without bacteremia. Seven patients with compromised immune systems (liver transplant, CD, stem cell transplant, cirrhosis) did not become bacteremic. ESBL is now routinely screened as part of donor selection. There is likely an under-recognition of infections because patients who undergo FMTs are medically complicated and infections cannot be clearly attributed to FMT.

Other potential risks include worsening the underlying disease process in question. As discussed above, a trial investigating FMT capsules in IBS patients demonstrated worse symptoms in the intervention group compared to placebo [39]. The FDA has released several statements to reduce the risk of infection as detailed below.

4.4 Risk in Inflammatory Bowel Disease

IBD patients have an increased risk of contracting CDI and infection can worsen disease progression. Refractory CDI is a commonly encountered scenario, and observational data suggests 33% higher incidence among IBD patients [60]. These patients are immunosuppressed because of malnutrition, medications, impaired mucosal barrier, and immune dysregulation, and are, therefore, at increased risk of iatrogenic infections that may occur even with the most rigorous stool screening. It is theoretically possible that FMT will exacerbate autoimmunity by further disrupting the microbiome. Studies demonstrate that FMT appears to be safe in this patient population. In one cohort of 67 patients at eight academic IBD centers who were followed for treatment of CDI as well as IBD activity at 3 months, FMT was successful in 80% of patients. IBD activity improved in 30% and worsened in 8%. Serious adverse events were rare including colectomy (1.4%), hospitalization for CDI (2.9%) and IBD flare (2.9%) [61]. A subsequent cohort of 78 patients noted an initial CDI cure rate of 77%, and 100% after repeat FMT. IBD flare was reported in only five patients, although two required colectomy and one diverting ileostomy [62]. These data about IBD specific effects cannot be extrapolated to patients without CDI because these patients, by definition, have marked dysbiosis that is different than unaffected IBD patients.

The potential risks of FMT are summarized in Table 2.

4.5 FDA Regulation

After advocacy from physicians and patient groups after initial reports detailing profound efficacy of FMT in the treatment of CDI, the FDA released guidelines in 2013. They recommended "enforcement discretion" regarding investigational new drug requirements to use FMT to treat CDI in refractory cases. These recommendations require informed consent by patients that FMT is an experimental therapy, but do not provide strict safety standards. After two patients contracted ESBL bacteremia in 2019, the FDA released stricter guidelines requiring donor screening with questions that address risk factors for colonization with multidrug resistant organisms as well as testing of donor stool and exclusion of positive samples. After further reports of transmission of enteropathogenic and Shiga toxin-producing *E. coli* (EPEC and STEC), the FDA has recommended further screening guidelines and disclosure of potential risks. The FDA has also mandated Covid-19 testing of stool donors. It is unknown whether and how the introduction of 2 approved products will impact the current FDA stance on enforcement discretion.

Table 2 Potential risks of FMT

Infection
Procedural complication
Worsening of underlying condition
Provider tolerability
Patient tolerability and adherence

5 Cost Effectiveness Studies

There have been several cost-effectiveness analyses that show improved efficacy and reduced cost with FMT in the treatment of recurrent CDI: (1) A U.S. decision analysis in 2014 comparing metronidazole, vancomycin, fidoxamicin and FMT (duodenal, enema, and colonoscopy) found that colonoscopic FMT dominated metronidazole and fidoxamicin (i.e. cheaper and more effective) and was a cost-effective alternative compared to vancomycin (ICER $17,016, less than willing-to-pay threshold of 50,000) [63]. (2) A Canadian Markov analysis evaluating metronidazole, oral vancomycin, fidaxamicin and FMT (duodenal, enema, and colonoscopy) found that colonoscopic FMT dominated all other strategies (i.e. cheaper and more effective) [64]. (3) A French study decision analysis that evaluated pulsed tapered vancomycin, fidaxamicin, and FMT (duodenal, colonoscopy, and enema) showed FMT via duodenal infusion and fidaxamicin dominated, leaving vancomycin, enema, and colonoscopy as options. They concluded that colonoscopy was not cost effective compared to enema with ICER €73,653/QUALY and assuming a willingness-to-pay threshold of €32,000/QUALY [65]. (4) An Australian Markov Model compared oral vancomycin with FMT (duodenal and colonoscopy); they noted no significant difference between duodenal vs colonoscopic methods, which both dominated vancomycin [66].

Of note, none of these trials included oral encapsulated FMT as there was not yet data demonstrating efficacy or non-inferiority of this route of administration [7]. There are no cost effectiveness analyses assessing FMT for treatment of any condition beyond recurrent CDI, nor are there any studies evaluating oral FMT. The cost of FMT is in constant flux in a changing regulatory environment and because of ongoing commercial endeavors. If FMT or a derivative product is validated as an Investigational New Drug through the FDA, the treatment cost may increase significantly.

6 Conclusions and Future Directions

Fecal microbial transplantation is a therapy that has made remarkable strides since it was rediscovered in the last decade. It is cost-effective for recurrent and refractory CDI with a cure rate over 90%. As our understanding of the role of the microbiome in disease continues to progress, manipulation of the flora through FMT will continue to become more sophisticated. In particular, safety, cost, and efficacy

Table 3 Cost effectiveness trials

Author	Healthcare system	Interventions	Findings
Konijeti et al.	United States	Metronidazole, vancomycin, fidoxamicin, FMT (duodenal, enema, and colonoscopy)	FMT dominated metronidazole and fidoxamicin, cost-effective alternative compared to vancomycin (ICER $17,016)
Lapointe-Shaw et al.	Canada	Metronidazole, oral vancomycin, fidaxamicin, FMT (duodenal, enema, and colonoscopy)	Colonoscopic FMT dominated all other strategies
Baro et al.	France	Pulsed tapered vancomycin, fidaxamicin, and FMT (duodenal, colonoscopy, and enema)	FMT via duodenal infusion and fidaxamicin were dominated. FMT via enema was cost effective alternative to pulsed-tapered vancomycin (ICER of €18,092/QALY) Colonoscopy was not cost effective compared to enema (ICER €73,653/QUALY)
Merlo et al.	Australia	Oral vancomycin with FMT (duodenal and colonoscopy)	No significant difference between duodenal vs colonoscopic methods, which both dominated vancomycin

are areas of focus, and indications may expand beyond CDI. Oral delivery may become increasingly adopted as a resource-saving measure in situations where diagnostic colonoscopy is not required. FDA regulations will help standardize and validate FMT based-therapies, but may increase cost to patients and the health system. Future studies of predictors of successful FMT may facilitate precision medicine whereby FMT is customized to individual patients for various indications. FMT has become an important treatment modality and remains an exciting area of research.

The various cost effectiveness trials for FMT are summarized in Table 3.

References

1. Zhang F, Luo W, Shi Y, Fan Z, Ji G. Should we standardize the 1,700-year-old fecal microbiota transplantation? Am J Gastroenterol. 2012;107(11):1755. https://doi.org/10.1038/ajg.2012.251. 1755; author reply p.1756.
2. Eiseman B, Silen W, Bascom GS, Kauvar AJ. Fecal enema as an adjunct in the treatment of pseudomembranous enterocolitis. Surgery. 1958;44(5):854–9. Accessed 23 Aug 2019. http://www.ncbi.nlm.nih.gov/pubmed/13592638.
3. Khoruts A. Fecal microbiota transplantation-early steps on a long journey ahead. Gut Microbes. 2017;8(3):199–204. https://doi.org/10.1080/19490976.2017.1316447.
4. Van Nood E, Vrieze A, Nieuwdorp M, et al. Duodenal infusion of donor feces for recurrent Clostridium difficile. N Engl J Med. 2013;368(5):407–15. https://doi.org/10.1056/NEJMoa1205037.
5. D'Haens GR, Jobin C. Fecal microbial transplantation for diseases beyond recurrent clostridium difficile infection. Gastroenterology. 2019;157(3):624–36. https://doi.org/10.1053/j.gastro.2019.04.053.
6. Khanna S, Assi M, Lee C, et al. Efficacy and safety of RBX2660 in PUNCH CD3, a phase III, randomized, double-blind, placebo-controlled trial with a Bayesian primary analysis for the prevention of recurrent Clostridioides difficile infection. Drugs. 2022;82(15):1527–38. https://doi.org/10.1007/s40265-022-01797-x.
7. Kao D, Roach B, Silva M, et al. Effect of Oral capsule- vs colonoscopy-delivered fecal microbiota transplantation on recurrent Clostridium difficile infection: a randomized clinical trial. JAMA. 2017;318(20):1985–93. https://doi.org/10.1001/jama.2017.17077.
8. Weingarden AR, Treiger O, Ulsh L, et al. Delivery of fecal material to terminal ileum is associated with long-term success of fecal microbiota transplantation. Dig Dis Sci. 2022;68:2006–14. https://doi.org/10.1007/s10620-022-07761-9.
9. Moayyedi P, Surette MG, Kim PT, et al. Fecal microbiota transplantation induces remission in patients with active ulcerative colitis in a randomized controlled trial. Gastroenterology. 2015;149(1):102–109.e6. https://doi.org/10.1053/j.gastro.2015.04.001.
10. Treatment Information—OpenBiome. https://www.openbiome.org/treatment-information. Accessed 24 Apr 2020.
11. Quraishi MN, Widlak M, Bhala N, et al. Systematic review with meta-analysis: the efficacy of faecal microbiota transplantation for the treatment of recurrent and refractory Clostridium difficile infection. Aliment Pharmacol Ther. 2017;46(5):479–93. https://doi.org/10.1111/apt.14201.
12. Vaughn BP, Vatanen T, Allegretti JR, et al. Increased intestinal microbial diversity following fecal microbiota transplant for active Crohn's disease. Inflamm Bowel Dis. 2016;22(9):2182–90. https://doi.org/10.1097/MIB.0000000000000893.
13. Vermeire S, Joossens M, Verbeke K, et al. Donor species richness determines Faecal microbiota transplantation success in inflammatory bowel disease. J Crohns Colitis. 2016;10(4):387–94. https://doi.org/10.1093/ecco-jcc/jjv203.
14. Kump P, Wurm P, Gröchenig HP, et al. The taxonomic composition of the donor intestinal microbiota is a major factor influencing the efficacy of faecal microbiota transplantation in therapy refractory ulcerative colitis. Aliment Pharmacol Ther. 2018;47(1):67–77. https://doi.org/10.1111/apt.14387.
15. Ott SJ, Waetzig GH, Rehman A, et al. Efficacy of sterile fecal filtrate transfer for treating patients with Clostridium difficile infection. Gastroenterology. 2017;152(4):799–811.e7. https://doi.org/10.1053/j.gastro.2016.11.010.
16. Theriot CM, Bowman AA, Young VB. Antibiotic-induced alterations of the gut microbiota Alter secondary bile acid production and allow for Clostridium difficile spore germination and outgrowth in the large intestine. mSphere. 2016;1(1):e00045–15. https://doi.org/10.1128/mSphere.00045-15.
17. Weingarden AR, Chen C, Zhang N, et al. Ursodeoxycholic acid inhibits Clostridium difficile spore germination and vegetative growth, and prevents the recurrence of Ileal Pouchitis associated with the infection. J Clin Gastroenterol. 2016;50(8):624–30. https://doi.org/10.1097/MCG.0000000000000427.
18. Kelly CR, Yen EF, Grinspan AM, et al. Fecal microbiota transplantation is highly effective in real-world practice: initial results from

the FMT National Registry. Gastroenterology. 2021;160(1):183–192.e3. https://doi.org/10.1053/j.gastro.2020.09.038.
19. McDonald LC, Gerding DN, Johnson S, et al. Clinical practice guidelines for Clostridium difficile infection in adults and children: 2017 update by the Infectious Diseases Society of America (IDSA) and Society for Healthcare Epidemiology of America (SHEA). Clin Infect Dis. 2018;66(7):e1–e48. https://doi.org/10.1093/cid/cix1085.
20. Kelly CR, Fischer M, Allegretti JR, et al. ACG clinical guidelines: prevention, diagnosis, and treatment of Clostridioides difficile infections. J Am Coll Gastroenterol. 2021;116(6):1124. https://journals.lww.com/ajg/Fulltext/2021/06000/ACG_Clinical_Guidelines__Prevention,_Diagnosis,.12.aspx.
21. Tariq R, Pardi DS, Bartlett MG, Khanna S. Low cure rates in controlled trials of fecal microbiota transplantation for recurrent Clostridium difficile infection: a systematic review and meta-analysis. Clin Infect Dis. 2019;68(8):1351–8. https://doi.org/10.1093/cid/ciy721.
22. Feuerstadt P, Harvey A, Yoho DS, Garcia-Diaz JB, Knapple WL, Bancke L. Retrospective analysis of the safety and efficacy of fecal microbiota, live-jslm (REBYOTA™) administered under enforcement discretion to patients with Clostridioides difficile infection. Open Forum Infect Dis. 2023;10:ofad171. https://doi.org/10.1093/ofid/ofad171.
23. Feuerstadt P, Louie TJ, Lashner B, et al. SER-109, an Oral microbiome therapy for recurrent Clostridioides difficile infection. N Engl J Med. 2022;386(3):220–9. https://doi.org/10.1056/NEJMoa2106516.
24. Allegretti JR, Kelly CR, Louie T, et al. S145 week 24 efficacy and safety data from PRISM3: a randomized, placebo-controlled trial evaluating CP101, an investigational orally administered microbiome therapeutic for the prevention of recurrent C. difficile Infection. Off J Am Coll Gastroenterol. 2021;116:S63. https://journals.lww.com/ajg/Fulltext/2021/10001/S145_Week_24_Efficacy_and_Safety_Data_from_PRISM3_.145.aspx.
25. Nelson RL, Suda KJ, Evans CT. Antibiotic treatment for *Clostridium difficile* -associated diarrhoea in adults. Cochrane Database Syst Rev. 2017;3:CD004610. https://doi.org/10.1002/14651858.CD004610.pub5.
26. Juul FE, Garborg K, Bretthauer M, et al. Fecal microbiota transplantation for primary *Clostridium difficile* infection. N Engl J Med. 2018;378(26):2535–6. https://doi.org/10.1056/NEJMc1803103.
27. Frank DN, St. Amand AL, Feldman RA, Boedeker EC, Harpaz N, Pace NR. Molecular-phylogenetic characterization of microbial community imbalances in human inflammatory bowel diseases. Proc Natl Acad Sci. 2007;104(34):13780–5. https://doi.org/10.1073/pnas.0706625104.
28. Rossen NG, Fuentes S, van der Spek MJ, et al. Findings from a randomized controlled trial of fecal transplantation for patients with ulcerative colitis. Gastroenterology. 2015;149(1):110–118.e4. https://doi.org/10.1053/j.gastro.2015.03.045.
29. Paramsothy S, Kamm MA, Kaakoush NO, et al. Multidonor intensive faecal microbiota transplantation for active ulcerative colitis: a randomised placebo-controlled trial. Lancet. 2017;389(10075):1218–28. https://doi.org/10.1016/S0140-6736(17)30182-4.
30. Costello SP, Hughes PA, Waters O, et al. Effect of fecal microbiota transplantation on 8-week remission in patients with ulcerative colitis. JAMA. 2019;321(2):156–64. https://doi.org/10.1001/jama.2018.20046.
31. Wang Y, Wiesnoski DH, Helmink BA, et al. Fecal microbiota transplantation for refractory immune checkpoint inhibitor-associated colitis. Nat Med. 2018;24(12):1804–8. https://doi.org/10.1038/s41591-018-0238-9.
32. Dai C, Liu W-X. Refractory immune checkpoint inhibitor-induced colitis improved by fecal microbiota transplantation: a case report. Inflamm Bowel Dis. 2022;28(3):e43–4. https://doi.org/10.1093/ibd/izab265.
33. Chen M, Liu M, Li C, et al. Fecal microbiota transplantation effectively cures a patient with severe bleeding immune checkpoint inhibitor-associated Colitis and a short review. Front Oncol. 2022;12:913217. https://doi.org/10.3389/fonc.2022.913217.
34. Amin R, Thomas AS, Wang Y. Fecal transplant sustained colitis remission on immunotherapy resumption. Ann Intern Med Clin Cases. 2022;1(10):e220490. https://doi.org/10.7326/aimcc.2022.0490.
35. Landy J, Walker AW, Li JV, et al. Variable alterations of the microbiota, without metabolic or immunological change, following faecal microbiota transplantation in patients with chronic pouchitis. Sci Rep. 2015;5(1):12955. https://doi.org/10.1038/srep12955.
36. Ford AC, Moayyedi P, Chey WD, et al. American College of Gastroenterology monograph on management of irritable bowel syndrome. Am J Gastroenterol. 2018;113(Suppl 2):1–18. https://doi.org/10.1038/s41395-018-0084-x.
37. Maqsood R, Stone TW. The gut-Brain Axis, BDNF, NMDA and CNS disorders. Neurochem Res. 2016;41(11):2819–35. https://doi.org/10.1007/s11064-016-2039-1.
38. Johnsen PH, Hilpüsch F, Cavanagh JP, et al. Faecal microbiota transplantation versus placebo for moderate-to-severe irritable bowel syndrome: a double-blind, randomised, placebo-controlled, parallel-group, single-Centre trial. Lancet Gastroenterol Hepatol. 2018;3(1):17–24. https://doi.org/10.1016/S2468-1253(17)30338-2.
39. Halkjær SI, Christensen AH, Lo BZS, et al. Faecal microbiota transplantation alters gut microbiota in patients with irritable bowel syndrome: results from a randomised, double-blind placebo-controlled study. Gut. 2018;67(12):2107–15. https://doi.org/10.1136/gutjnl-2018-316434.
40. Holvoet T, Joossens M, Jerina B, et al. 617 - fecal microbiota transplantation in irritable bowel syndrome with predominant abdominal bloating: results from a double blind, placebo-controlled clinical trial. Gastroenterology. 2018;154(6):S-130. https://doi.org/10.1016/S0016-5085(18)30860-6.
41. Holster S, Lindqvist CM, Repsilber D, et al. The effect of allogenic versus autologous fecal microbiota transfer on symptoms, visceral perception and fecal and mucosal microbiota in irritable bowel syndrome. Clin Transl Gastroenterol. 2019;10(4):e00034. https://doi.org/10.14309/ctg.0000000000000034.
42. Ianiro G, Eusebi LH, Black CJ, Gasbarrini A, Cammarota G, Ford AC. Systematic review with meta-analysis: efficacy of faecal microbiota transplantation for the treatment of irritable bowel syndrome. Aliment Pharmacol Ther. 2019;50(3):240–8. https://doi.org/10.1111/apt.15330.
43. El-Salhy M, Hatlebakk JG, Gilja OH, Kristoffersen AB, Hausken T. Efficacy of faecal microbiota transplantation for patients with irritable bowel syndrome in a randomised, double-blind, placebo-controlled study. Gut. 2020;69(5):859–67. https://doi.org/10.1136/GUTJNL-2019-319630.
44. Bajaj JS, Kassam Z, Fagan A, et al. Fecal microbiota transplant from a rational stool donor improves hepatic encephalopathy: a randomized clinical trial. Hepatology. 2017;66(6):1727–38. https://doi.org/10.1002/hep.29306.
45. Keshavarzian A, Green SJ, Engen PA, et al. Colonic bacterial composition in Parkinson's disease. Mov Disord. 2015;30(10):1351–60. https://doi.org/10.1002/mds.26307.
46. Lai S-W, Liao K-F, Lin C-L, Sung F-C. Irritable bowel syndrome correlates with increased risk of Parkinson's disease in Taiwan. Eur J Epidemiol. 2014;29(1):57–62. https://doi.org/10.1007/s10654-014-9878-3.
47. Sampson TR, Debelius JW, Thron T, et al. Gut microbiota regulate motor deficits and Neuroinflammation in a model of Parkinson's dis-

ease. Cell. 2016;167(6):1469–1480.e12. https://doi.org/10.1016/j.cell.2016.11.018.
48. Kang D-W, Adams JB, Coleman DM, et al. Long-term benefit of microbiota transfer therapy on autism symptoms and gut microbiota. Sci Rep. 2019;9(1):5821. https://doi.org/10.1038/s41598-019-42183-0.
49. Saus E, Iraola-Guzmán S, Willis JR, Brunet-Vega A, Gabaldón T. Microbiome and colorectal cancer: roles in carcinogenesis and clinical potential. Mol Asp Med. 2019;69:93–106. https://doi.org/10.1016/J.MAM.2019.05.001.
50. Gopalakrishnan V, Spencer CN, Nezi L, et al. Gut microbiome modulates response to anti-PD-1 immunotherapy in melanoma patients. Science. 2018;359(6371):97–103. https://doi.org/10.1126/SCIENCE.AAN4236.
51. Vrieze A, Van Nood E, Holleman F, et al. Transfer of intestinal microbiota from lean donors increases insulin sensitivity in individuals with metabolic syndrome. Gastroenterology. 2012;143(4):913–916.e7. https://doi.org/10.1053/j.gastro.2012.06.031.
52. Kootte RS, Levin E, Salojärvi J, et al. Improvement of insulin sensitivity after lean donor feces in metabolic syndrome is driven by baseline intestinal microbiota composition. Cell Metab. 2017;26(4):611–619.e6. https://doi.org/10.1016/j.cmet.2017.09.008.
53. Zipursky JS, Sidorsky TI, Freedman CA, Sidorsky MN, Kirkland KB. Patient attitudes toward the use of fecal microbiota transplantation in the treatment of recurrent Clostridium difficile infection. Clin Infect Dis. 2012;55(12):1652–8. https://doi.org/10.1093/cid/cis809.
54. Kahn SA, Rubin DT. When subjects violate the research covenant: lessons learned from a failed clinical trial of fecal microbiota transplantation. Am J Gastroenterol. 2016;111(11):1508–10. https://doi.org/10.1038/ajg.2016.153.
55. Klasco R. What is a fecal transplant, and why would I want one? New York: New York Times; 2019.
56. Zipursky JS, Sidorsky TI, Freedman CA, Sidorsky MN, Kirkland KB. Physician attitudes toward the use of fecal microbiota transplantation for the treatment of recurrent Clostridium difficile infection. Can J Gastroenterol Hepatol. 2014;28(6):319–24. https://doi.org/10.1155/2014/403828.
57. Quraishi MN, Segal J, Mullish B, et al. National survey of practice of faecal microbiota transplantation for Clostridium difficile infection in the UK. J Hosp Infect. 2017;95(4):444–5. https://doi.org/10.1016/j.jhin.2016.10.023.
58. Surawicz CM, Brandt LJ, Binion DG, et al. Guidelines for diagnosis, treatment, and prevention of Clostridium difficile infections. Am J Gastroenterol. 2013;108(4):478–98. https://doi.org/10.1038/ajg.2013.4.
59. DeFilipp Z, Bloom PP, Torres Soto M, et al. Drug-resistant E coli bacteremia transmitted by fecal microbiota transplant. N Engl J Med. 2019;30:2043–50. https://doi.org/10.1056/NEJMoa1910437.
60. Razik R, Rumman A, Bahreini Z, McGeer A, Nguyen GC. Recurrence of Clostridium difficile infection in patients with inflammatory bowel disease: the RECIDIVISM study. Am J Gastroenterol. 2016;111(8):1141–6. https://doi.org/10.1038/ajg.2016.187.
61. Fischer M, Kao D, Kelly C, et al. Fecal microbiota transplantation is safe and efficacious for recurrent or refractory Clostridium difficile infection in patients with inflammatory bowel disease. Inflamm Bowel Dis. 2016;22(10):2402–9. https://doi.org/10.1097/MIB.0000000000000908.
62. Tabbaa OM, Aboelsoud MM, Mattar MC. Long-term safety and efficacy of fecal microbiota transplantation in the treatment of *Clostridium difficile* infection in patients with and without inflammatory bowel disease: a tertiary care Center's experience. Gastroenterol Res. 2018;11(6):397–403. https://doi.org/10.14740/gr1091.
63. Konijeti GG, Sauk J, Shrime MG, Gupta M, Ananthakrishnan AN. Cost-effectiveness of competing strategies for management of recurrent Clostridium difficile infection: a decision analysis. Clin Infect Dis. 2014;58(11):1507–14. https://doi.org/10.1093/cid/ciu128.
64. Lapointe-Shaw L, Tran KL, Coyte PC, et al. Cost-effectiveness analysis of six strategies to treat recurrent Clostridium difficile infection. PLoS One. 2016;11(2):e0149521. https://doi.org/10.1371/journal.pone.0149521.
65. Baro E, Galperine T, Denies F, et al. Cost-effectiveness analysis of five competing strategies for the management of multiple recurrent community-onset Clostridium difficile infection in France. PLoS One. 2017;12(1):e0170258. https://doi.org/10.1371/journal.pone.0170258.
66. Merlo G, Graves N, Brain D, Connelly LB. Economic evaluation of fecal microbiota transplantation for the treatment of recurrent Clostridium difficile infection in Australia. J Gastroenterol Hepatol. 2016;31(12):1927–32. https://doi.org/10.1111/jgh.13402.

Future Directions in the Microbiome

Mark Pimentel

In this book, a prolific group of authors have attempted to summarize the field of research in the human gut microbiome. A recent search of the literature for the term "human gut microbiome" produces 17,969 references, and this increases every day. The literature has run far ahead of any clinician's ability to stay 'in-the-know' regarding the microbiome. More importantly, how does one assemble such a body of work into actionable clinical treatments at this time? No single book or work of medical literature can summarize such a body of work completely. The intent of this book is to synthesize important areas of research and clinical focus in the gut microbiome, and provide an up-to-date reference for clinicians seeking to expand their knowledge.

In this chapter, the intent is to summarize the main findings and discuss the future potential of the collective work.

1 The Myriad of Microbes

In Chapter "The Players within the Intestinal Microbiome" we learned of the myriad of microbes in the gut. These include bacteria, fungi, parasites, archaea, and bacteriophages. This "soup" of microbes has tremendous metabolic potential, and has evolved in collaboration with us as the host organism. A wise infectious disease expert once said, "microbes have been around a lot longer than humans…maybe we evolved as a vehicle for them?". This is an interesting concept and perhaps to some extent is true. Based on existing research, most of this collaboration between host and microbes appears to be mutually beneficial, but it can go wrong.

One of the challenges of the twenty-first century is to understand the complex interactions within the microbiome. As indicated, there are number of different groups of microbes that are competing for niches in the microbiome. Even now, there is no true definition of a 'normal microbiome', as humans exhibit similarities but are also ultimately unique in our microbial compositions. The complexity is like a combination lock comprising types of organisms, balances in the relative abundance of each organism, changes in these balances between the small intestine and the colon, host interactions (we produce antimicrobial proteins such as defensins), the inflammatory cascade, diet, exercise, geography, culture, genetics, and so much more.

One problem already encountered in the study of the microbiome is the question, "what diseases are associated with the microbiome?". These days, it can seem from the literature as though the microbiome is associated with all diseases. Everyone is studying the microbiome in their individual areas of research and finding "differences". A known statistical concept is that if you look at 100 items, 5 will be significantly different between groups, due to statistical chance alone. When comparing thousands of microbes and their relative abundances, this is only augmented further. Surely the microbiome does not cause everything. This makes sifting through the literature ever more difficult for the uninformed clinician.

2 We Understand Infection

For centuries, clinicians have tuned into the role of infection in human health and disease. Diseases like the plague and tuberculosis have been scourges to humanity. These are simpler concepts that were initially described by Koch's postulates [1]. One organism equals one disease. Ridding a patient of this pathogenic organism leads to resolution of the illness. For the microbiome, this is not the case. Since the publication of the Human Microbiome Project (HMP) [2], there are few if any examples of a single organism - single disease effect. Rather, the new era is in "communities".

M. Pimentel (✉)
Medically Associated Science and Technology (MAST) Program, Cedars-Sinai Medical Center, West Hollywood, CA, USA

Karsh Division of Gastroenterology and Hepatology, Department of Medicine, Cedars-Sinai Medical Center, Los Angeles, CA, USA
e-mail: pimentelm@cshs.org

M. Pimentel et al. (eds.), *Clinical Understanding of the Human Gut Microbiome*, https://doi.org/10.1007/978-3-031-46712-7_10

But we can still learn from infectious diseases. In Chapter "Gut Microbes: The Gut-Brain Connection" we learn of the amazing effects a single organism can have on the host. One example is post-infectious irritable bowel syndrome (IBS). A single bout of gastroenteritis caused by, for example, *Campylobacter jejuni*, can lead to a chronic illness, IBS [3]. This does not fulfill Koch's postulates in the traditional sense since the disease occurs long after *C. jejuni* is gone from the body, as the illness remains in its shadow. Complex research has now shown that exposure to a toxin (cytolethal distending toxin, CdtB) during gastroenteritis leads to autoimmunity to vinculin, and this cascade results in small intestinal bacterial overgrowth (SIBO), which constitutes a marked change in the microbiome [4]. In this example, a single event has a long-term effect on the microbiome. Antibodies to CdtB and vinculin can now be measured, and can definitively diagnose IBS with diarrhea. Furthermore, this provides for future opportunities to treat IBS. Could removing this antibody cure IBS? This needs urgent exploration. There must be other, similar examples in clinical medicine that have yet to be identified.

3 "Functional" Bowel Disease

It is ironic that 20 years ago, "functional" bowel disease was considered to be a "diagnosis of exclusion". A diagnosis of irritable bowel syndrome (IBS) was given to patients only when all other causes of altered bowel function and pain were excluded. This enigmatic disease is now the protype for microbiome diseases. In fact, our enhanced understanding of post-infectious IBS and SIBO has completely changed how IBS is managed. It could be argued that, similar to *H. pylori* and peptic ulcer disease, up to 60% of IBS is caused by a disturbed small intestinal microbiome, for which an antibiotic may be of benefit. This represents a marked change for this disease.

What IBS has also taught us is that, despite the Human Microbiome Project's definition of the "gut microbiome", stool is <u>not</u> representative of the gut. In fact, the human small intestinal microbiome is remarkably unique and cannot be understood from stool analyses [5]. Furthermore, gases generated in the small intestine have been studied via breath testing in attempts to understand the small intestinal microbiome. In this area of research, hydrogen and methane have been very important measurable analytes. Methane was recently discovered to be associated with, and to play a causal role in, patients with constipation [6, 7]. This may be proven to be one of the few recent single organism - single disease relationships in modern medicine. However, the future is here with a new breath test that also measures the last gas produced by gastrointestinal microbes, hydrogen sulfide (H_2S) [8]. It is now recognized that H_2S correlates with diarrhea and even with inflammation in the gut. This new breath test (simultaneously measuring H_2, CH_4, and H_2S) is poised to be an important research and clinical tool for the future.

4 The Gut-Brain Axis

To think that the brain (at such a distance from the gut) would be so affected by the microbiome seemed unlikely until recently. However, it has been known for decades that hepatic encephalopathy is due to the effects of microbial byproducts on the brain in patients with underlying liver disease. The effects of hepatic encephalopathy are profound. Other extreme examples of neurological effects of the microbiome include tetanus due to *Clostridium tetani* and Guillaime-Barre syndrome due to previous *C. jejuni* infection. There must also be more subtle effects on nervous system function, and understanding these is obviously complex. The greatest areas to watch for in the future of microbiome research are Parkinson's disease, depression, and perhaps autism, as all of these conditions have synchronously recognized changes in gut function along with, or in the case of Parkinson's (associated with constipation) decades before, the illness is recognized. One or more of these areas could in time be shown to represent a cause-and-effect relationship. This area needs to be followed closely by clinicians, as shown in Chapter "Gut Microbes: The Gut-Brain Connection".

5 Obesity and Diabetes

Metabolic disorders are among the first to have been studied in attempts to understand the importance of the microbiome in contributing to a human condition. The extent to which the microbiome solely or directly causes obesity is unknown, as it is recognized that metabolic disorders such as obesity are multifactorial in nature (discussed in Chapter "Changes in the Gut Microbiome as seen in Diabetes and Obesity"). However, it is interesting to see the importance of methanogens in this relationship. Methanogens slow the gut [9]. A slower gut could rationally allow a greater time for energy harvest from food [10]. Understanding these relationships more clearly is being accomplished in labs today, and this is an area with a high likelihood of yielding future treatments in clinical medicine.

6 Gut Inflammation

Our instinct in medicine is to equate bacteria with "inflammation", based on our understanding of the term "infection". However, that understanding is shifting. While the word "probiotic" is now a cliché to imply "good" bacteria, there are beneficial effects of microbes. As explored in Chapter "Fecal Transplant: The Benefits and Harms of Fecal

Microbiota Transplantation", there are microbes that have anti-inflammatory properties. The area of IBD begs the question not of which microbe(s) are causing disease, but perhaps which microbes are missing that keep the gut immune system in check? The importance of IBD research in the microbiome is the incredible work demonstrating a synergistic effort to connect host genetics to the microbiome. It is also an example of the possibility that IBD symptoms and phenotypes are not due to bacteria, but rather due to fungi such as *Malassezia restricta*.

7 Bugs as Drugs

As initially introduced above, the term "probiotics" implies that the bacteria are "good". To be frank, there are no "good" or "bad" bacteria. If bacterial populations are in their correct compositions and relative abundances, health is achieved. Finding the common probiotic *Lactobacillus acidophilus* growing from the blood stream or on a heart valve is not "good". Bacteria residing where they belong in the right abundance is "good". If the last decade has taught us one thing, is that it is the compositions of microbial communities that are good, usually not a single species. This premise has been the basis for fecal transplant.

The concept of using bacteria as treatments is the basis for a "bugs as drugs" approach. This may take many forms in the future. The use of probiotics is one form. As seen in Chapter "The Role of the Microbiome in Inflammatory Bowel Disease", the entire microbiome can be used as a fecal transplant. But bacteria may be manipulated in the future in more complex ways. For example, bacteria can be genetically engineered to produce specific product or chemical compound. When that bacteria is then introduced into a human being, it produces something that treats or ameliorates the disease the patient is suffering from. There are emerging technologies that have the potential to address several diseases using this type of technology.

8 Drugs from Bugs

Another future approach in microbiome research is also emerging. If science can identify a specific bacterial product that is beneficial, this product can be isolated and used as a treatment.

9 Drugs for Bugs

Finally, another and very promising future approach to the microbiome is to target the properties of the microbiome. For example, if a bacteria, or more likely a group of bacteria with related functions, produces an agent or protein that is harmful to humans, since the metabolic pathways of bacteria are so different from those of humans, blocking this pathway with a drug could help treat the human condition or disease with minimal side effects for the human host. This is a recognized small molecule approach to medicine.

10 In Conclusion

There is still a great deal of excitement surrounding the importance and potential of the human microbiome. The great challenge of our time is to understand what constitutes a "normal" microbiome. Some argue that with modern medicine and antibiotics, probiotics, and food preservatives, we may never know what the "real" microbiome in normal humans is or was supposed to be. There is something slightly worrying in this thinking. However, we are quickly understanding how important the gut microbiome is for our health and well-being. Understanding this complex relationship will begin to educate us and thus yield a better future for our food supply, antibiotics selection, probiotic use, and many other important aspects of human health. This understanding also brings tremendous promise for the treatment of many diseases.

References

1. R K. Die Aetiologie der Tuberkulose. Mitt Kaiser Gesundh. 1884;2:1–88.
2. Turnbaugh PJ, Ley RE, Hamady M, et al. The human microbiome project. Nature. 2007;449:804–10.
3. Klem F, Wadhwa A, Prokop LJ, et al. Prevalence, risk factors, and outcomes of irritable bowel syndrome after infectious enteritis: a systematic review and meta-analysis. Gastroenterology. 2017;152:1042–1054.e1.
4. Morales W, Triantafyllou K, Parodi G, et al. Immunization with cytolethal distending toxin B produces autoantibodies to vinculin and small bowel bacterial changes in a rat model of postinfectious irritable bowel syndrome. Neurogastroenterol Motil. 2020;32:e13875.
5. Leite GGS, Weitsman S, Parodi G, et al. Mapping the segmental microbiomes in the human small bowel in comparison with stool: a REIMAGINE study. Dig Dis Sci. 2020;65:2595.
6. Triantafyllou K, Chang C, Pimentel M. Methanogens, methane and gastrointestinal motility. J Neurogastroenterol Motil. 2014;20:31–40.
7. Kunkel D, Basseri RJ, Makhani MD, et al. Methane on breath testing is associated with constipation: a systematic review and meta-analysis. Dig Dis Sci. 2011;56:1612–8.
8. Lin E, Chua KS, Pichetshote N, et al. Measurement of hydrogen sulfide during breath testing correlates to patient symptoms. Gastroenterology. 2017;152:S205–6.
9. Park YM, Lee YJ, Hussain Z, et al. The effects and mechanism of action of methane on ileal motor function. Neurogastroenterol Motil. 2017;29:1.
10. Barlow GM, Lin EA, Mathur R. An overview of the roles of the gut microbiome in obesity and diabetes. In: Bagchi D, Nair S, editors. Nutritional and therapeutic interventions for diabetes and metabolic syndrome. 2nd ed. Amsterdam: Elsevier; 2018. p. 65–91.

Index

M. Pimentel et al. (eds.), *Clinical Understanding of the Human Gut Microbiome*, https://doi.org/10.1007/978-3-031-46712-7

GPSR Compliance

The European Union's (EU) General Product Safety Regulation (GPSR) is a set of rules that requires consumer products to be safe and our obligations to ensure this.

If you have any concerns about our products, you can contact us on ProductSafety@springernature.com

In case Publisher is established outside the EU, the EU authorized representative is:

Springer Nature Customer Service Center GmbH
Europaplatz 3
69115 Heidelberg, Germany

Batch number: 10731528

Printed by Printforce, the Netherlands